<u>SECTIONS</u>

ELECTROCHEMICAL ASPECTS OF THE HYDROMETALLURGY OF GOLD

By

Garth Louis Putnam

CONTENTS

A. INTRODUCTION

1. HISTORICAL:

The history of gold metallurgy is older than that of any other metal. As its high specific gravity (19.3) made possible its recovery from alluvial deposits by such primitive methods as washing in wooden vessels, the noble metal was known to the ancients at least prior to 4000 B.C.

The next development in the metallurgy of gold was the introduction of amalgamation. Pliny (A.D. 23-79) in Volume 35 of his treatise on Natural History (1) refers to the extraction of gold by crushing the ore and then amalgamating; and this method is still a very important one today. As clearly shown by Professor A. F. Taggart (2) in his Handbook of Ore Dressing, amalgamation is one of our most economical methods, and when part of the gold is recoverable in this way, it is nearly always most profitable to amalgamate before subjecting the ore to flotation or cyanidation.

Two thousand years later, following the introduction of the amalgamation process, and as the chemistry of gold began to be more intensely developed, aqueous chlorine solutions were applied (1848) to the extraction of gold from roasted sulfide ore and to the tailings from the amalgamation processes. Next, in less than forty years (1887) appeared a more widely applicable chemical method, that of MacArthur, Forrest, and Forrest,[1] usually referred to as the cyanide method or as the MacArthur-Forrest process.

A major share of the credit for the industrial development of the cyanide process belongs to John Van Nostrand Dorr. The vast economic importance of his processes and equipment was pointed out at the presentation of the 1941 Perkin Medal to Dr. Dorr (3). In North America today, Dorr processes and equipment, many of which were originally designed for the cyanide extraction of gold, are used in 95 percent of the gold industry, 93 percent of the copper industry, 93 percent of the lead industry, 85 percent of the zinc industry, 100 percent of the nickel industry, 100 percent of the aluminum industry, and 15 percent of the iron industry (3).

2. PURPOSE OF THIS WORK:

An excellent review of the cyanide process as practiced at the present time was published recently by Neil Munro of The Dorr Company (4). We quote directly from his paper:

> Of the gold produced in Canada, and valued at $166,000,000 in 1938, about 75 percent, valued at $120,000,000, was extracted by

1. Many writers, who have evidently failed to examine the original communications, refer to the process as the MacArthur-Forrest process. There were two Forrests (19).

> the use of the cyanide method. In its application cyanide salts valued at $1,500,000 and lime valued at $150,000 were used....
>
> Concentrations of cyanide range from 1/40 percent to 1/10 percent potassium cyanide or equivalent; concentrations of lime ranging from 1/200 percent to 1/10 percent calcium oxide provide the alkalinity required and aid in the settlement of ground solids....
>
> Some forms of elemental carbon, and minerals containing copper, arsenic, antimony, cobalt, nickel, and chromium destroy the effective dissolving power of the cyanide solution and give rise to increased cost and tailing loss unless means to counteract their effect are used.... <u>If more were known about the chemistry of the main dissolution and precipitation reactions, and the effects of the deleterious compounds mentioned above on these reactions, means of counteracting their effect could be applied in a more rational manner than the rule-of-thumb methods used in most plants</u>. (Our underlining.)

Munro also gives flow-sheets for several modern cyanide mills. In a typical instance, 75 percent of the gold content of the ore is dissolved during grinding in a cyanide solvent; most of the remaining gold is dissolved in cyanide solution contained in three agitators. <u>It should be noted that vigorous agitation of the pulp (ground ore + solvent) is used throughout the process, and that the size of the agitators for a given mill capacity must necessarily depend on the time required to dissolve the gold particles</u>.

One does not have to study Dorr's book (5) in detail to see that the primary emphasis is on equipment rather than on chemistry. The problem which we have set for ourselves is to find means of decreasing the time required to dissolve gold through the application of chemical rather than mechanical means. While the facilities at our disposal were limited, we have tried to attack the problem of improving the cyanide process in the manner in which Dorr would have done it: "Dr. Dorr ... had an unerring ability to pick out the most needed thing to be done and to solve the problem of doing it." (3) Bearing in mind that perhaps the most important problem facing the cyanide industry today is to economically decrease the time required for the dissolution of the gold, we have attempted by this present work to obtain a general picture of the ways by which this object might be accomplished.

B. COLORIMETRIC DETERMINATION OF GOLD

INTRODUCTION:

One difficulty that immediately presented itself in our study of the various gold extraction processes was the confused state of the literature pertaining to a <u>rapid method</u> for the determination of gold.[1] The colorimetric test, which many authorities look upon as standard, is the stannous chloride "purple of Cassius" test.[2] Bugbee, whose textbook[3] is widely used in colleges and universities, including the Columbia School of Mines, gives the following directions for applying the test to cyanide solutions (<u>8</u>):

> Pour 1000 ml. of cyanide solution into a wide-necked bottle; to this add 3 drops of a saturated solution of lead acetate, and shake the mixture. Add 0.75 gram of zinc dust and shake again. Add 20 ml. of saturated cyanide solution and shake for 30 seconds. Pour into an evaporating dish and allow the dust to settle. Pour off the solution, put the dish on a spirit lamp, and dissolve the zinc dust with about 10 ml. of aqua regia; boil this until the bulk is reduced to a volume of about 4 ml. Allow the chloride of lead to settle, and then pour the solution into a test tube. Cool the solution in the tube, and add 2 drops of a solution of stannous chloride. If gold is present to the extent of 0.02 dwt. per ton of the original cyanide solution, a very slight coloration will be perceived in the liquid; 0.03 shows a slight yellow; 0.04 a slight pinkish yellow; 0.06 a strong pink; and 0.08 the purple of Cassius.

With this method the amount of gold is estimated from the shade, or tint, of color; and it may also be observed that the volumes of stannous chloride and lead acetate are given in terms of drops of unspecified size.

REDUCTION OF AURIC CHLORIDE TO COLLOIDAL GOLD:

The first phase of our work consisted of a study of several chemical methods for preparing colloidal gold. Nearly all of the methods which we tried have been proposed as qualitative or as colorimetric tests for gold (<u>9</u>, <u>10</u>, <u>11</u>). A solution of auric chloride

1. For examples of our application of colorimetric methods, please see Tables 11 and 19 of the following sections.
2. The product resulting from the reaction of stannous chloride with gold halides is commonly referred to as "colloidal gold." According to Professor A. W. Thomas (<u>6</u>), this is the oldest example of a metal having been obtained in the colloidal state. The test was discovered by Andreas Cassius in 1663 (<u>6</u>). While this sol has the color of colloidal gold, its properties, according to Zsigmondy and Thiessen (7), are predominantly those of colloidal stannic acid.
3. The widespread acceptance of Bugbee's text indicates its standing in the field of gold assaying. We quote his work extensively because we have the very highest respect for it.

was prepared by dissolving C.P. precipitated gold in aqua regia and evaporating the gold solution almost to dryness (10). For each trial of Table 1, 10 ml. of a solution containing 5 mg. of gold per liter were used. The color of the liquid was observed through a column about 3 inches deep.

When 1 ml. of reagent was added to 10 ml. of the solution containing 5 mg. of gold per liter, reagents which gave no color within 10 minutes were: 3 M potassium iodide, 2 M potassium iodide-3 M potassium hydroxide, 1 M hydrogen peroxide, saturated acetylene-0.5 M potassium hydroxide, saturated benzaldehyde, 3 M formaldehyde, 0.5 M sodium thiosulfate, 1 M ferrous sulfate, 0.1 M sodium acid sulfite, 0.1 M sodium sulfite-0.1 M potassium hydroxide. The fact that these weaker reducing agents gave no color indicates that the speed of reduction of gold ions in dilute solutions may depend to a large extent on the oxidation-reduction potential of the reagent, and may also help to explain why sea water is able to retain small amounts of gold in solution. In some cases, at least, the addition of a strong reducing agent causes color development in a solution to which a weaker reducing agent has been added some time previously (Table 1, alkaline formate trial).

The experiments of Table 1 indicated that the stannous chloride, alkaline formaldehyde, and Nessler's reagent (K_2HgI_4-KI and KOH) tests were the most sensitive. Nessler's reagent, unfortunately, gives a yellow color with traces of ammonia, and no practicable method could be found for preventing this interference. Moreover, the shades and color intensities of both the Nessler's reagent and the alkaline formaldehyde tests were not reproducible.

THE PURPLE OF CASSIUS TEST:

Favorable preliminary results having been obtained, the stannous chloride method was studied more in detail.

Reagents. The reagents were prepared as follows:

1. Stannous Chloride, 0.25 M. Into a 30-ml. test tube were placed 22.6 g. of C.P. stannous chloride crystals ($SnCl_2 \cdot 2H_2O$) and 2.33 ml. of concentrated (11.2 M) hydrochloric acid. On heating almost to boiling, the stannous chloride dissolved and the solution became clear within one or two minutes. After maintaining at the boiling point for two or three minutes, the solution was cooled to 40°-50°, poured into 300 ml. of distilled water, and made up to a final volume of 400 ml. It was found necessary to follow this procedure closely in order to obtain a clear solution of low acidity. When the reagent became turbid, it was filtered; and reagents more than 8 hours old were not used. This reagent was 0.25 M (0.50 N) with respect to stannous chloride, and less than 0.07 M with respect to hydrochloric acid.

Table 1

Preliminary Experiments with Various Gold Tests

Reagent	Mls. of Reagent Used	Remarks
1 M H_2O_2 - 1 M KOH	1, 2 and 5	After ten minutes, very pale blue. Intensity gradually increased on standing.
0.06 M K_2HgI_4-0.06 M KI and 3.6 M KOH	1, 2 and 5	Order of addition and relative volumes of reagents varied. Colors intense but not reproducible. 0.01 mg./liter of ammonia interferes.
0.09 M resorcinol	1	Dark green. Nitrite ion and oxidizing agents interfere.
0.09 M resorcinol-0.3 M KOH		Immediate formation of reproducible blue-violet color. Traces of chlorine and several other oxidizing agents give similar color. Traces of nitrite ion give intense yellow.
Satd. benzaldehyde-0.5 M KOH	1	Pale violet.
3.3 M formaldehyde-2 M KOH	1	Intense violet. Ten minutes standing required for full color development. Intensity and shade not reproducible.
0.11 M glucose-0.2 M KOH	1	Gradual development of purple color. Not as intense as formaldehyde.
0.28 M butyraldehyde-0.2 M KOH		Pale violet. Not as intense as glucose test.
Satd. hydrogen sulfide	1	Very pale yellow.
1 M potassium formate-2 M KOH	1	Slow development of pale blue color. After permitting to stand five minutes and then adding formaldehyde, color became as intense as that produced by formaldehyde-KOH.
0.1 M $SnCl_2$	1	Immediate development of intense yellow color to full intensity. Reproducible. Over a wide range of stannous chloride concentrations, intensity of color not a function of stannous chloride concentration.

2. Stannous Chloride, Saturated. This was prepared according to Bugbee (8), who gives these directions: "Make a saturated solution of the C.P. salt in 1 part of hydrochloric acid and 15 parts of distilled water. Place a few shavings of tin in the bottle to keep in the stannous form."

3. Bromine Water, Saturated. A 300-ml. bottle was partly filled with distilled water and a large excess of C.P. bromine added. The bottle was shaken vigorously and filled with distilled water whenever the reagent was depleted.

4. Gold Solutions. a. Auric Bromide, 500 Mg. of Gold per Liter. C.P. precipitated gold, which had been dried at a temperature such that the color and luster of massive gold were absent, was used. (On igniting at a temperature above about 400°, precipitated gold acquires a metallic luster and dissolves with difficulty.) Using a watch glass, 125 mg. of gold powder were weighed and transferred to a 250-ml. beaker, using 5 ml. of saturated bromine water to wash the small amount of adhering dust from the watch glass. Agglomerated particles of gold powder were crushed with a stirring rod and 10 ml. of bromine water added, 0.5 to 2 ml. at a time; between each addition the beaker was covered with a watch glass, to prevent loss of bromine, and swirled. The gold bromide solution was then warmed to about 50° in order to remove most of the excess bromine and then poured into a 250-ml. volumetric flask. The undissolved gold remaining in the beaker was treated with bromine water, 2 ml. at a time, until dissolved, less than 20 ml. being required. The gold bromide solution was warmed to 50°, transferred to the 250-ml. volumetric flask, and made up to the mark with distilled water. The total time required to dissolve the gold was less than 30 minutes. This reagent contains 500 mg. of gold per liter as auric bromide, and is stable for several months.

b. Aurous Cyanide, 381 Mg. of Gold Per Liter. To a 150-ml. gas-washing bottle were added 99.9 mg. of C.P. precipitated gold powder and 100 ml. of a solution containing 5 grams of sodium cyanide per liter. After drawing air through the solution for 24 hours, all but 4.7 mg. of the gold had dissolved; therefore 95.2 mg. went into solution. The volume was made up to 250 ml. and the solution contained 381 mg. of gold per liter.

c. Gold Chloride, Aqua Regia Solution. Using 20 ml. of aqua regia, 47.8 mg. of precipitated gold were dissolved; the solution was evaporated to a volume of about 2 ml. and then made up to a volume of 250 ml. with distilled water. The concentration of gold was therefore 191.2 mg. per liter.

5. Lead Acetate, Saturated. This reagent was a saturated solution of the C.P. salt in distilled water (8).

6. Sodium Cyanide, Saturated. An excess of C.P. sodium cyanide was shaken vigorously with distilled water (8).

7. Chlorine Water. A stream of chlorine was generated by the action of hydrochloric acid on calcium hypochlorite, the gas

being washed with water before being absorbed in distilled water. The concentration of free chlorine was determined by pipetting 10 ml. into an excess of potassium iodide and titrating with standard sodium thiosulfate solution (12).

8. Lime Water Saturated. A clear, saturated solution was prepared by shaking a large excess of calcium hydroxide with distilled water and decanting.

Procedure. Comparison of Colors. Colors were compared with 50-ml. Nessler tubes having a maximum variation in length of 2.5 percent. As it was found advantageous, from the standpoint of maintaining a constant tint of color, to have approximately the same concentrations of gold in both standard and unknown, the standard and unknown test solutions were diluted until the color intensities of 50-ml. portions of the stannous chloride-gold solutions were approximately equal. In the cases of the experiments given in Table 8, the dilutions were made in intervals of 20 ml.; in all other cases the dilutions were made in intervals of 50 ml. Final comparison of the colors was made by pouring a portion of the more highly colored solution from the Nessler tube into a graduated cylinder or other clean vessel, until the colors matched. The depths of the fluids in the Nessler tubes were then measured with a ruler or aluminum bar scale prepared for the purpose.

Precautions. Colloidal gold solutions and solutions of gold halides have a marked tendency to form purple-colored films on glassware. This film is readily soluble in chlorine water. When working with gold solutions we made a practice of washing all glassware with chlorine water before using. To remove stannic oxides from glassware, concentrated hydrochloric acid was used. Solutions containing tin salts should not be allowed to stand overnight in glassware, as hydrolysis takes place and the stannic oxides, on aging, become practically insoluble in concentrated hydrochloric acid.

Hydrolysis of Stannous Chloride. Preliminary experiments (Table 1) having shown that low acid concentrations were desirable when making the stannous chloride test, experiments were made on the effect of sodium chloride on the hydrolysis of dilute stannous chloride solutions. When the solutions hydrolyze, it becomes impossible to compare the colors with any degree of accuracy.

In making the trials of Table 2, freshly prepared 0.5 N stannous chloride reagent was added to 25 ml. of dilute sodium chloride solution, and the solutions rapidly mixed. The time required for the appearance of perceptible turbidity was noted, giving the results of Table 2.

Table 2 proves that sodium chloride accelerates the formation of turbidity in dilute stannous chloride solutions, such as we have when making the colorimetric test for gold. Moreover, Table 2 indicates that the tendency for stannous chloride solutions to become turbid can be reduced by lowering the stannous chloride concentration. It is possible that the electropositive sodium and stannous ions act as precipitants for electronegative colloidal stannic oxides present in the solution (13).

Table 2

Effect of Sodium Chloride on the Time Required for the Appearance of Turbidity in Stannous Chloride Solutions

	Time Required for Appearance of Turbidity with Stannous Chloride Concentration of: 0.002 N	0.010 N	0.029 N
No NaCl Added	48 hours[a]	48 hours[a]	48 hours[a]
0.01 N NaCl	16 hours[b]	16 hours[b]	16 hours[b]
0.05 N NaCl	80 seconds	20 seconds	10 seconds
0.10 N NaCl	40 seconds	20 seconds	2 seconds
0.50 N NaCl	40 seconds	5 seconds	1 second

a. These trials were somewhat turbid after 48 hours, but there was no precipitation. Precipitation had taken place in the cases of all of the other trials.
b. After 16 hours there were no precipitates with these trials or with the trials containing no sodium chloride, but there were precipitates with all of the other trials.

Effect of Acid Concentration on Tint of Color. In order to determine the effect of acid concentration on the tint or shade of color, 40 ml. of a solution containing 5 mg. of gold, as bromide, per liter were added to each of 10 50-ml. Nessler tubes, and the solutions made up to the mark with 1.57 N or 11.2 N hydrochloric acid and distilled water. All solutions therefore contained 4 mg. of gold per liter. The Nessler tubes were closed with a rubber stopper which had been washed with chlorine water, and inverted twice. Using a burette, 1 ml. of 0.25 M stannous chloride was added to each tube, which was immediately closed with a rubber stopper and inverted twice. The preparation of all Nessler tubes of each series was completed within three minutes after the addition of the stannous chloride reagent to the first tube of the series. One or two groups of ten trials each were made for each acid concentration. After 15 minutes' standing, one of the tubes of each series was taken as the standard and the rest of the tubes were compared with it. The results are given in Table 3.

Table 3

Effect of Acid Concentration on Tint of Color

Series No.	Conc. of HCl in Colorimetric Medium N	No. of Trials in Series	Average Percent Deviation From Mean of Series[a]	Maximum Percent Deviation From Mean of Series[a]	Tint of Color After Fifteen Minutes' Standing
1	0.002	20	5.1	10.6	Yellow
2	0.08	10	3.8	8.4	Light Brown
3	0.16	20	4.4	12.9	Tan with Faint Purple Component
4	0.32	10	6.1	10.2	Tan with Distinct Purple Tinge
5	0.64	10	5.3	13.6	Purple

a. Please see text.

The data of Table 3 indicated that the acid concentration was an important factor affecting the tint of color in the stannous chloride test. The values for the "Percent Deviation From Mean" in Table 3, however, have little bearing on the reproducibility of the test under ordinary analytical conditions, as it is almost impossible to prepare standards which have exactly the same ages and compositions as the unknowns. The figures on "Percent Deviation from Mean" are included for the purpose of showing the order of magnitude of our experimental error.

When making a series of analyses, it is apparent that it would be difficult and time-consuming to have to use standards which were of the same age as the unknowns. Moreover, the factor of the rate of color development has a very important bearing on the accuracy and reproducibility of the test, and for this reason was studied.

Solutions containing 2.5 mg. of gold per liter and 5.0 mg. of gold per liter were prepared by diluting the gold bromide reagent. For each trial 0.100 mg. of gold in the form of the dilute bromide solution was added to a 50-ml. Nessler tube. The required amount of 1.57 N or 11.2 N hydrochloric acid was then added with a burette, the solution made up to the mark of the Nessler tube with distilled water, and inverted twice using a rubber stopper that had been washed in chlorine water. The gold concentration with each trial was 2 mg. per liter.

One ml. of 0.5 N stannous chloride was then rapidly added with a burette to the Nessler tube, and the tube inverted once, the time of the start of the trial being taken as the time when the tube was inverted. The addition of the stannous chloride and the inversion of the tube required less than ten seconds. After standing ten minutes, 2.5 ml. of solution were withdrawn from the Nessler tube. Using another Nessler tube, a second trial identical in all respects to the previous trial, was started, and the time required for the colors to match within the limits of visual error recorded. Three trials were made with each acid concentration, giving the results of Table 4.

The "Percent Deviation from Mean" shows the probable order of magnitude of our experimental error.

The results of Table 4 prove that color development takes place far more rapidly in the weak acid solutions. Moreover, as the time required for the development of the color to practically full intensity is much less at low acid concentrations, the results of Table 4 indicate that under ordinary conditions of analysis, more reproducible results would probably be obtained at low acid concentrations.

Regarding the stability of the colloidal solutions, we have never observed any precipitation of gold from the yellow solutions, except in those cases where a flocculent precipitate of stannic oxides or oxy-chlorides carried down the yellow colloidal gold. With the purple solutions, especially when strongly acid, purple precipitates which are neither flocculent nor voluminous form within a few hours and the solutions become colorless. With the yellow solutions in which conditions are such that hydrolysis of the tin salts does not take place, the solutions may generally be preserved without change for at least a week, and when precipitation does finally take place it is preceded by a change of color from yellow or brown to purple.

Moreover, in connection with the effect of acid, a very interesting fact was observed which to our knowledge has never been recorded in the literature: By adding hydrochloric acid to the yellow solutions, so that the solutions are made 2 N to 6 N with respect to hydrochloric acid, the color changes immediately to blue or purple. This color change has been observed with solutions containing up to 300 mg. of gold per liter. With solutions containing 2 to 20 mg. of gold per liter, the addition of hydrochloric acid causes the solutions to become colorless for a period ranging from a fraction of a second to several seconds, after which a blue or purple color develops.

Effect of Stannous Chloride Concentration. In order to determine the effect of stannous chloride concentration on the gold test, a solution was prepared which contained 10 mg. of gold as bromide per liter, 30 mg. of chlorine per liter, and 40 mg. of hydrochloric acid per liter. The chlorine was added in order to simulate analytical conditions and to make certain that all of the gold was present in the auric state; an excess of strong oxidizing agent must be present in gold solutions to prevent photochemical reduction to the aurous or metallic states. It is difficult, if not impossible, to remove all of the free halogen from an auric halide solution without decomposing a part of the auric compound. The experiments on the effect of stannous chloride concentration differed from those previously given in that the concentration of free halogen was slightly greater.

Table 4

Effect of Acid Concentration on Development of Color

Expt. No.	HCl Conc. N	Seconds Time, After Start of Second Trial, Required for Colors to Match	Percent Deviation from Mean	Shade of Color After Twenty to Thirty Minutes' Standing[a]
1	0.002	30	5.4	Yellow
2	0.002	40	26.2	Yellow
3	0.002	25	21.2	Yellow
Mean	0.002	31.7	17.6	Yellow
4	0.08	90	22.8	Yellow
5	0.08	80	9.1	Yellow
6	0.08	50	31.8	Yellow
Mean	0.08	73.3	21.2	Yellow
7	0.16	360	10	Light brown
8	0.16	360	10	Light brown
9	0.16	480	20	Light brown
Mean	0.16	400	13.3	Light brown
10	0.64	900	9.8	Purple
11	0.64	840	2.4	Purple
12	0.64	720	12.2	Purple
Mean	0.64	820	8.1	Purple
13	5.0[b]	1200	15.5	Dark blue
14	5.0[b]	1380	2.8	Dark blue
15	5.0[b]	1680	18.3	Dark blue
Mean	5.0[b]	1420	12.2	Dark blue

a. With the 0.64 N and 5.0 N HCl solutions there was a distinct difference in shade of color between samples 20 minutes and 30 minutes old.

b. A time of more than 180 seconds was required before any visible coloration appeared.

The stannous chloride reagent was prepared by dissolving 22.6 g. of C.P. stannous chloride crystals ($SnCl_2 \cdot 2H_2O$) in 2.33 ml. of concentrated hydrochloric acid, as indicated in the section on Reagents. This solution, cooled to about 40°, was poured into 100 ml. of water and the volume made up to 200 ml. Reagents of other stannous chloride concentrations were prepared by diluting this 0.50 M (1.0 N) solution. In the cases of the trials with 0.137 M stannous chloride in the colorimetric media, it was necessary to filter the stannous chloride reagent in order to remove all turbidity.

In making a series of trials, 25 ml. of the solution containing 10 mg. of gold per liter were added to each of twelve 50-ml. Nessler tubes. After adding the gold solution, the required volume of stannous chloride was added with a 1-ml. graduated pipette or with a 50-ml. burette, the burette being used when the volume of reagent was greater than 1 ml. The solution was then made up to the mark of the tube with distilled water, the tube closed with a rubber stopper that had been washed in chlorine water, and inverted twice, these operations being completed as rapidly as possible. After standing for fifteen minutes, comparisons of the relative color intensities were made as indicated in the section on Comparison of Colors. The results are given in Table 5.

With the exception of Trials 4, 5, and 6, which were tan to purple-brown, all trials were yellow. The probable reason for the low results of Trials 1, 2, and 3 was that the amount of stannous chloride was not sufficient to combine with all of the gold and excess oxidizing agents.

Table 5 proves that the shade of color and intensity of color are fairly constant over a wide range of stannous chloride concentrations (approximately 0.0014 M to 0.14 M).

Effect of Gold Concentration on Intensity of Color. The effect of gold concentration on intensity of color was next determined, the solutions being prepared as follows:

The required volume of gold bromide solution, prepared by diluting the reagent containing 500 mg. of gold per liter, was added to each 50-ml. Nessler tube, and sufficient distilled water to give a volume of 40 to 45 ml. was then added. The tubes were closed with a rubber stopper which had been washed in chlorine water, and inverted twice. Using a burette, 0.61 ml. of the 0.25 M stannous chloride reagent was then added to each tube, corresponding to a stannous chloride concentration in the colorimetric medium of 0.003 M and a hydrochloric acid concentration of 0.001 M. The volume was then rapidly made up to the mark with distilled water, the tube stoppered and inverted three times. After ten minutes' standing, the colorimetric comparisons were begun, the gold concentrations of all solutions, including the standards, being diluted to 2 mg. per liter. (Please see Comparison of Colors.) Approximately twenty-five minutes, from the time of starting the trials, were required to complete the tests of each series.

Table 5

Effect of Stannous Chloride Concentration on Determination of Gold in a Solution Containing 2 Mg. of Gold Per Liter

Trial No.	Molarity of Stannous Chloride in Colorimetric Medium	Ml. of Stannous Chloride Reagent Used	Molar Conc of Stannous Chloride Reagent	Apparent Recovery of Gold[a] %	Percent Deviation From Mean
1	0.00014	0.27	0.025	35.6	0.7
2	0.00014	0.27	0.025	29.4	5.5
3	0.00014	0.27	0.025	39.7	4.8
Mean	0.00014	0.27	0.025	34.9	3.7
4	0.00055	1.10	0.025	128	0.3
5	0.00055	1.10	0.025	117	10.7
6	0.00055	1.10	0.025	138	10.3
Mean	0.00055	1.10	0.025	127.7	7.1
7	0.00137	2.74	0.025	118	5.3
8	0.00137	2.74	0.025	105	7.7
9	0.00137	2.74	0.025	115	2.3
Mean	0.00137	2.74	0.025	112.7	5.1
10	0.00275	5.50	0.025	100[a]	4
11	0.00275	5.50	0.025	114	10
12	0.00275	5.50	0.025	98	6
Mean	0.00275	5.50	0.025	104	6.7
13	0.00275	5.50	0.025	100[a]	4.7
14	0.00275	5.50	0.025	102	2.7
15	0.00275	5.50	0.025	112	7.3
Mean	0.00275	5.50	0.025	104.7	4.9
16	0.0275	5.50	0.25	98	3.7
17	0.0275	5.50	0.25	100	1.7
18	0.0275	5.50	0.25	107	5.3
Mean	0.0275	5.50	0.25	101.7	3.6
19	0.055	11.0	0.25	107	2
20	0.055	11.0	0.25	105	0
21	0.055	11.0	0.25	103	2
Mean	0.055	11.0	0.25	105	1.3
22	0.137	13.7	0.50	100	9.3
23	0.137	13.7	0.50	88	2.7
24	0.137	13.7	0.50	78	12.7
Mean	0.137	13.7	0.50	90.7	8.2

a. Trial 10 taken as the standard for Trials 1 to 12, inclusive; Trial 13 taken as the standard for Trials 13 to 24, inclusive.

Table 6 indicates that by using the low-acid stannous chloride test, the gold concentration can be estimated within about 12 percent, even though the standard has a concentration very different from that of the unknown.

Table 6

Effect of Gold Concentration on Intensity of Color

Test No.	Mg. of Gold Added Per Liter of Colorimetric Medium	Mg. of Gold Found Per Liter of Colorimetric Medium	Percent Error
1	2.00	1.92	-4.0
2	2.00	1.75	-12.5
3	2.00	1.70	-15.0
Mean	2.00	1.79	-10.5
4	5.00	4.35	-13.0
5	5.00	4.75	-5.0
6	5.00	4.72	-5.6
Mean	5.00	4.61	-7.9
7	10.0	10.0[a]	0.0
8	10.0	10.5	5.0
9	10.0	9.7	-3.0
Mean	10.0	10.1	0.7
10	20.0	21.4	7.0
11	20.0	20.6	3.0
12	20.0	22.0	10.0
Mean	20.0	21.3	6.7
13	10.0	10.0[a]	0.0
14	10.0	9.8	-2.0
15	10.0	10.5	5.0
Mean	10.0	10.1	1.0
16	50.0	58.0	16.0
17	50.0	53.0	6.0
18	50.0	57.5	15.0
Mean	50.0	56.2	12.3
19	100	106	6.0
20	100	117	17.0
21	100	110	10.0
Mean	100	111	11.0

a. No. 7 was taken as the standard for Tests 1 to 12, inclusive; No. 13 was the standard for Tests 14 to 21, inclusive.

Colorimetric Determination of Gold in Synthetic Leaching Solutions. The tests with the low-acid stannous chloride test having proved its superiority to the standard test in which the acid concentration is both very high and uncontrolled, experiments were made with a synthetic leaching solution. A stock solution was first prepared which had the composition:

1. NaCN --------- 10 g./L. as free NaCN.
2. Fe ----------- 0.5 g./L. as $K_4Fe(CN)_6$.
3. Cu ----------- 0.5 g./L. as $Na_2Cu(CN)_3$.

The sodium cuprocyanide was prepared by adding sodium cyanide to copper sulfate solution until the precipitate which first formed had completely dissolved.

For each trial 50 ml. of the above stock solution, 500 ml. of saturated lime water, varying amounts of gold cyanide solution (see Reagents), and distilled water to make a total volume of one liter, were used. The concentrations of the various constituents were therefore:

1. Free NaCN ------------------ 0.5 g./L.
2. Fe ------------------------- 0.025 g./L.
3. Cu ------------------------- 0.025 g./L.
4. CaO ------------------------ 0.6 g./L.

A solution of this composition is typical of those encountered in cyanidation practice (5).

Trials were first made by the method given by Bugbee (8). (Please see Introduction of this section for exact quotation of directions given by Bugbee.) The directions given by Bugbee are substantially identical with those given by Snell (9), Treadwell (10), Yoe (11), Scott and Furman (14), and many others.[4] On evaporating the aqua regia solution of the zinc-lead-gold precipitate, the volume was reduced to 2 to 4 ml.; it was not found possible to control the volume to any greater degree than this. Before addition of the saturated stannous chloride reagent, the volume was made up to 4 to 5 ml. with distilled water. Our results are given in Table 7.

The results of Table 7 show that all of the solutions, regardless of the gold content, have practically the same shade or tint of color when the directions of Bugbee (8) are followed. In this connection it will be recalled (please see Introduction to this section) that the alleged variation in tint of color with variation in the gold concentration is the basis of the colorimetric determination as given in the standard analytical texts. If differences in the shade of color are obtained when this gold test is made, these differences must be ascribed to variations in the heretofore uncontrolled acid concentrations rather than to changes

4. The directions generally call for three "drops" of lead acetate and two "drops" of stannous chloride. The size of a drop, as is well known, depends on the surface tension of the solution and the size of the capillary; we found that a 0.6 mm. external diameter capillary gave 100 drops of distilled water per ml., while a 6 mm. glass tube gave eleven drops per ml. In following the directions, 0.04 ml. was taken as equal to one drop.

in the gold concentrations. Bugbee's method gives no indication of the actual gold content, and we therefore believe it is fair to say that the Purple of Cassius test, as known heretofore, has little or no quantitative significance.

Table 7

Experiments with the Purple of Cassius Gold Test

Trial No.	Dwt. of Gold per Metric Ton of Cyanide Solution	Mg. of Gold per Liter of Cyanide Solution	Mg. of Gold per Liter of Colorimetric Medium[a]	Tint of Color Found	Tint of Color According to Bugbee
1	0.02	0.0311	6	Blue	Very slight coloration
2	0.02	0.0311	6	Blue	Very slight coloration
3	0.03	0.0467	9	Violet	Slight yellow
4	0.03	0.0467	9	Tan[b]	Slight yellow
5	0.04	0.0622	12	Purple	Pinkish yellow
6	0.04	0.0622	12	Violet	Pinkish yellow
7	0.06	0.0934	19	Purple	Strong pink
8	0.06	0.0934	19	Purple	Strong pink
9	0.08	0.124	25	Purple	Purple
10	0.08	0.124	25	Purple	Purple

a. Assuming that all of the gold was recovered in the precipitation step. The concentration ratio was 200:1 if the volume in the test tube was 5 ml.

b. On evaporation of the aqua regia solution, the volume was reduced to about 1 ml.

Colorimetric Determination of Gold with the Low-Acid Stannous Chloride Test. We next tried our modification of the stannous chloride test, in which the acid concentration is controlled. Moreover, we altered the precipitation conditions slightly in order to insure complete precipitation of the gold. Our procedure was that of Bugbee (8), modified as indicated below:

One liter of the cyanide solution was poured into a 4-liter beaker, 0.12 ml. of saturated lead acetate solution was added, and the precipitate which formed dissolved by stirring. To the solution was added 0.75 gram of zinc dust, and the solution swirled. 20 ml. of saturated sodium cyanide and 20 ml. of concentrated ammonia solution (spec. grav. 0.90)[5] were added and the solution swirled for 5 minutes.[6] All but about 50 ml. of the solution was

5. We have experimentally found that ammonia promotes the precipitation of gold by zinc dust when copper is present (Section D). When copper is present, gold is precipitated with difficulty by zinc dust, and this problem, frequently encountered in cyanide practice (15), has not yet been satisfactorily solved.

6. The time of swirling the solution was increased from 30 seconds to 5 minutes in order to insure complete precipitation of the gold.

decanted from the precipitate, which was transferred to a 250-ml. beaker and then to a funnel containing an 11 cm. No. 2 Whatman filter paper, the precipitate being washed to the apex of the filter paper. After thoroughly washing the precipitate with distilled water, dilute nitric acid (please see Reagents) was cautiously added until all of the precipitate, with the exception of the gold particles, had dissolved. The filter paper was washed with distilled water, the black particles being washed to the apex of the filter paper. A 100-ml. beaker was then placed under the funnel, and 10 ml. of saturated bromine water passed through the filter paper, 1 ml. at a time, over a period of 15 minutes. During the dissolution of the gold, the funnel was kept covered with a watch glass in order to prevent loss of bromine. The filter was then washed with 8 ml. of water, and the filtrate heated to about 45° C on a hot plate.[7] The beaker was removed from the hot plate and allowed to stand for 40 minutes, during which time most of the free bromine escaped from the solution. The solutions were then made up to volumes of 18 ml., placed in 50-ml. Nessler tubes, and 2 ml. of 0.25 M stannous chloride reagent added to each solution. Immediately after the addition of the stannous chloride, the tubes were closed with a rubber stopper that had been washed in chlorine water, and the tubes inverted twice. The gold standards, prepared by diluting the stock solution of gold bromide to 1.73 mg. per liter, were treated in a similar manner. After standing for more than 10 minutes, the relative color intensities of the tubes were determined by the method indicated under Comparison of Colors.

The colors, before dilution, varied from light yellow to light brown. In no instance was a purple or blue color obtained.

Table 8 shows that by controlling the acidity, the stannous chloride test can be placed on an at least semi-quantitative basis. There are no cupellation losses with our method; and special apparatus, such as muffle furnaces and assay balances, are dispensed with. The errors in weighing with an assay balance are generally of the order of 0.01 mg. (16), corresponding to a 25 percent error with 0.04 mg., and the substitution of filtration for decantation (17), probably reduces the errors due to loss of gold when separating the gold particles from the solution. When the amount of gold to be determined is less than 0.04 mg., the low acid-stannous chloride test is probably superior to the gravimetric assay procedure with respect to accuracy and speed.

SUMMARY[8] OF SECTION B:

As the result of our experimental study of the stannous chloride test for gold, we find the following new facts:

1. When the acid concentration is maintained below about 0.05 N,

7. Low results are frequently obtained when the solutions are heated above 50° C.

8. Our references to Professor Bugbee's book should not be construed as indicating a desire on our part to be critical of his work. We consider Professor Bugbee's text to be the best in its field.

Table 8

Determination of Gold in Cyanide Solutions by Means of the Low-Acid Stannous Chloride Test

Trial No.	Dwt. of Gold Added per Metric Ton of Cyanide Solution	Mg. of Gold Added per Liter of Cyanide Solution	Dwt. of Gold Found per Ton[a]	Percent Error
1	0.00	0.00	0.0013	
2	0.00	0.00	0.000	
Mean	0.00	0.00	0.001	
3	0.02	0.0311	0.017	15
4	0.02	0.0311	0.014	30
Mean	0.02	0.0311	0.0155	22.5
5	0.03	0.0467	0.030	0
6	0.03	0.0467	0.032	6.6
Mean	0.03	0.0467	0.031	3.3
7	0.04	0.0622	0.050	25
8	0.04	0.0622	0.043	7.5
Mean	0.04	0.0622	0.047	16.3
9	0.06	0.0934	0.061	1.7
10	0.06	0.0934	0.069	15
Mean	0.06	0.0934	0.065	8.4
11	0.08	0.124	0.071	11
12	0.08	0.124	0.071	11
Mean	0.08	0.124	0.071	11

a. Corrected for blank (Trials 1 and 2).

the tint of color which is obtained with the stannous chloride test for gold is not purple but is yellow to brown.

2. The tint of color which is obtained with the stannous chloride test for gold is primarily a function of the acid concentration.

3. The gold concentration of a solution cannot be determined from the tint of color obtained by the Purple of Cassius test, at least by visual observation.

4. When a yellow or brown solution, prepared by the stannous chloride method and containing gold in the colloidal state, is treated with sufficient hydrochloric acid to bring the acid concentration to 2 N to 6 N, the solution changes from the relatively stable yellow form to the unstable purple form. With gold

concentrations of 2 to 20 mg. per liter, there is a distinct time interval during which the solution is colorless.

5. The time required for the development of the color to full intensity is less at low concentrations than at high concentrations of hydrochloric acid, and the test is therefore probably more reproducible at low concentrations of hydrochloric acid.

6. The addition of sodium chloride to 0.002 N to 0.029 N stannous chloride, in amount sufficient to bring the sodium chloride concentration to 0.01 N to 0.50 N, greatly increases the tendency for the solutions to become turbid, and this may be due to precipitation of electronegative colloids from the solutions. We conclude that the salt concentrations should be reduced to the lowest practicable values when making the stannous chloride test.

7. Over the range of 0.0014 M to 0.14 M stannous chloride, the tint and intensity of color with the low-acid stannous chloride test are not appreciably affected by change in the stannous chloride concentration.

8. Gold may be more accurately and more precisely determined by our low-acid stannous chloride test than by the widely used and heretofore generally accepted high-acid stannous chloride test.

C. CHLORINATION IN THE PRESENCE OF CHLORIDES

METHODS:

Introduction. Having developed an improved method for the colorimetric determination of gold (Section B), we made a study of several gold solvents. Three general methods for determining the efficiencies[1] of gold solvents were considered:

1. Determination of the loss in weight of uniform gold particles when subjected to the solvents under uniform conditions.
2. Determination of the amount of gold dissolved by the solvent by analysis of the solvent.
3. Determination of the time required for the solvent to dissolve gold particles having uniform compositions and dimensions.

Regarding Method 1, it would be necessary to use a micro-balance for the work, as the cost of the gold would otherwise be prohibitive. The time requirement for each test would be large, and the work would therefore be limited in scope. During each test extreme care would have to be taken to have adequate and uniform circulation, and yet avoid all contamination of the solvent;[2] temperature and other variables, including the hydrogen sulfide content of the air that is bubbled through the solution, would have to be controlled. Because of these facts, we did not use Method 1.

Method 2 does not necessarily involve the use of an assay balance, and when the colorimetric method of Section B is employed, the time required per test is not great. This method was applied when we made the first tests with halogen solvents (Table 11).

It was not long, however, before a variation of Method 3 was adopted, which was considered superior to either Method 1 or 2 for our purpose. By far the greater proportion of the results of this dissertation were obtained by measuring the time required for the solvent to dissolve gold leaf.

Apparently the gold leaf solution test was first used in 1857 by M. Faraday (18), who wrote as follows concerning the action of potassium cyanide on gold leaf:

1. We define the *efficiency* of a solvent as being inversely proportional to the time required for the solvent to completely dissolve uniform gold particles under the experimental or industrial conditions. Although the word *activity* was formerly used in this connection, the term has since been adopted by physical chemists engaged in thermodynamic studies, and consequently our use of the term would be ambiguous.

2. In Section G, Figure 7, we show that 1 mg. of sulfide ion per liter increases by a factor of 6 the time required for a cyanide solution to dissolve gold leaf, and the deleterious action of such minute traces of sulfide ion has heretofore been unknown. Consequently the use of sulfur-containing rubber in contact with the alkaline cyanide solutions would not be permissible. Some investigators, not aware of the deleterious action of such minute traces of sulfur, give imposing tables of 4-place data on the rates at which gold dissolves in cyanide solutions, but say nothing about the presence of rubber tubes in the apparatus, or about precautions taken to avoid the introduction of hydrogen sulfide.

> Air-voltaic circles are formed in these cases, and the gold is dissolved almost exclusively under their influence. When one piece of gold leaf was placed on the surface of a solution of cyanide of potassium, and another, moistened on both sides, was placed under the surface, both dissolved; but twelve minutes sufficed for the solution of the first, whilst about twelve hours were required for the submerged piece. In weaker solutions, and with silver also, the same results were obtained: from 60 to 100 fold as much time being required for the disappearance of the submerged metal as for that which floating, was in contact both with air and solvent. (Our underlining.)

Clearly, the experiments of Faraday, made thirty years before the discoveries of MacArthur, Forrest, and Forrest (19), anticipated their work, and this fact should be recognized.

Reagents. All reagents were of C.P. grade. The methods of preparing and standardizing many of the reagents are given in Section B. Concentrations of sodium cyanide were determined by the standard method involving titration with silver nitrate, using potassium iodide as the indicator (20). The bromine reagents were standardized by pipetting 5 ml. into an excess of potassium iodide solution and titrating with sodium thiosulfate (12).

Regarding the sodium cyanide reagent, it was necessary to make certain that the amount of sulfur in the form of sulfide ion was less than the amount (0.1 mg. per liter -- please see Section G) which has a deleterious action on the dissolving of gold. Experiments as follows were made:

a. To 100 ml. of a solution containing 42.5 grams of sodium cyanide per liter were added 0.10 g. of crystalline sodium sulfide ($Na_2S \cdot 9H_2O$) and 0.50 g. of lead nitrate ($Pb(NO_3)_2$). The solution was then neutralized to a faint pink with phenolphthalein, or a pH of about 8, and distilled into a 0.1 M sodium carbonate solution. The distillate had a concentration of 14.2 grams of sodium cyanide per liter, and did not give a sulfide test with lead nitrate (21).

b. One hundred ml. of a solution containing 42.5 grams of sodium cyanide per liter were distilled from lead nitrate, as in (a) above, but the addition of sodium sulfide was omitted. The distillate contained 26.2 grams of sodium cyanide per liter.

All solutions were diluted to 1.0 grams of sodium cyanide per liter with 0.05 M sodium carbonate-0.05 M sodium bicarbonate solution. The time periods required for the reagents to dissolve gold leaf were then determined in duplicate experiments, all trials being made simultaneously. The results with the various reagents were as follows:

1. Solution prepared from the solution containing 42.5 grams of sodium cyanide per liter (no distillation or addition of lead nitrate); 180 and 200 seconds.
2. Solution prepared from the sodium sulfide-lead nitrate-sodium cyanide distillate; 210 and 200 seconds.

3. Solution prepared from the lead nitrate-sodium cyanide distillate; 180 and 190 seconds.

No. 2 of the above experiments shows that the lead nitrate treatment removes all of the sulfide ion, while No. 3 shows that the reagent from which all traces of sulfide ion had been removed had the same efficiency as the unpurified chemical (No. 1). Moreover, colorimetric tests for sulfur (21) showed that the laboratory chemical contained less than 4 mg. of sulfide ion per kg. of sodium cyanide. Accordingly, a reagent containing 1.0 gram of sodium cyanide per liter would have less than 0.004 mg. of sulfur, in the form of sulfide, per liter; and this is below the range at which sulfide ion is deleterious.

Regarding the gold leaf, the material which we used was the 23-carat, XX grade product manufactured by the W. H. Kemp Company, 350 Hudson Street, New York, N. Y. 23-carat gold alloys have a nominal composition of 95.8 percent gold. On dissolving a leaf, weighing 15.2 mg., in aqua regia and diluting, a clear solution was obtained, the test being made under conditions that would have detected 0.5 percent of silver in the gold leaf. The principal impurity was shown to be copper by addition of ammonia. The mean of separate determinations of the weights and areas of three gold leaves gave a weight of 0.206 mg. per sq. cm.; and if we assume that the specific gravity is the mean (18.9) of that indicated by a composition of 95.8 percent gold (spec. grav. 19.3) and 4.2 percent copper (spec. grav. 8.9), the thickness may be calculated to be 11 millionths of a cm., corresponding to 0.11 microns or 4 millionths of an inch.

Reproducibility of the Gold Leaf Solution Test. In order to determine the reproducibility of the gold leaf solution test, uniform squares, 3 x 3 mm., were prepared by placing a gold leaf between two sheets of paper and cutting with a scissors. Several No. 3 cork stoppers, which had not been used previously, were then wet with water and placed in a covered 250-ml. beaker. By removing the paper from one side of the gold leaf, and placing the gold leaf in contact with the bottom (smaller plane face) of the moistened cork, the gold leaf adhered and retained its original square shape. All corks with adhering gold leaf were kept in a covered 250-ml. beaker until just before use, as it is very important that the corks do not dry. To each of five 10-ml. test tubes were added 5 ml. of the reagent being tested, and the test tubes were stoppered with corks having adhering squares of gold leaf. All test tubes were then vigorously shaken by hand at a uniform rate of about 3 vibrations per second. The time of solution was taken as the time at which the particles of dissolving gold leaf became invisible.

Regarding the air supply, a large excess was present over that required to dissolve the gold leaf; a square of leaf 0.5 cm. on edge[3] weighs about 0.05 mg. and therefore requires about 0.002 mg. of oxygen for its solution according to the equation (22)

$$O_2 + 4\ Au + 8\ NaCN + 2\ H_2O \longrightarrow 4\ NaAu(CN)_2 + 4\ NaOH.$$

Five ml. of air corresponds to 1.2 mg. of oxygen, or a 600-fold excess of oxygen. Yet there is less possibility of the introduction of hydrogen sulfide into the solution than would be the case if a vigorous stream of air were passed into the solution.

In Table 9 are given some results on the reproducibility of the gold leaf solution test, five trials being made simultaneously.

Table 9

Reproducibility of the Gold Leaf Solution Test

Trial No.	Composition of Solution	Seconds Time Required to Dissolve Gold Leaf	Percent Deviation from Mean
1	0.5 g./L. NaCN	210	12.5
2	0.5 g./L. NaCN	270	12.5
3	0.5 g./L. NaCN	260	8.3
4	0.5 g./L. NaCN	240	0
5	0.5 g./L. NaCN	255	6.2
6	0.5 g./L. NaCN	225	6.4
7	0.5 g./L. NaCN	240	0
8	0.5 g./L. NaCN	270	12.5
9	0.5 g./L. NaCN	220	8.3
10	0.5 g./L. NaCN	225	6.2
Mean	0.5 g./L. NaCN	240.5	7.3
11	3.5 g./L. Br_2	165	12.7
12	3.5 g./L. Br_2	180	4.8
13	3.5 g./L. Br_2	225	19.0
14	3.5 g./L. Br_2	210	11.1
15	3.5 g./L. Br_3	225	19.0
16	3.5 g./L. Br_2	170	10.0
17	3.5 g./L. Br_2	195	3.2
18	3.5 g./L. Br_2	165	12.7
19	3.5 g./L. Br_2	180	4.8
20	3.5 g./L. Br_2	195	3.2
Mean	3.5 g./L. Br_2	189	10.1

The results of Table 9 show that under the experimental conditions the average deviation of the results from the mean is about 10 percent. Further discussion of this table is given at the conclusion of Table 10.

3. This is taken as the basis for calculation because we used squares of this size in many of the tests which will be described later.

We then made tests in order to determine whether the cork had any effect on the time required for the solvents to dissolve gold leaf. Three 60-ml. glass-stoppered bottles were used, and comparative tests made by replacing the glass stoppers with corks. Twenty-five ml. of reagent were used per test, other details being similar to those of Table 9. The results are given in Table 10.

Table 10

Effect of Cork on Time Required for the Dissolution of Gold Leaf

Trial No.	Seconds Time Required for Dissolution of Gold Leaf with			
	0.50 g./L. NaCN Solution		1.7 g./L. Cl_2 Solution	
	Glass Stopper	Cork Stopper	Glass Stopper	Cork Stopper
1	200	220	80	78
2	235	210	70	84
3	210	240	81	77
Mean	215	223	77	80

The results of Tables 9 and 10 show clearly that fairly reproducible results can be obtained by the gold leaf solution test. Among the several outstanding advantages of the test are the following:

1. No foreign materials other than cork and glass come in contact with the solution. In this respect the test simulates cyanidation practice, in which the cyanide solutions are in contact with silicates and wooden vessels, as well as with the gold (5).

2. Under similar conditions, glass-stoppered bottles give practically the same results as do cork-stoppered bottles (Table 10), and within the limits of experimental error the cork is without effect.

3. Vigorous agitation is employed, as is generally the case in cyanidation practice (4).

4. A series of 5 to 12 trials can be made at one time under identical conditions, and the relative efficiencies of the solvents for dissolving gold leaf can be established within 10 to 20 percent.

5. The gold is in a uniform condition in all tests, and, like nearly all native gold (5), contains copper.

6. The cost of the gold is not prohibitive. At the present time a book of 25 leaves (8.6 x 8.6 cm.) costs about $1.25.[4]

7. The time required per test is less than with any other method. One can therefore study a larger number of solvents and obtain a more complete picture of the essential variables than would otherwise be possible.

4. Eimer and Amend Company, New York, N. Y.

Because of its desirable characteristics, the test was used in nearly all of our subsequent work. Among the well-known and established facts concerning cyanidation which we have confirmed by the gold leaf test, and which we have reported in this dissertation, may be mentioned:

1. Calcium hydroxide retards the dissolution of gold in cyanide solutions (23). See Tables 15 and 24. As far as we know, there are no adequate explanations for the behavior of calcium hydroxide, as much larger concentrations of sodium hydroxide and of calcium sulfate have practically no deleterious effect (23).

2. In the presence of calcium hydroxide, soluble lead compounds cause further retardation of the dissolution of gold (24). See Tables 14 and 16. As far as we know, this fact has not been explained from a theoretical standpoint.

3. The presence of lead in solutions of pH approximately 10 is not deleterious and is often beneficial (25). See Tables 13 and 19. We believe that our interpretation represents the first satisfactory explanation of this fact (Section D).

4. Traces of mercury salts accelerate the dissolution of gold in cyanide solutions (26). See Tables 18 and 19. While our results agree with those of Dowsett and John (26), we do not agree with their interpretation of the results.

5. Sulfide ion, when present to the extent of more than 5 mg. per liter, retards the dissolution of gold (65). See Tables 27 and 28.

6. Potassium ferricyanide accelerates the dissolution of gold in cyanide solutions (28). See Tables 24 and 25.

7. Copper ammonia cyanide is an efficient solvent for gold (29). See Table 20.

With the gold leaf method it is possible to make several hundred tests per day, and in our first experiments we studied practically every known solvent for gold. The results which we present in this dissertation are based almost entirely on findings which are new or which bear on contradictory information recorded in the literature. We believe, however, that the seven well-known facts in the cyanidation of gold, listed above and described later in connection with new material, establish the usefulness and reliability of the gold leaf test, which has never, as far as we know, given results which are in qualitative disagreement with any established fact.

REVIEW OF THE CHLORINATION PROCESS:

Fifty years ago, when chlorine was prepared by chemical instead of electrochemical action and was worth 35 cents instead of 1.75 cents per lb. (30), gold ores were treated on an extensive scale by roasting and subsequent leaching with aqueous solutions of chlorine (31). Recoveries of 95 to 98 percent, with chlorine consumptions of 2 lbs. per ton of ore, were typical (31), the gold-bearing leaching solutions being reduced with ferrous sulfate or hydrogen sulfide, and the precious metal content recovered by sedimentation. Later, with the introduction of the cyanide process, discovered by Elsner (32) and Faraday (18), and advocated by MacArthur, Forrest, and Forrest (19), the chlorination process was soon displaced, abandoned, and forgotten. We believed, however, that developments since 1888 merited a reconsideration of the chlorination process. Cyanidation does have serious disadvantages, such as:

1. Oxidized copper minerals form stable complexes which in lime solutions are valueless as gold solvents. Because of the high reagent consumption, ores containing as little as 0.2 percent of oxidized copper are frequently not amenable to cyanidation (5).

2. Since a basic leach solution is required, all iron sulfates and free acids must be neutralized with lime. The cost of the lime is often an important item (4).

3. The gold particles dissolve very slowly. In order to secure a reasonable recovery, the ores must, in some instances, be leached for 3 to 10 days in order to secure reasonable recovery. Mechanical agitation is often used as a means of accelerating the dissolution of the gold particles (4, 5).

4. The sodium cyanide consumed in the process represents a major cost item (4). At the present time the price of sodium cyanide, in cents per lb., is about eight times that of chlorine (30).

Because of these factors we believed the chlorine process to be worthy of investigation. Our experiments soon indicated that chlorides accelerate the dissolution of pure gold and silver-free gold alloys, and we believe this is a new discovery. Although chloride solutions have been extensively applied to the leaching of silver and silver-gold ores, the processes were based entirely on the well-known solubility of silver chloride in concentrated chloride solutions (33). It was also known that gold would dissolve more rapidly in the presence of "nascent chlorine" (34), and the "nascent chlorine solutions generally contained chloride ion; the chlorine was generated in the presence of the ore by chemical reaction (33). Nevertheless, as far as we know, no one has ever shown that the addition of chlorides to a chlorine solution will decrease the time required for gold particles to dissolve in the solvent.

The prior art of the chlorination of gold ores is summarized in several of the older metallurgical treatises, including those of Schnabel (35), Aaron (36), Eissler (34), Greenawalt (33), and Kustel (37). Pertinent quotations from these works are as follows:

Greenawalt, page 374:

> ... and if the ore contains silver ... the chlorine may be used in connection with a chloride solution, as is frequently done in extracting the precious metals by the Plattner process.

Schnabel, page 870:

> With the object of extracting gold as well as silver, Roëzner and Paterä have suggested the use of brine saturated with chlorine, but this method has not come into common use.

Schnabel, page 996:

Chlorination proceeds especially well when the chlorine is generated in revolving vessels themselves from chloride of lime and sulfuric acid, as it has a powerful chloridizing action on gold in the native state.

Eissler, page 237:

The Von Pateras and Roeszner process, which is based on the solubility of chloride of silver in concentrated salt solution....

Eissler, page 241:

Gold is more readily attacked by nascent chlorine than by chlorine held in solution in water.

Kustel, page 130:

The object of this [Patera and Roeszner] process is the extraction of gold and silver together.

Efforts to locate the original communications of Roeszner and Patera have not been successful, no mention of them or citations to the literature being found which would assist in locating the original articles. The discrepancies in spelling the names Roeszner and Patera are given as they occur in the texts. It is possible that the latter inventors never published an account of their work.

We have made as thorough a search of the literature, including the patent literature, as time would permit and have found no record of any experiments which would indicate or prove that the addition of a chloride or a bromide to a solution of the free halogen will accelerate the corrosion of gold or of silver-free gold alloys.

THE ACTION OF VARIOUS SOLVENTS ON PURE GOLD:[5]

The action of various solvents on pure gold was first studied, using 24-carat gold foil, 0.1 mm. thick, obtained from Eimer and Amend Company, New York City. The purity of 24-carat gold is generally greater than 99.9 percent (39). The sodium cyanide reagents were prepared by diluting a stock solution of C.P. sodium cyanide, the solution being analyzed within 48 hours before each test by the standard method involving titration with silver nitrate, using a potassium iodide indicator (40). The chlorine reagents were prepared and standardized by the method given in Reagents, Section B (12). The bromine reagents were likewise standardized by reacting with an excess of potassium iodide and titrating with sodium thiosulfate, using a starch indicator.

The sea water was taken near Bay Shore, Long Island, during an incoming tide; this location was chosen because the sea water had no evidence of pollution, unlike other coastal waters in the vicinity of New York City. The salt concentration was found by evaporation to be 29.43 grams per liter, which indicates that the sea water was diluted to a slight extent with fresh water.

5. These results were not obtained by the gold leaf method, as we developed the gold leaf method after the results of Table 11 were completed. Nearly all of the data given in Table 11, however, were later verified by the gold leaf method (Tables 12 and 13).

Table 11

Milligrams of Gold Dissolved in Five Minutes Per Square Centimeter of Total Surface Area of Gold Foil

Trial No.	Reagent	Mg. of Gold Dissolved per Sq.Cm. Surface Area in Five Minutes
	Sodium Cyanide Solutions	
1	2.5 G. of NaCN per Liter	0.21
2	0.5 G. of NaCN per Liter	0.23
3	0.1 G. of NaCN per Liter	0.03
	Chlorinated Sea Water	
4	0.059 G. of Cl_2 per Liter	0.013
5	0.059 G. of Cl_2 per Liter + 2.5 G. of H_2SO_4 per Liter	0.65
6	0.059 G. of Cl_2 per Liter + 7.0 G. of H_2SO_4 per Liter	0.69
7	0.118 G. of Cl_2 per Liter	0.58
8	0.118 G. of Cl_2 per Liter + 2.5 G. of H_2SO_4 per Liter	1.35
9	1.18 G. of Cl_2 per Liter	8.98
10	1.18 G. of Cl_2 per Liter + 7.0 G. of H_2SO_4 per Liter	9.07
	Miscellaneous Chlorine Solutions (Distilled Water)	
11	0.437 G. of Cl_2 per Liter	0.32
12	0.437 G. of Cl_2 per Liter + 0.5 G. of H_2SO_4 per Liter	0.76
13	0.437 G. of Cl_2 per Liter + 8 G. of H_2SO_4 per Liter	2.66
14	0.437 G. of Cl_2 per Liter + 10 G. of H_2SO_4 per Liter	0.31
15	2.0 G. of Cl_2 per Liter	1.68
16	2.0 G. of Cl_2 per Liter + 0.5 G. of Br_2 per Liter	1.62
17	2.0 G. of Cl_2 per Liter + 2.0 G. of $CaCl_2$ per Liter	3.38
18	2.0 G. of Cl_2 per Liter + 8.0 G. of HCl per Liter	16.8
19	2.0 G. of Cl_2 per Liter + 20 G. of Na_2SO_4 per Liter	1.64
20	2.0 G. of Cl_2 per Liter + 20 G. of NaF per Liter	1.75
	Bromine Solutions	
21	30 G. of Br_2 per Liter	9.18
22	30 G. of Br_2 per Liter + 20 G. of NaCl per Liter	80
23	30 G. of Br_2 per Liter + 25 G. of KBr per Liter	185

Before making the tests, the gold foil was thoroughly cleaned with pumice soap, washed with water, heated to a red heat, cooled and cut into squares 2 x 2 mm., all squares having a constant area within about 10 percent. For each trial of Table 11 one or more squares, which had not been used in previous tests, were placed in a 10-ml. test tube with 5 ml. of the solvent under investigation, and the test tube corked and shaken by hand until about 0.05 mg. of gold had gone into solution, the relative efficiencies of the solvents being approximately known from preliminary tests. Five to ten trials were made at one time, and the rate of shaking was about two vibrations per second. At the conclusion of the test, the solvent was decanted from the gold foil, and gold content determined by the low acid-stannous chloride test of Section B.

The data of Table 11 were interesting from these standpoints:

1. The results indicated that under the experimental conditions chloride ion greatly accelerated the dissolution of gold in chlorine and bromine solutions.

2. Sodium fluoride and sodium sulfate did not accelerate the dissolution of gold in chlorine solutions.

3. Except in very dilute solutions, where hypochlorite formation is an important factor,[6] sulfuric acid did not accelerate the dissolution of the gold.

4. The dissolution of the gold, under the experimental conditions, proceeded much more rapidly with the halogen solvents than with the cyanide solvents.

The values obtained with sodium cyanide agree within 20 percent with the values given by Barsky and co-workers (23). Further discussion is reserved for the conclusion of this section.

Table 12

Time Required for Sodium Chloride-Chlorine Solutions to Dissolve Gold Leaf

Grams of Sodium Chloride per Liter of Solvent	Seconds Time Required to Dissolve Gold Leaf With Solvents Having Chlorine Concentrations of 3.46 G./L.	1.73 G./L.	0.691 G./L.	0.346 G./L.	0.173 G./L.
0.000	26	80			
1.09	22	--	420		
2.19	17	40	---	650	
4.58	10	20	55	232	
8.77	7	15	39	105	480
17.5	6	12	30	70	180
35.1	5	9	24	65	105

THE ACTION OF VARIOUS SOLVENTS ON GOLD LEAF:

We next tried the gold leaf solution test, the procedure being similar to that used when determining the reproducibility of the test (Tables 9 and 10). The only change that was made was that the squares of gold leaf were 5 x 5 mm. instead of 3 x 3 mm. As our

6. The equilibrium constant of the reaction $Cl + H_2O \rightarrow H^+ + Cl^+ + HClO$ is 4.84×10^{-4} at 25° C (41).

principal object was to determine the effect of the sodium chloride, the trials of each series had identical chlorine concentrations, the sodium chloride concentration being varied. The three to seven trials of each series were made simultaneously. The results are given in Table 12 and in Figures 1 and 2, and are similar to the results obtained in other series.

Discussion of Table 12 is reserved for the conclusion of this section.

In order to confirm Trials 21 and 22 of Table 11, in which it was shown that sodium chloride accelerates the dissolution of gold in bromine water, several experiments were made by the gold leaf method (Table 13). The procedure was similar to that used in determining the reproducibility of the gold leaf test, the squares of gold leaf being 5 x 5 mm. All solvents had bromine concentrations of 2.0 grams per liter.

Table 13

Effect of Sodium Chloride on the Time Required for the Dissolution of Gold Leaf in Bromine Solution

Trial No.	Concentration of Sodium Chloride in Grams per Liter	Seconds Time Required for Solution of Gold Leaf
1	0	1120
2	5	870
3	10	625
4	30	310
5	100	80

DISCUSSION OF RESULTS:

The results obtained by two entirely different methods are in substantial agreement (Tables 11, 12, and 13), and prove quite conclusively that the addition of sodium chloride to chlorine or bromine solutions accelerates the dissolution of pure gold and of gold leaf, while sulfates and fluorides (Table 11) have practically no effect. This effect has not been recognized heretofore.

Among the possible explanations for the action of chloride ion may be mentioned:

1. Formation of a Stable Auric Ion. If the accelerating action were due to the formation of a stable auric complex, we would expect that sulfate and fluoride ions, which generally form more stable complexes than do chloride ions, would be effective in promoting the dissolution of gold. Unfortunately, lack of information as to the stability of the sulfate and fluoride auric complexes prevents us from checking this point, but we have seen (Table 11) that sulfate and fluoride ions have little or no effect.

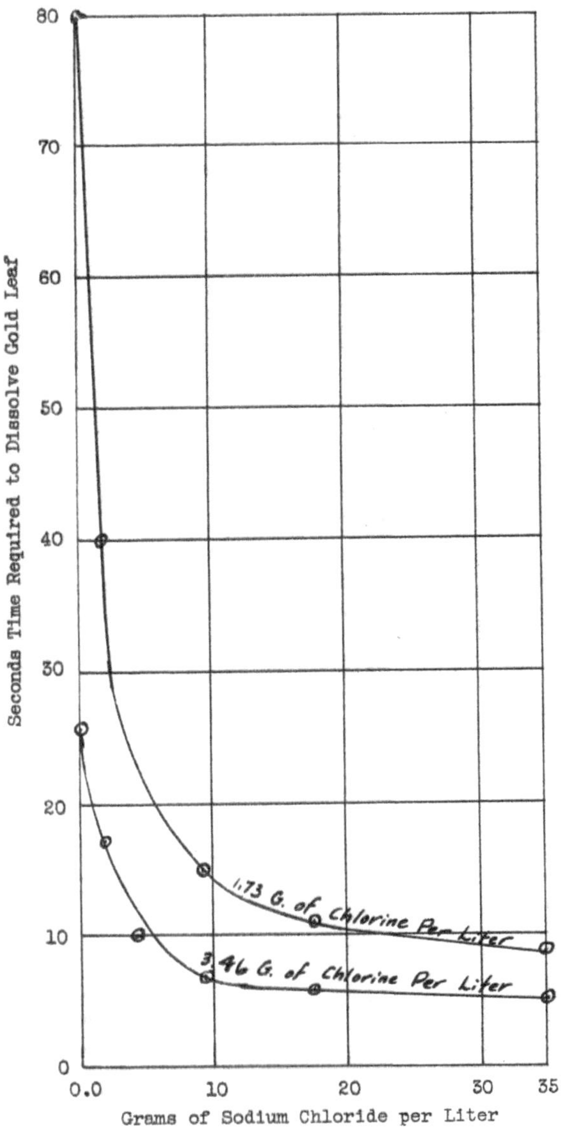

Fig. 1. Effect of Sodium Chloride on Time Required for Chlorine Solutions to Dissolve Gold Leaf

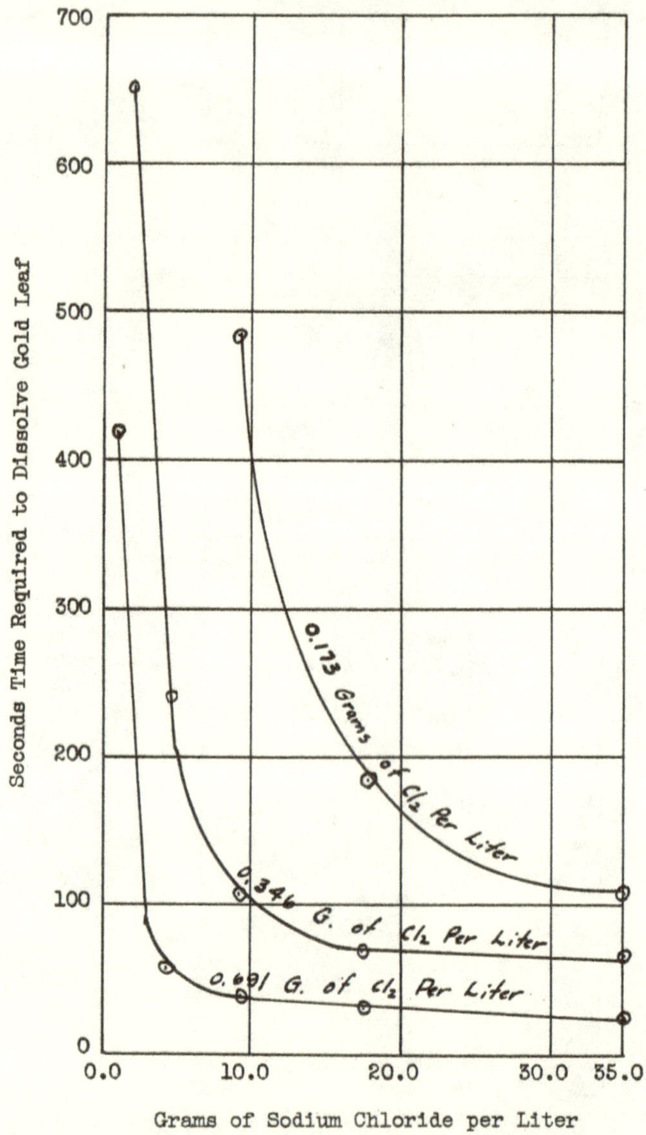

Fig. 2. Effect of Sodium Chloride on Time Required for Chlorine Solutions to Dissolve Gold Leaf

However, it may be shown in another way that the formation of a stable auric complex does not explain the behavior of chloride ion. If the formation of auric ion were responsible, moderate concentrations of chloride should be nearly as effective as high concentrations. The reaction is (42): 2 Au + 3 Cl_2 + 2 NaCl --> 2 $NaAuCl_4$, and the thermodynamics are such that only very negligible concentrations of free chlorine and chloride ion are in equilibrium with the auric complex, as is apparent from the electrode potentials (42)

$$4\ Cl^- + Au = AuCl_4^- + 3\ e^-;\ E^\circ = -1.00\ v.$$

$$2\ Cl^- = Cl_2 + 2\ e^-;\ E^\circ = -1.358\ v.$$

and dissociation constant (42)

$$AuCl_4^- = Au^{+++} + 4\ Cl^-;\ K = 5 \times 10^{-22}.$$

Accordingly, if the molar ratio NaCl:Cl_2 were 2:3, sufficient sodium chloride would be present in the film to combine with all of the chlorine to form the complex auric ion, and the maximum effect should be obtained with the 2:3 ratio. But the trials of Table 12 prove that this is not the case with any of the five chlorine concentrations tested. The molar ratio, NaCl:Cl_2, of 2:3 corresponds to a weight ratio of 0.55:1, and when the weight concentration of sodium chloride exceeds 0.55 of the chlorine weight concentration, the further addition of sodium chloride should have no effect. Reference to Table 12, however, shows that with every one of the chlorine concentrations tested, the addition of sodium chloride beyond the weight ratio of 0.55:1 progressively decreases the time required to dissolve the gold leaf.

While it is agreed that a reaction will not take place unless there is a decrease in free energy, we do not believe that the magnitude of the decrease in free energy has much to do with the time required to reach equilibrium. Getman and Daniels write (43):

> It may be stated at the start that there is no apparent relation between reaction rate and chemical affinity. For example, the polymerization of nitrogen dioxide to nitrogen tetroxide is a mild reaction involving only 14,600 calories per mol, and yet it takes place so rapidly that its rate has not yet been measured. On the other hand, a violent reaction like the union of hydrogen and oxygen to give a mol of water, involving a free energy decrease of 56,560 calories, proceeds with immeasurably slow velocity at room temperature.... The equilibrium constant tells nothing of the absolute velocities.

2. Suppression of Hypochlorite Formation. If this hypothesis were correct, the addition of chlorides to bromine solutions should have no effect on the dissolution of gold (Table 13 and trials 21 and 22 of Table 11), and sulfuric acid should be as

effective as sodium chloride (Trials 5, 6, 8, 10, and 14 of Table 11). As Table 11 shows, however, such is not the case except in dilute solutions in sea water, and this exception may be accounted for by the well-known buffer capacity of sea water (44). For these reasons we reject the hypothesis of hypochlorite formation.

3. The Solvent Action of Sodium Chloride for Aurous Chloride and Bromide. If we assume that the first reaction between chlorine and gold is

$$2\ Au + Cl_2 \rightarrow 2\ AuCl,$$

and that the over-all rate of reaction is controlled by the rate at which the insoluble aurous chloride film is removed from the gold surface by the reactions

$$AuCl + Cl_2 + Cl^- \rightarrow AuCl_4^-, \text{ and}$$

$$AuCl + Cl^- \rightarrow AuCl_2^-,$$

it is believed that the results are satisfactorily explained. Thus Lengfeld (45) has shown that aurous chloride is insoluble in water but is soluble in sodium chloride. The $AuCl_2^-$ is later oxidized to $AuCl_4^-$, but this is not a rate-controlling reaction. As Ostwald (46) said: "In the last analysis, diffusion determines the velocity of chemical change."

CONCLUSIONS OF SECTION C:

Our results are believed to warrant the following conclusions:

1. It is experimentally shown for the first time that the dissolution of pure gold and of gold leaf in chlorine water is promoted by relatively high concentrations of sodium chloride.

2. It is experimentally shown for the first time that the dissolution of pure gold and of gold leaf in bromine solutions is promoted by relatively high concentrations of sodium chloride.

3. Chlorinated sea water is an excellent solvent for pure gold.

4. The results are best explained on the basis that the primary reaction product is insoluble, and that the rate of removal of this reaction product from the gold surface is the rate-controlling factor in the reaction.

D. CONTACT AGENTS IN THE CYANIDE PROCESS

INTRODUCTION:

Although cyanide solvents have been used extensively for the recovery of gold from ores and concentrates, the normal rate of solution of the precious metal as determined by Barsky[1] (23) and others (5, 28) and checked by ourselves,[2] is less than about 3 mg. per sq. cm. per hour (23), corresponding to a corrosion depth of approximately 1.5 microns (0.00006 inches) per hour. Although this slow rate may be increased to some extent in certain cases, as, for example, when the gold is partially embedded in iron pyrites (47), the leaching operation is almost invariably the "bottle-neck" of the cyanide process.

In the operation of cyanide mills it is frequently necessary for the leaching solvent to remain in contact with the ores for 50 to 200 hours in order to dissolve all of the gold particles (4, 5). Not only does this slow solution rate increase the capital costs, but the cyanide consumption is also increased due to vaporization of hydrocyanic acid and secondary reactions with minerals such as malachite ($CuCO_3 \cdot Cu(OH)_2$), pyrrhotite (FeS), and stibnite (Sb_2S_3), which very frequently are associated with the gold. Accordingly, it is not surprising that numerous attempts have been made to accelerate the dissolution of gold in the ores. One of the more successful procedures which we discovered in our preliminary tests was the addition of salts of "contact" metals to the cyanide solvent. Among these metal salts found by us to decrease the time required to dissolve gold leaf are the salts of lead, bismuth, and thallium. The literature on this subject may be summarized as follows:

1. Lead Salts: According to Hayden (25), the use of lead compounds in large-scale operations at the Morro (Brazil) gold mine has permitted increased gold recoveries due to the buffering action of lead hydroxide and control thereby of the pH. But Hayden overlooked the valuable contact agent effect. Beyers (28) found that lead compounds render gold passive when the potassium cyanide

1. Using highly refined methods and apparatus in which temperature and flow of solution past the surface of the noble metal were kept constant, Barsky and co-workers (23) have proven that the rate of solution of gold and its silver alloys in pure sodium cyanide is controlled by the rate of diffusion of oxygen to the metal-solution interface. Barsky also proved that the rate of solution of gold in sodium cyanide is constant over long periods of time, and that lime decreases the rate of solution. Other workers (27) have shown that sodium carbonate, di-sodium phosphate, and sodium hydroxide have a negligible effect on the rate of solution.

2. Over the range of 0.25 to 5.0 grams of sodium cyanide per liter, the rate of solution of gold is very nearly constant, if the pH of the solution be maintained above 10 (23, 20), and calcium hydroxide and other deleterious compounds are absent. We have verified this with the gold leaf test.

concentration is below 0.08 percent. And Leaver, Woolf and Jackson (24) have reported that lead salts are very deleterious in lime-sodium cyanide solvents. Our own research soon disclosed that these earlier investigators were not aware of the true function of the lead salts and the best operating conditions.

2. Mercury Salts. The use of 5 to 10 mg. per liter of mercuric salts as oxidizing agents in the cyanide process was recently patented by Dowsett and John (26). The specification and the unpublished communications given in the file wrapper of the patent state that a mercuric oxycyanide is formed which oxidizes the gold and is thereby converted to mercurous cyanide. When mercuric chloride is added in excess of 50 to 100 mg. per liter, the gold particles become covered with mercury and are then protected from attack by the cyanide solution (26). Mercury salts then retard rather than accelerate the solution of gold. On the other hand, Keith [cited by Beyers (28)] found that mercury salts to the extent of 50 mg. of mercury per liter of solution increased the rate of dissolution of gold. Keith (1890) believed that the mercury acted as an oxygen carrier by being continuously re-dissolved by the cyanide solution and precipitated by the gold.

3. Bismuth and Thallium Salts. As far as we have been able to determine, no one has ever shown that these salts will accelerate the dissolution of gold under any conditions.

We looked into this contact effect more carefully and record below our observations on the behavior of various metal salts on the time required to dissolve gold leaf. While others have reported at best a negligible increase in the solution rate of gold, we find on the contrary that there are several metal salts, in concentrations of 1 to 40 mg. per liter, which decrease the time required for cyanide solutions to dissolve gold leaf to 0.5 to 0.032 of the time required in the absence of these metal salts. In the interpretation of our results we present an electrochemical theory which, we believe, satisfactorily accounts for the observations made.

LEAD SALTS AS CONTACT AGENTS:

As indicated in Section C above (Tables 9 and 10), a very simple and relatively precise method for comparing the effectiveness of gold solvents is to measure the time required to dissolve gold leaf, a method employed by Michael Faraday in 1857. It was shown that sulfide ion was absent from our cyanide solutions, and precautions necessary in making the test were outlined. It was pointed out that gold leaf, like native gold, contains small amounts of copper and silver, and that the results are therefore comparable with what one might expect in cyaniding gold ores. The procedure used in the tests of this section was similar to that of Section C, the gold leaf squares being 0.5 cm. on edge.

Effect of Lead Nitrate Concentration. A solution was prepared which contained, per liter, 1.0 grams of sodium cyanide, 1.06

grams of sodium carbonate, and 0.84 grams of sodium bicarbonate. The pH of this strongly buffered[3] solution is 9.9 (62). The results are recorded in Table 14 and are plotted in Figure 3.

Table 14

Lead Nitrate as a Contact Agent in Buffered Sodium Cyanide Solution

Mg. of Lead Nitrate, $Pb(NO_3)_2$, per Liter of Cyanide Solution - - -	50	10	5	1	0.5	0.1	0.0
Seconds Time Required to Dissolve Gold Leaf -	25	30	35	70	98	120	180

Referring to Table 14, the amount of cyanide solution used in each test was 5 ml. and the weight of gold leaf per test was 0.051 mg. as determined by direct weighing of a leaf 8.6 x 8.6 cm. In the second to the last test of Table 14 we used 0.1 mg. of lead nitrate per liter of cyanide solution, and the time required to dissolve the gold leaf was reduced more than 30 percent. In the 5 ml. of cyanide solution used in the test there were 0.0005 mg. of lead nitrate. Accordingly, 0.0005 mg. of lead nitrate will accelerate the dissolution of 0.051 mg., or over 100 times its weight of gold. The observed acceleration cannot be attributed to the oxidizing power of the lead ion. We shall discuss this more fully below.

Effects of Calcium Hydroxide and of Sodium Carbonate on the Contact Effect of Lead Nitrate. A solution of 1.0 gram of sodium cyanide in one liter of saturated lime water was prepared. The pH of this solution may be calculated to be about 12.7 (63). The results obtained when lead nitrate was added to this saturated solution of calcium hydroxide were as follows:

Table 15

Effect of Lead Nitrate on the Dissolution Velocity of Gold Leaf in Saturated Calcium Hydroxide Solutions of Sodium Cyanide

Mg. of Lead Nitrate per Liter of Solution - - - - - -	500	100	10	0
Seconds Time Required to Dissolve Gold Leaf - - - -	1800	1260	800	800

3. The Elsner equation (32) for the dissolution of gold in sodium cyanide is:

$$4\ Au + 2\ H_2O + O_2 + 8\ KCN \longrightarrow 4\ KAu(CN)_2 + 4\ KOH.$$

Consequently, the pH of an unbuffered cyanide solution dissolving gold does not remain constant. The sodium carbonates have no appreciable effect on the rate of dissolution of gold (27).

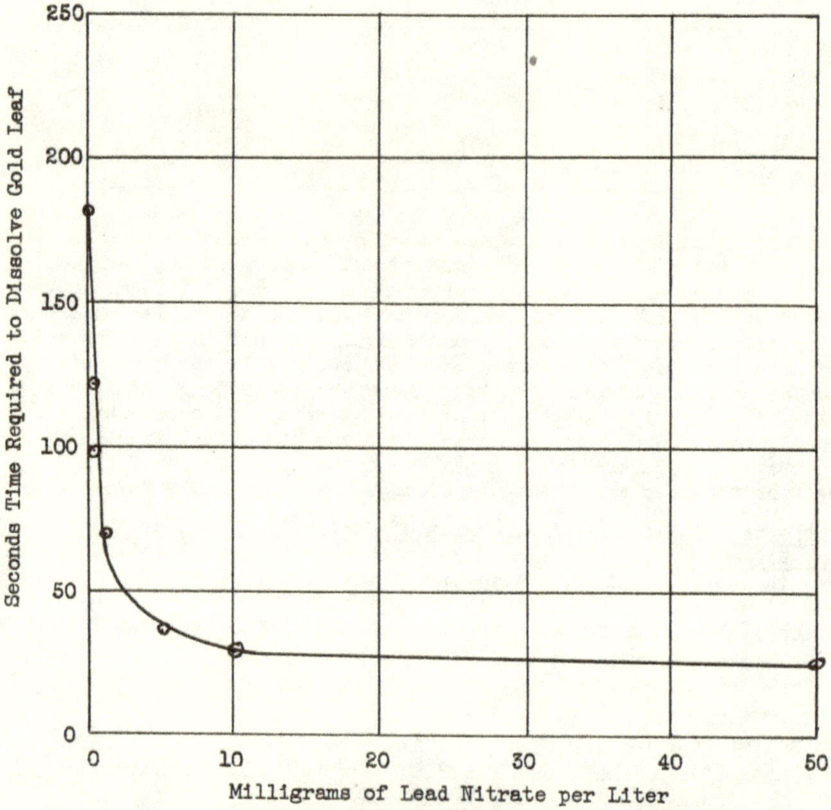

Fig. 3. Effect of Lead Nitrate on Time Required to Dissolve Gold Leaf in Sodium Carbonate Solutions Containing 1.0 Gram of Sodium Cyanide per Liter

We find (Table 15) that in lime-water solutions the addition of lead nitrate does not accelerate the dissolution of gold leaf. Quite the contrary, the lead compound increases the inhibiting effect of calcium hydroxide (23). This is in accord with the reports of Leaver, Woolf, and Jackson (24) concerning the deleterious effect of soluble lead compounds in lime solutions.

The effect of increasing the sodium cyanide concentration in a saturated lime-water solution containing 10 mg. of lead nitrate per liter was next studied. (Table 16).

Table 16

Effect of Sodium Cyanide Concentration on Velocity of Gold Leaf in Saturated Lime-Water Solutions Containing 10 Mg. of Lead Nitrate per Liter

Grams of Sodium Cyanide per Liter of Solution - - - -	1.0	2.0	5.0
Seconds Time Required to Dissolve Gold Leaf - - - -	780	420	165

It is evident (Table 16) that the higher the cyanide concentration, the faster the gold leaf dissolves.

The experiments presented in Table 17 were made under conditions similar to those of Table 16, with the exception that the cyanide solutions were 50 percent saturated with respect to calcium hydroxide. The results of Table 16 and 17 are plotted in Figure 4.

Table 17

Effect of Sodium Cyanide Concentration on the Dissolution of Gold Leaf in 50 Percent Saturated Lime-Water Solutions Containing 10 Mg. of Lead Nitrate per Liter

Grams of Sodium Cyanide per Liter of Solution - - - -	1.0	2.0	5.0
Seconds Time Required to Dissolve Gold Leaf - - - -	640	240	90

Table 17 also proves that when the sodium cyanide concentration is increased in the lime-water solution containing lead nitrate, the time required to dissolve the gold leaf is decreased.

Effect of Sodium Cyanide Concentration on Dissolution of Gold Leaf in Sodium Carbonate Solutions. Tests were next made with a solution of somewhat lower pH (Table 18). All solutions were 0.02 M

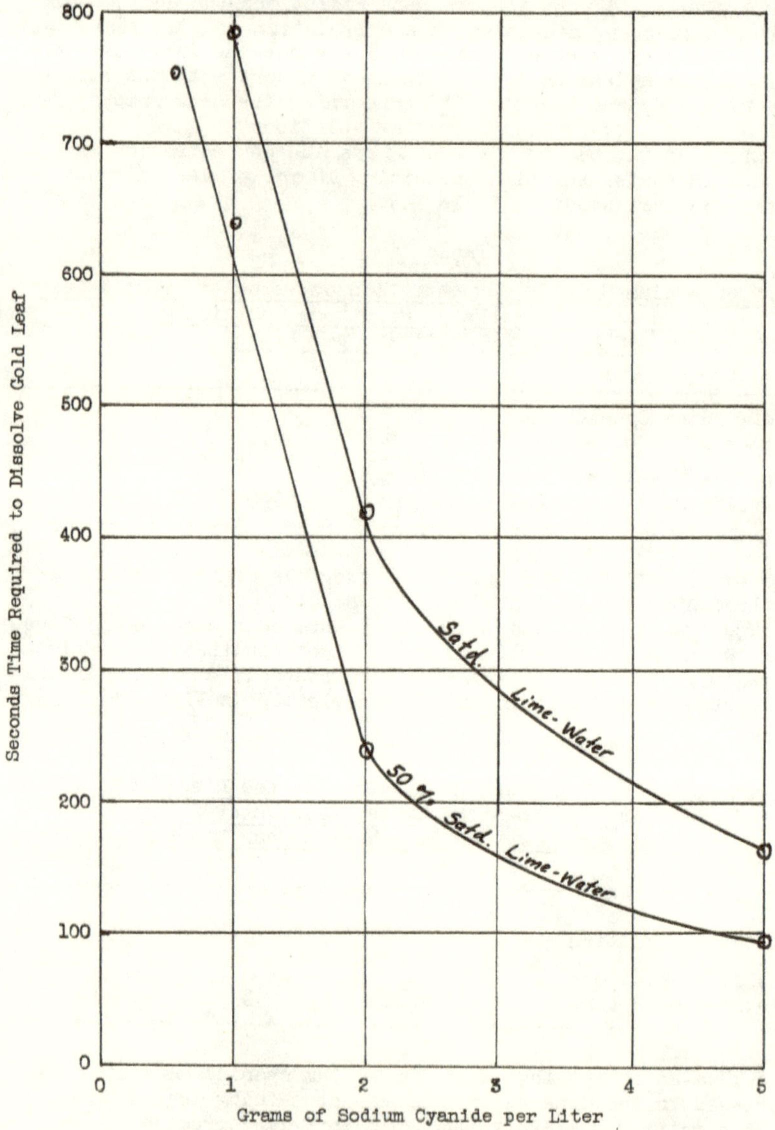

Fig. 4. Effect of Sodium Cyanide Concentration on Time Required to Dissolve Gold Leaf in Lime-Water Containing 10 Mg. of Lead Nitrate per Liter

(2.12 G./L.) with respect to sodium carbonate, and to alternate trials 10 mg. of lead nitrate were added per liter of solution. The pH of 0.02 M sodium carbonate is approximately 11.5 (62).

Table 18

Lead Nitrate as a Contact Agent: Effect of Cyanide Concentration on the Dissolution of Gold Leaf in Sodium Carbonate - Lead Nitrate Solutions

Composition of Solution	Seconds Time Required To Dissolve Gold Leaf
0.2 G. of NaCN per Liter - - - - - - - - - - - - - - -	220
0.2 G. of NaCN + 10 Mg. of Pb $(NO_3)_2$ per Liter - - - -	130
0.5 G. of NaCN per Liter - - - - - - - - - - - - - - -	210
0.5 G. of NaCN + 10 Mg. of $Pb(NO_3)_2$ per Liter - - - -	100
1.0 G. of NaCN per Liter - - - - - - - - - - - - - - -	200
1.0 G. of NaCN + 10 Mg. of $Pb(NO_3)_2$ per Liter - - - -	50

Table 18 shows that when lead nitrate is absent, change in the sodium cyanide concentration has little effect on the dissolution of gold leaf, and this is in accord with the classic experiments of MacLaurin (20). In the presence of lead nitrate, however, the sodium cyanide concentration has a very marked effect on the time required to dissolve gold leaf. Our theory of the action of lead nitrate and other contact agents satisfactorily accounts for this extraordinary behavior (see below) of sodium cyanide.

Effect of the pH of the Cyanide Solution. Although the accelerating effect of lead compounds had been definitely established in the case of sodium carbonate-sodium cyanide solutions, the results with the lime-water cyanide solutions (Tables 15 and 16) required further study to account for the absence of the accelerating effect due to lead. It may be calculated that the solutions of Tables 15 and 16 had a pH ranging between 12.3 and 12.6 (63), and we suspected that the high pH was to blame. Tests with sodium hydroxide in place of lime-water supported this suspicion; lead nitrate had no accelerating effect in sodium hydroxide-sodium cyanide solutions. We established an upper limit in pH of about 12 beyond which there was a retardation instead of an acceleration in the dissolution of gold leaf in solutions containing 1.0 gram of sodium cyanide per liter. The more dilute (less than 2 grams per liter) solutions of sodium cyanide are the only solutions which are of interest from an industrial viewpoint.

With this upper limit of a pH of 12 in mind, we made a series of experiments with strongly buffered cyanide solutions of lower pH, the values being estimated from tables compiled by Kolthoff and Rosenblum (62). That the buffering agents used have no effect on the dissolution of gold has been shown by other workers (23, 27, 28, 48) and has been confirmed by us with the gold leaf solution method. For reasons that follow,[4] the use of buffering agents was chosen in preference to other methods:

1. Measurements with the antimony electrode are vitiated by the presence of traces of metals which the antimony is capable of displacing. As little as 0.1 mg. of copper per liter, for example, may be a source of great error (59).

2. In the absence of buffering agents, the pH of a cyanide solution dissolving gold rises as the gold dissolves, in accordance with the Elsner equation:

$$4\ Au + 8\ KCN + O_2 + 2\ H_2O \rightarrow 4\ KAu(CN)_2 + 4\ KOH.$$

3. Because of the high pH of our solutions and the presence of sodium salts, we did not believe the glass electrode method to be applicable for our purpose. (Please see footnote on page 51). The glass electrode is also very inaccurate in concentrated salt solutions of low pH (60, 61).

Table 19

Dissolution of Gold Leaf in Buffered Cyanide Containing Lead Nitrate and Mercuric Chloride

Concentrations, in Grams per Liter, of Salts in Buffered Solutions Containing 1.0 Gram of Sodium Cyanide per Liter	Estimated pH (62)	Seconds Time Required to Dissolve Gold Leaf in the Presence of 10 Mg. of $Pb(NO_3)_2$ per Liter	Seconds Time Required to Dissolve Gold Leaf in the Presence of 10 Mg. of $HgCl_2$ per Liter
10.6 g. Na_2CO_3 + 8.4 g. $NaHCO_3$	9.9	55	80
3.71 g. H_3BO_3 + 2.97 g. Na_2CO_3	9.0	85	60
4.95 g. H_3BO_3 + 1.06 g. Na_2CO_3	8.5	140	140
12.4 g. H_3BO_3 + 1.06 g. Na_2CO_3	8.0	130	140
21.4 g. Na_2HPO_4 + 2.76 g. KH_2PO_4	7.6	270	90

Table 19 indicates that lead is effective over the pH range of 8.0 to more than 9.9. Without lead nitrate additions, as recorded in Tables 15 and 19, the dissolution time was 180 to 220 seconds. Tables 15, 16, and 17 show that the upper pH limit with 1.0 gram of sodium cyanide per liter is about 12.

4. For further discussion of the difficulties involved in determining the pH of cyanide solutions, please see Section E.

MERCURY COMPOUNDS AS CONTACT AGENTS:

When 5 mg. of mercury, as mercurous nitrate or mercuric chloride, were added to a liter of a solution containing 1 gram of sodium cyanide, gold leaf dissolved in less than 60 seconds as against 180 seconds without mercury. The presence of 40 grams of sodium hydroxide per liter, 20 grams of sodium carbonate per liter, or of 1.7 grams of calcium hydroxide per liter did not prevent mercury compounds from accelerating the dissolution of gold leaf. The effective pH range for mercury was found to be 7.6 (Table 20) to more than 14, and the effective concentration of mercury was found by us to be from less than 0.1 mg. to more than 50 mg. per liter.

Dowsett and John (26) have reported an accelerating effect due to mercury compounds; but Dowsett and John were not interested in the pH of the solutions and give an entirely different interpretation of their results. Our interpretation is given at the conclusion of this section. (Please see Theory of Contact Agents.)

BISMUTH AND THALLIUM COMPOUNDS AS CONTACT AGENTS:

Salts of bismuth and thallium were likewise found to have a marked accelerating effect. The results of Table 20 were obtained with a solution containing 1.0 gram of sodium cyanide per liter.

Table 20

Salts of Bismuth and Thallium as Contact Agents

Mg. of Salt Added per Liter of Solution	Seconds Time Required to Dissolve Gold Leaf in Satd. $Ca(OH)_2$	Seconds Time Required to Dissolve Gold Leaf in 0.2 M Na_2CO_3
No salts of bismuth or thallium	780	210
TlCl, 40 Mg./L.	90	150
TlCl, 300 Mg./L.	60	45
$TlCl_3$, 40 Mg./L.	60	45
$TlCl_3$, 300 Mg./L.	25	45
$Bi(NO_3)_3$, 30 Mg./L.	50	90
$Bi(NO_3)_3$, 300 Mg./L.	50	90

Table 20 indicates that bismuth and thallium are effective contact agents or cyanide accelerators.

It goes without saying that many other metal salts have been investigated for their contact effect. The salts of only four metals, lead, bismuth, mercury, and thallium have been found effective.

THEORY OF CONTACT AGENTS:

Julian and co-workers (48) and Christy (47) have shown that gold particles which are embedded in iron pyrites dissolve at faster rates than do gold particles which are not in contact with iron pyrites. This behavior is attributed to air depolarization over the entire surface of the pyrites-gold couple (47), and this observation suggested to us that a similar action may be involved with cyanide accelerators. Christy (47) has shown that lead, bismuth, and mercury are more noble than gold in 0.065 percent potassium cyanide solutions. We believe that the four contact metals, lead, bismuth, mercury, and thallium, are displaced from solution by the gold, thereby facilitating oxygen depolarization and increasing the solution rate of the gold in accordance with the well-known equation which is commonly attributed to Elsner (32):

$$O_2 + 4\ Au + 8\ KCN + 2\ H_2O \rightarrow 4\ KAu(CN)2 + 4\ KOH. \quad (1)$$

The action involved with our four contact metals or cyanide accelerators is analogous to that of copper sulfate in promoting the dissolution of electrolytic zinc in hydrochloric acid. Instead of diffusion of hydrogen away from the metal surface, diffusion of atmospheric oxygen to the gold surface is involved.

That our interpretation is plausible may be shown by considering the changes in electrode potentials which take place when the hydrogen ion and cyanide ion concentrations are varied. The discussion which follows is based on data compiled by Latimer (42):

A. Equilibrium Constants

$$HCN \rightarrow H^+ + CN^-;\ K = 4 \times 10^{-10}$$

$$H_2O \rightarrow H^+ + OH^-;\ K = 1 \times 10^{-14}$$

$$Au(CN)_2^- \rightarrow Au^+ + 2\ CN^-;\ K = 5 \times 10^{-39}$$

$$PbO\ (red) + H_2O \rightarrow Pb + 2\ OH^-;\ K.S.P. = 5.5 \times 10^{-6}$$

$$Pb\ (OH)_2 \rightarrow Pb^{++} + 2\ OH^-;\ K.S.P. = 2.8 \times 10^{-16}$$

B. Electrode Potentials

$$2\ OH^- + Pb \rightarrow PbO\ (red) + H_2O + 2\ e^-;\ E = 0.578.$$

$$2\ CN^- + Au \rightarrow Au(CN)_2 + e^-;\ E = 0.60.$$

$$3\ OH^- + Bi \rightarrow BiOOH + H_2O + 3\ e^-;\ E = 0.46.$$

$$OH^- + Tl \rightarrow TlOH + e^-;\ E = 0.3445.$$

$$4\ CN^- + Hg \rightarrow Hg(CN)_4^{--} + 2\ e^-;\ E = 0.37.$$

First we shall consider the effect of pH on the cyanide ion concentration. The hydrolysis reaction is:

$$Na^+ + CN^- + H_2O \rightarrow Na^+ + OH^- + HCN. \quad (2)$$

or, omitting the sodium ion,

$$CN^- + H_2O \rightarrow OH^- + HCN. \quad (3)$$

From Latimer:

$$\frac{(H^+)(CN^-)}{HCN} = 4 \times 10^{-10}, \text{ and} \quad (4)$$

$$(H^+)(OH^-) = 10^{-14} \quad (5)$$

Dividing (4) by (5):

$$\frac{(CN^-)}{(HCN)(OH^-)} = 4 \times 10^{4} \quad (6)$$

From equation (6):

If pH = 14, $CN^-/HCN = 3 \times 10^4$.

If pH = 13, $CN^-/HCN = 4 \times 10^3$.

If pH = 10, $CN^-/HCN = 4$.

If pH = 9, $CN^-/HCN = 0.4$.

If pH = 8, $CN^-/HCN = 0.04$.

From the above equations we see that the cyanide ion concentration depends on the pH. Moreover, these calculations demonstrate why it is that the addition of lime to sodium cyanide solutions prevents loss of the cyanide due to vaporization of hydrocyanic acid; and the well-known fact that cyanide solutions of pH below about 8 have little dissolving power for gold is theoretically substantiated (5).

We shall now calculate the electrode potentials of gold and of lead in cyanide solvents, in order to see whether it is theoretically possible for gold to displace lead ion in such solvents.

The aurous ion concentration of a solution which is 1.0 M with respect to aurocyanide ion and 1.0 M with respect to cyanide ion will be:

$$\frac{(Au^+)(CN^-)^2}{Au(CN)_2} = 5 \times 10^{-39}. \quad (7)$$

$$Au^+ = 5 \times 10^{-39} \quad (8)$$

In actual practice, however, the solutions will be more dilute. We shall assume that the sodium cyanide concentration is 0.02 M (1 gram per liter) instead of 1.0 M, and that the aurocyanide concentration is 5×10^{-5} M (0.01 gram of gold per liter) instead of 1 M. Under these conditions,

$$\frac{(Au^+)(2 \times 10^{-2})^2}{5 \times 10^{-5}} = 5 \times 10^{-39}, \text{ and} \quad (9)$$

$$Au^+ = 6.25 \times 10^{-40}. \quad (10)$$

According to Latimer, the electrode potential of metallic gold in a solution in which both the cyanide ion and aurocyanide ion are at unit activity is 0.60 volt, so that in 0.02 M sodium cyanide solutions having a pH above 10 and a gold concentration of 5 x 10 M, the electrode potential of gold is:

$$E = 0.60 + \frac{RT}{nF} \ln \frac{5 \times 10^{-39}}{6.25 \times 10^{-40}} = 0.60 + 0.05 = 0.65 \text{ v.} \quad (11)$$

We shall now calculate the electrode potential of lead, assuming that the red modification of lead oxide is the stable phase. Lead hydroxide is the lead compound in equilibrium with our cyanide solutions, but as the data cited by Latimer show (see above) there is very little difference between the solubility products of lead hydroxide and the red modification of the lead oxide. The error involved in assuming that the stable phase is lead oxide will be less than 0.02 v., if the solubility product data are correct. As for the formation of cyanide complexes with lead ion, we can find no evidence either from the literature or from our own experiments that lead ion is capable of forming a stable complex cyanide ion, and we shall therefore assume that the presence of sodium cyanide (0.02 M) causes no change in the concentration of lead ion. Moreover, we shall assume that the lead ion concentration is governed by the hydroxyl ion concentration and the solubility product of lead oxide.

According to the data cited by Latimer, the electrode potential of lead in a solution saturated with lead oxide (red) and containing hydroxyl ion at unit activity is 0.578 v. (see above). The lead ion concentration in such a solution will be:

$$Pb^{++} = 5.5 \times 10^{-16}. \quad (12)$$

In a solution of pH = 13, the lead ion concentration will be:

$$Pb^{++} = \frac{5.5 \times 10^{-16}}{(0.1)^2} = 5.5 \times 10^{-14}, \quad (13)$$

and the electrode potential of lead will be:

$$E = 0.573 - \frac{RT}{nF} \ln \frac{5.5 \times 10^{-14}}{5.5 \times 10^{-16}} = 0.52. \quad (14)$$

Accordingly, at a pH of 13 and in a 0.02 M sodium cyanide solution saturated with lead oxide (red) and 5×10^{-5} M with respect to aurocyanide ion, the electrode potential of gold is 0.65 - 0.52 = 0.13 volt *more base* than lead, and in such solutions metallic gold should displace lead ion, *if the lead metal, as produced by the reaction, is in a massive, strain-free condition*.

Now let us see what happens when the pH is reduced to 10. When this is done, the lead ion concentration becomes:

$$Pb^{++} = \frac{5.5 \times 10^{-16}}{(10\text{-}4)^2} = 5.5 \times 10^{-8}, \quad (15)$$

corresponding to an electrode potential, for lead, of:

$$E = 0.573 - \frac{RT}{nF} \ln \frac{5.5 \times 10^{-8}}{5.5 \times 10^{-16}} = 0.341. \quad (16)$$

At a pH of 10, the driving force which tends to displace lead ion is therefore increased from 0.13 volt (pH = 13) to 0.65 - 0.34 = 0.31 volt. It seems very probable that the metallic lead produced by the reaction of gold with lead ion has an electrode potential which is more electropositive (Lewis and Randall notation) than massive, annealed lead, and that unless the driving force of the reaction is at least 0.10 volt, no displacement of the lead ions by the gold can take place. However that may be, the preceding calculations prove that it is theoretically possible for gold to displace lead ion and that lead should be more effective at a pH of 10 than at a higher pH, in direct confirmation of our experimental data. Similar reasoning applies in the case of mercury, bismuth, and thallium salts, although in these cases the tendency for gold to displace the metal ion is much greater, as shown by the electrode potentials cited by Latimer, and the effect of pH is comparatively unimportant, as shown by our data (Tables 19 and 20).

The above explanation gives us insight into the seemingly anomalous behavior of lead salts. The gold electrode potential is made less noble by increase in the cyanide ion concentration (Equation 7), and the ability of the gold to displace the lead ions is thereby increased. As we have seen (Tables 16 and 17), lead nitrate, even in the presence of calcium hydroxide (pH = 12.3 to 12.7), becomes effective when the sodium cyanide concentration is increased. Moreover, this hypothesis shows why it is that lead salts are not effective in dilute cyanide solutions of high pH (Tables 15, 16, and 17). As the hydroxyl ion concentration increases, the lead ion concentration decreases; metallic gold can no longer displace lead ion from solution; and lead salts then have no accelerating effect on the solution of metallic gold in cyanide solvents.

SUMMARY OF SECTION D:

A large number of metal salts were tested for their effect on the time required for sodium cyanide solutions to dissolve gold leaf.

Salts of the four metals, lead, bismuth, mercury, and thallium were found to greatly decrease the time required for sodium cyanide solutions to dissolve gold leaf. In some instances less than 1 part per million (1 mg. per liter) of metal salt had a definite effect.

This beneficial effect is apparently limited to the four metals. In the case of lead nitrate, no beneficial effect is observed if the pH of the cyanide solution is above a definite value which depends upon the sodium cyanide concentration.

In theoretically interpreting the behavior of the contact agents, it is shown that their effect in accelerating the dissolution of the gold is probably due to the galvanic couple gold-lead, gold-mercury, gold-bismuth, or gold-thallium, the gold in each case being the less noble metal. Gold goes into solution, and the lead or other metal is precipitated. That gold is less noble than lead in sodium cyanide solutions of pH = 10 is confirmed by calculations based on published single potential data.

A lower limiting concentration of contact metal ion is necessary to bring about the accelerating effect. Thus, if the pH of the cyanide solution is above 11.5 to 14, the lead ion concentration is too low to be effective. Similarly, the effect disappears if the cyanide ion concentration is too low, as the gold electrode potential thereby becomes more noble than that of lead or other contact metal.

E. COPPER AMMONIA CYANIDE PROCESS

INTRODUCTION:

We next turned our attention to the copper ammonia cyanide process. While this method was successfully applied prior to 1905 in several large scale operations (29, 50), it apparently is not in use today (51). Not only is the literature concerning the method of a meager character, but modern metallurgical treatises either omit mention of the process or confine the description to one or two sentences (5, 52). Our interest in this process was prompted by the facts that it is applicable to gold ores containing copper (53), that it might be used with telluride ores which ordinary cyanide solutions do not attack (5), that the price of ammonia has greatly declined in recent years, and that ammoniacal solvents have been successfully applied to the extraction of copper from its ores (54).

The copper ammonia cyanide process consists in leaching the ore with a solvent containing ammonia, a cupric compound, and an alkali or alkaline earth cyanide (53, 55). The ammonia is reported to act as a protection[1] against the cyanicide,[2] copper (29). Since a plausible explanation for the action of the ammonia has never been given and the chemical composition of the gold-dissolving constituent is unknown, it may be concluded that the theory of the copper ammonia cyanide process is not well understood.

PRELIMINARY EXPERIMENTS:

The experimental methods were very similar to those used in testing the action of chlorides in chlorine solutions (Section C). Gold leaf of thickness 0.11 microns was dissolved in the solvents.

To a cyanide solution containing 1 gram of sodium cyanide per liter was added an amount of pulverized malachite[3] which was greater than the amount which would dissolve in the cyanide solution; the power of the cyanide solution to dissolve gold leaf was completely destroyed. A similar result was obtained with C.P. copper sulfate. That copper oxides, hydroxides, carbonates, and

1. According to Edquist (56), a cupric cyanide is formed which is a vigorous oxidizing agent. This, however, does not explain the very important role of the ammonia; the reaction product of a cyanide with an excess of a cupric compound has no solvent action for gold unless ammonia is present (51, 53, 56).

2. The term *cyanicide* is in very general use and means any substance which destroys the power of the cyanide solutions to dissolve gold. Thus, when an ore contains arsenic, antimony, or oxidized copper minerals, it is said to contain a *cyanicide*. The phrase "protection against the cyanicide" is commonly used to explain empirically known facts, and gives a fair indication of the 1941 status of the theory of cyanidation.

3. Malachite has the composition $CuCO_3 \cdot Cu(OH)_2$. The mineral used in our work was obtained from Ward's Natural Science Establishment, Rochester, New York. It was found to contain 58.2 percent copper, using the sodium cyanide titration method (57); the theoretical composition is 57.6 percent copper.

sulfates are cyanicides has been known almost from the inception of the cyanide process (5, 29, 58), and has been explained by the fact that copper forms a very stable complex with cyanide ions (58). Our experiments with gold leaf showed that sodium chloride, sodium thiocyanate, and many other compounds did not restore the activity of the sodium copper cyanide solutions. We confirmed the fact, however, that ammonia produces a remarkable increase in the gold-dissolving power of the solution (58), and discovered that many organic compounds which contain amine groups, such as ethylene diamine, urea, and ethyl amine, would also restore the power of the solution to dissolve gold leaf. As the organic compounds had no apparent advantages over ammonia, we confined our tests to ammonia.

EFFECT OF SODIUM CYANIDE CONCENTRATION ON TIME REQUIRED TO DISSOLVE GOLD LEAF:

A solution was prepared which contained 22 grams of ammonia per liter and 50 grams of crystalline copper sulfate ($CuSO_4 \cdot 5 H_2O$) per liter. Into 5 ml. portions of this solution were pipetted varying amounts of a sodium cyanide solution containing 50 grams of sodium cyanide per liter, and the periods of time necessary for the solvents to dissolve gold leaf determined (Table 21). With all solutions the amount of copper was several times that required to form the cyanide complex, the exact composition of which is unknown. The measurements were made in duplicate at the room temperature of 30°C.

Table 21

Effect of Sodium Cyanide Concentration on Time Required to Dissolve Gold Leaf in Copper Ammonia Cyanide Solutions

Sodium Cyanide Concentration, Grams per Liter	0.2	0.5	1.0	2.0	5.0	10.0
Seconds Time Required to Dissolve Gold Leaf	135	90	45	45	45	45

As sodium cyanide solutions having concentrations of 0.5 to 1.0 grams per liter all required at least 180 seconds to dissolve gold leaf (please see Table 9), cyanide in the form of copper ammonia cyanide may in some instances be a more efficient[4] solvent than in the form of sodium cyanide.

The results obtained by adding sodium cyanide to malachite solutions in ammonia were similar to those obtained with copper sulfate (Table 21). The malachite solutions were somewhat less efficient,

4. Please see Section C for definition of "efficiency."

but this was probably due to the presence of small amounts of ferrous sulfates or carbonates (58).

EFFECT OF VARIOUS BASES AND SALTS ON THE STABILITY OF THE SOLUTIONS:

Jarman and Brereton (58) have found that an excess of lime prevents copper compounds from dissolving in ammonia solutions, and recommended the use of lime in the process. As they did not state whether this affects the stability of the solvent, lime and other agents which alter the pH[5] were studied.

Table 22

Stability of Copper Ammonia Cyanide Solutions

Substances Added to the Solution	Seconds Time Required to Dissolve Gold Leaf after 5 hrs.	30 hrs.	100 hrs.
Nothing added - - - - - - - - - - - - - - -	200	250	450
50 g. silica sand - - - - - - - - - - - - -	180	250	280
0.5 g. ammonium sulfate - - - - - - - - - -	180	180	180
0.5 g. ammonium sulfate + 50 g. of sand -	180	180	240
1.0 g. of sodium bicarbonate + 1.0 g. of sodium carbonate (Na_2CO_3) - - - - - - -	160	200	650
1.0 g. of sodium bicarbonate + 1.0 g. of sodium carbonate + 50 g. of sand - - -	180	210	800
1.0 g. of sodium carbonate - - - - - - - -	180	260	830
1.0 g. of sodium carbonate + 50 g. of sand	180	280	750
1.0 g. of calcium hydroxide - - - - - - - -	240	600	2650
1.0 g. of calcium hydroxide + 50 g. of sand - - - - - - - - - - - - - - - - - -	180	480	3120

To 50 ml. portions of a solution containing 10.2 grams of copper (57), originally in the form of malachite; 26.8 grams of ammonia per liter, determined by titration with hydrochloric acid and methyl orange indicator; and 0.5 gram of sodium cyanide per liter

5. We do not give the actual pH values of the solutions because this would have necessitated a detailed investigation to find a reliable method for the pH measurement. The antimony electrode may give erroneous results in the presence of 1 mg. per liter of copper (59). Regarding the glass electrode, it was found by one of us (60) that the acid concentrations of concentrated salt solutions, as determined by this method, may be *high* by factors at least as great as 890 (89,000 percent error); and Gardiner and Sanders (61) report enormous deviations in concentrated sodium chloride solutions. It is well known that the glass electrode is practically useless in very basic solutions, such as calcium hydroxide solutions (61). Regarding colorimetric methods, there is the salt error to consider; ammonia and cyanides react with many organic compounds; and our solutions were highly colored due to the ammono-cupric ion present. With the hydrogen electrode there would be the possibility of copper or traces of other metals plating out on the electrode.

(20), were added various substances. Duplicate trials were made in which 50 grams of pulverized sand were added. The solutions were kept at room temperature, approximately 22°C, in glass-stoppered bottles and protected from light.

Table 22 indicates that calcium hydroxide, which was present in more than sufficient amount to saturate the solution slowly, destroys the active constituent. If the solubility of calcium hydroxide is 1.7 grams per liter and the degree of ionization is 0.90 (63), and the dissociation constant of water is 1×10^{-14}, the pH of a saturated calcium hydroxide solution is 12.7, which is much higher than the pH of any of the other solutions (62) of Table 22.

PRECIPITATION OF GOLD FROM COPPER AMMONIA CYANIDE SOLUTIONS:

One of the major difficulties in the cyanide leaching of ores containing soluble copper is the difficulty in precipitating the gold (64). When the usual zinc dust method is used, the zinc particles become covered with a thin film of copper or brass (65) and are then no longer capable of displacing the precious metal.[6] Indeed, Bertrand Hunt, the first person to apply the copper ammonia cyanide method on a tonnage scale, found it necessary to use aluminum as the precipitating agent (48).

Table 23

Effect of Ammonia Concentration on Percent of Gold Precipitated by Zinc Dust

Time of Contact with Zinc Dust	Percent of Gold Content Precipitated in the Presence of 0.486 G./L. of Ammonia	1.22 G./L. of Ammonia	2.43 G./L. of Ammonia
5 minutes - - - - - - - - - -	3	50	95
20 minutes - - - - - - - - - -	10	80	100
60 minutes - - - - - - - - - -	20	90	100

It was recently found by one of us that the reaction between zinc and ammoniacal copper solutions is markedly influenced by the ammonia concentration. An adherent, yellow brass coating is obtained at certain ammonia concentrations, and a non-adherent, dark-red copper deposit at other concentrations (65). The effect of ammonia on the precipitation of gold from cupriferous cyanide solutions was therefore studied. A solution was prepared which contained, per liter, 0.5 g. of sodium cyanide, 0.486 g. of ammonia, 0.5 g. of copper originally in the form of malachite, and 5 mg. of gold.

6. That this is true is known to everyone engaged in the art, but the reason for this peculiar behavior is not well understood.

This solvent had a pale blue color, indicating the presence of an excess of copper over that required to form the copper cyanide complex. For each trial 50 ml. of this solution were stirred with 0.50 g. of zinc dust and the gold content of the precipitate determined by the low acid-stannous chloride test (Section B). The results are presented in Table 23.

Table 23 indicates that ammonia, when present to the extent of 1.2 grams per liter, was effective in activating the zinc dust.

SUMMARY OF SECTION E:

The copper ammonia cyanide method offers an attractive method for treating oxidized copper-gold ores. If ammonia is present, copper compounds may actually enhance the ability of cyanide solutions to dissolve gold.

The presence of relatively large concentrations of soluble cupric compounds is not deleterious. An excess of calcium hydroxide is undesirable, as it slowly destroys the gold-dissolving constituent.

Ammonia, when present to the extent of more than 1.2 gram per liter, was found to promote the precipitation of gold by zinc dust, thereby overcoming one of the principal objections to copper.

F. THE FERRICYANIDE METHOD FOR THE EXTRACTION OF GOLD

INTRODUCTION:

That ferrocyanide ion is invariably present in gold leaching solutions is well known (5), and it is also well known that ferricyanides accelerate the dissolving of gold in alkali cyanide solutions (67). The purpose of our work was to find an economical method for converting ferrocyanides to ferricyanides, in the presence of sodium cyanide, and to stabilize the resulting solutions.

EFFECT OF POTASSIUM FERRICYANIDE ON THE DISSOLUTION OF GOLD LEAF:

Variable Potassium Ferricyanide Concentration. The test for the effectiveness of the cyanide solvents was the gold leaf solution test described in detail in Section C. In the experiments of Table 24 the solvent was a saturated lime-water solution containing 0.2 gram of sodium cyanide per liter, plus added variable amounts of potassium ferricyanide. We used saturated lime-water because lime-water is the base generally used in cyanidation.

Table 24

Effect of Potassium Ferricyanide Concentration on the Dissolution of Gold Leaf in Saturated Lime-Water Solutions Containing 0.2 Gram of Sodium Cyanide per Liter

Grams of $K_3Fe(CN)_6$ per Liter - - - -	None	0.1	0.2	0.5	1.0
Seconds Time Required to Dissolve Gold Leaf - - - - - - - -	720	250	180	100	80

The results of Table 24 are plotted in Figure 5.

Under the conditions indicated above, potassium ferricyanide greatly accelerates the dissolution of gold leaf. With solutions containing 0.2 gram of sodium cyanide per liter, the addition of 1.0 gram of potassium ferricyanide per liter decreases the time required to dissolve gold leaf by factors of 1/3 to 1/8. That ferricyanide ion at a concentration of less than 0.1 gram per liter is effective in overcoming the deleterious effect of calcium hydroxide is new discovery. It will be recalled that lead as a contact agent (Section E above) was ineffective in saturated lime-water solutions (Tables 17 and 18).

Effect of Sodium Cyanide Concentration. The time required for sodium cyanide solutions to dissolve gold leaf was determined, the potassium ferricyanide concentration being 1.0 gram per liter and

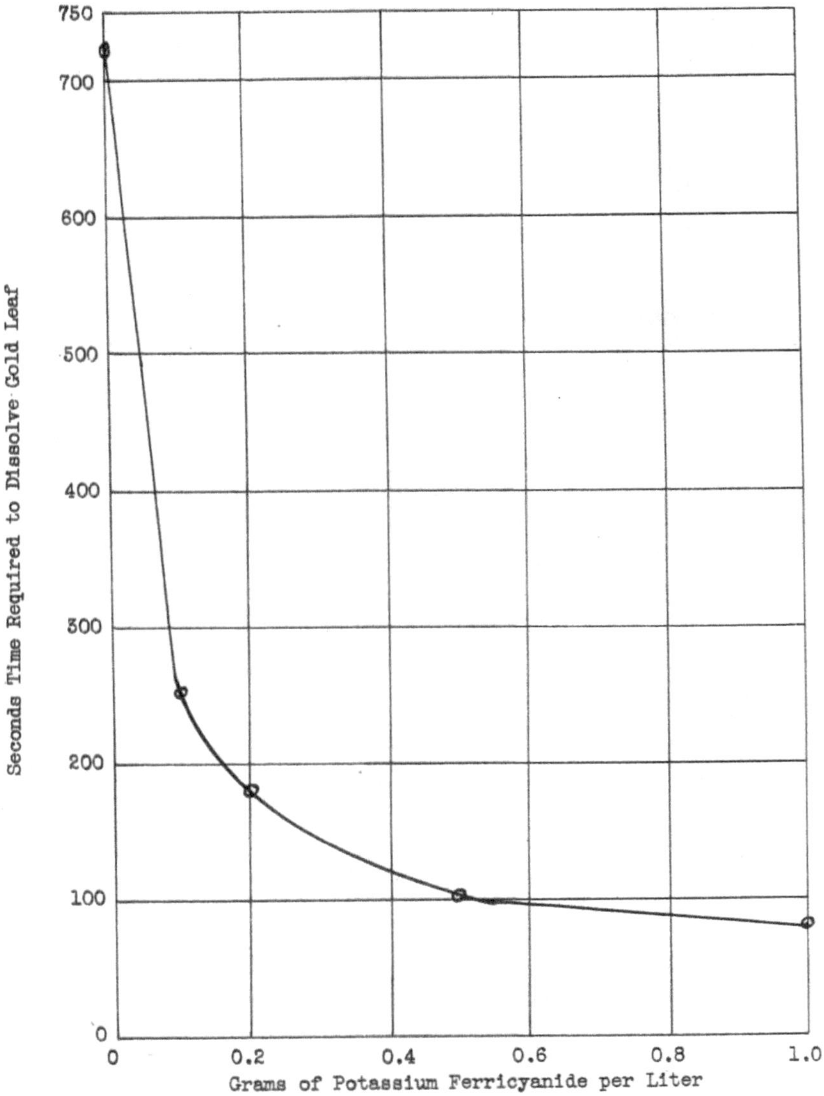

Fig. 5. Effect of Potassium Ferricyanide Concentration on Time Required to Dissolve Gold Leaf in Saturated Lime-Water Containing 0.2 Gram of Sodium Cyanide per Liter

the sodium cyanide concentration being varied. As in the preceding series the solutions were saturated with calcium hydroxide. All trials were made simultaneously (Table 25).

Table 25

Effect of Sodium Cyanide Concentration in Solutions Containing 1.0 Gram of Potassium Ferricyanide per Liter

Grams of Sodium Cyanide per Liter	0.5	0.2	0.5	1.0	2.0	5.0
Seconds Time Required to Dissolve Gold Leaf	240	80	65	45	40	40

The data of Table 25 are plotted in Figure 6.

The results of Table 25 indicate that sodium cyanide solutions which are more dilute than those generally used in cyanidation have appreciable solvent power for gold leaf if potassium ferricyanide is present. Comparison of the results of Table 25 with Tables 17 and 18 indicates that ferricyanide ion is especially beneficial in the presence of calcium hydroxide.

STABILITY OF SOLUTIONS:

The results of Tables 24 and 25 were interesting, but on attempting to apply the ferricyanide-cyanide solvents to gold ores, we were surprised to find that within a few hours the solutions lost their characteristic yellow color and much of their solvent power for gold leaf. In order to determine the nature of the underlying reaction, we added crystalline potassium ferricyanide to calcium hydroxide-sodium cyanide of known strength, permitted the solutions to stand until decolorized, and then determined the free cyanide (20). It was found that approximately two mols of potassium ferricyanide were reduced to the colorless state per mol of sodium cyanide destroyed. An equation which accounts for this reaction is:

$$NaCN + 2\ K_3Fe(CN)_6 + 2\ KOH \longrightarrow NaCNO + 2\ K_4Fe(CN)_6 + H_2O.$$

Moreover, it was soon apparent to us that there was a constituent present in the ores, even in some containing 98 percent or more of silica, which catalyzed the decomposition of the ferricyanide-cyanide solutions. A large number of tests finally revealed that soluble copper was responsible for this destruction of the ferricyanide. The presence of even 5 mg. of copper per liter of solution caused a marked increase in the rate at which the solvents lost their yellow color (ferricyanide ion is intensely yellow) and their power to dissolve gold leaf.

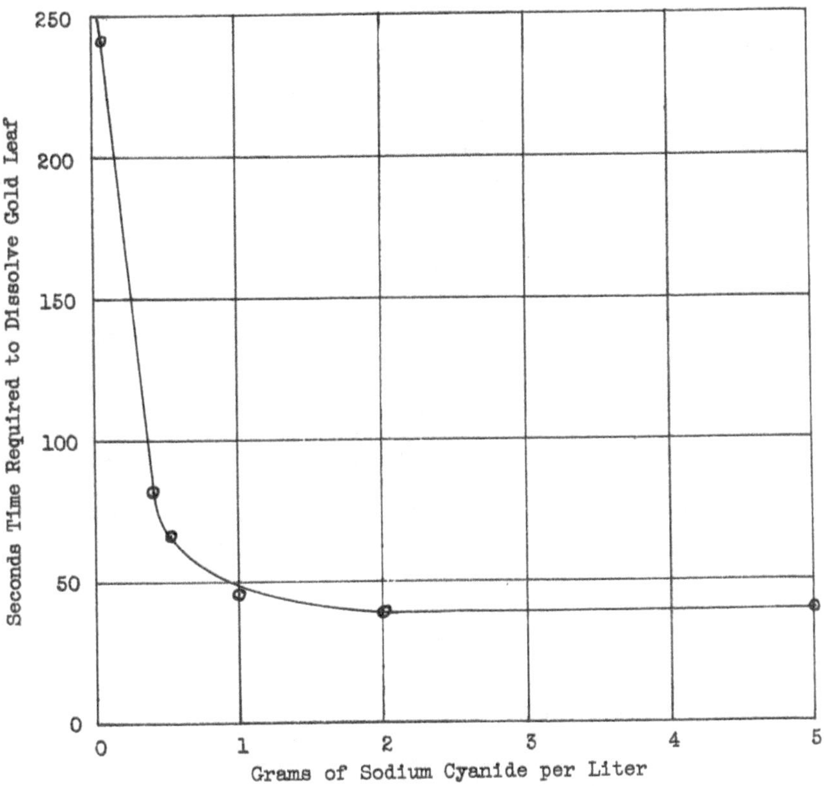

Fig. 6. Effect of Sodium Cyanide Concentration on Time Required to Dissolve Gold Leaf in a Saturated Lime-Water Solution Containing 1.0 Gram of Potassium Ferricyanide per Liter

It was also discovered that the complex cyanide of copper is a vigorous catalyst for the reaction between sodium arsenite and potassium ferricyanide. An extremely sensitive test for traces of copper in cyanide solutions consists in adding to the cyanide solution an equal volume of a freshly prepared solution containing 1.0 gram of potassium ferricyanide and 1.0 gram of dibasic sodium arsenite (Na_2HAsO_3) per liter. When as little as 3.5 mg. of copper per liter of cyanide solution is present, decolorization of the ferricyanide takes place within one second; with 0.35 mg. per liter, a time of 4.5 minutes is required; and with no copper present the solution retains a distinct yellow color for at least four hours. This simple test for copper is a new discovery on our part, and is useful for rapidly determining whether or not an ore can be treated by the ferricyanide-sodium cyanide process.

PREPARATION OF FERRICYANIDES:

Having found what we think is the most important variable affecting the stability of the cyanide-ferricyanide solutions, we next investigated feasible methods for introducing ferricyanide ion into the cyanide solutions. The most interesting and promising method seemed to be the oxidation of the ferrocyanide ion which is normally present in metallurgical cyanide solutions. It is well known that the soluble iron salts and basic ferrous oxides present in gold ores react with sodium cyanide to form ferrocyanide ion, and that the concentration of the latter builds up due to constant recirculation of the solvent (5). Accordingly, if one desired to use ferricyanides in cyanidation, all that would in general be necessary would be to find a method for economically converting the ferrocyanides of the leaching solvent to ferricyanides.

In order to study the conversion of ferrocyanides to ferricyanides, it was necessary to have some method for determining ferricyanide ion in the presence of cyanide and ferrocyanide ions. This is by no means a simple matter. We were not able to find in any of the standard analytical texts either a volumetric or a gravimetric method suitable for this purpose. Fortunately, however, potassium ferricyanide solutions have an intense yellow color, and the intensity of this yellow color is not affected by the constituents normally present in the cyanide solutions. It was possible, therefore, to estimate the amount of ferricyanide by colorimetric comparison in Nessler tubes. As a check, the efficiencies of the solvents were determined by the gold leaf test, the amount of copper present in the gold leaf being too small to affect the results.

Various methods for converting potassium ferrocyanide to ferricyanide were tried, including photochemical oxidation with atmospheric oxygen and direct sunlight, calcium hypochlorite, hydrogen peroxide, potassium permanganate, and ozone. Although fairly efficient at low pH, photochemical oxidation gave rise to excessive decomposition of the sodium cyanide present in the solution.

Photochemical oxidation of the ferrocyanide is probably an important factor in the operation of many cyanide mills, and to our knowledge has never been taken into account in reports on the chemistry of the cyanide process. Very frequently metallurgical cyanide solvents have a distinct yellow color due to the presence of ferricyanides, but that their formation is due to photochemical action has never been reported.[1] Upon introducing hypochlorites and bromates into the solution we found that these oxidizing agents destroyed the free cyanide content of the solution, but did not oxidize the ferrocyanides. Fortunately the opposite was true with potassium permanganate and ozone. Both of these reagents preferentially attacked the ferrocyanide. Thus, for example, a saturated calcium hydroxide solution containing 1.02 grams of potassium ferrocyanide and 0.511 gram of sodium cyanide per liter was treated with the volume of 0.12 N potassium permanganate theoretically necessary to effect the oxidation of the ferrocyanide according to the equation:

$$3\ K_4Fe(CN)_6 + KMnO_4 + 2\ H_2O = 3\ K_3Fe(CN)_6 + MnO_2 + 4\ KOH. \quad (1)$$

After adding the permanganate and filtering, it was found by silver nitrate titration (20) that the sodium cyanide concentration was 0.486 gram per liter, and by colorimetric comparison with Nessler tubes it was found that the potassium ferricyanide concentration was 0.91 gram per liter, the results being corrected for the 2 percent increase in volume caused by addition of the permanganate. The time required for the solvent to dissolve gold leaf was 35 seconds.

In order to demonstrate the applicability of the permanganate oxidation method, saturated lime-water solutions of sodium cyanide and potassium ferrocyanide were oxidized with the volume of 0.12 N potassium permanganate theoretically required according to Equation (1). Sodium cyanide was determined by the standard silver nitrate method (20), the concentration of potassium ferrocyanide stock solution used in making up the reagents was determined by titrating with potassium permanganate (20), and the ferricyanide concentration determined by comparing the color with that of ferricyanide solutions of known concentration, using the balancing tube method and standards of similar ferricyanide concentration. The results are presented in Table 26.

Table 26 indicates that potassium ferrocyanide can be quantitatively oxidized to ferricyanide in the presence of sodium cyanide and lime-water, and this is a new discovery on our part. Furthermore, it must be noted that all of the ferricyanide solutions of Table 26 were tested by the gold leaf method and found to have much greater solvent efficiencies for gold leaf than the original unoxidized solutions.

1. The yellow color is said to be due to ferrocyanides (Ref. 5, page 242). This is, however, impossible, as reduced ferrocyanide solutions are colorless. Oxidation of a part of the ferrocyanide content to ferricyanide must be involved.

Table 26

Permanganate Oxidation of Ferrocyanide in the Presence of Sodium Cyanide

Initial Conc. of $K_4Fe(CN)_6$ Grams per Liter	Grams of $K_4Fe(CN)_6$ per Liter Converted to $K_3Fe(CN)_6$[a]	Initial Conc. of NaCN in Grams per Liter	Final Conc. of NaCN in Grams per Liter[b]
0.556	0.54	0.120	0.117
0.556	0.51	0.240	0.229
0.556	0.54	2.40	2.11
1.11	1.10	0.240	0.226
1.11	1.10	2.40	2.30

a. Values may be 5 to 15 percent high due to presence of suspended manganese dioxide.

b. Corrected for the 1 to 2 percent increase in volume caused by addition of permanganate solution.

DISCUSSION:

1. Regarding the cost of the ferricyanide-permanganate method, one mol of permanganate is required per three mols of ferrocyanide (Equation 1). In order to oxidize the ferrocyanide content of one ton of cyanide solvent containing 0.2 gram of potassium ferrocyanide per liter or the equivalent amount of sodium ferrocyanide, 0.0573 lb. of potassium permanganate is theoretically necessary. At the present market price of 19.3 cents per lb. (30), the cost of the permanganate would be only 1.11 cents per ton of solution. This is but a very small fraction of the cost of the sodium cyanide normally present in one ton of cyanide solvent.

2. Potassium ferrocyanide has no effect on the dissolution of gold leaf.

3. Potassium permanganate has been used as an oxidizing agent in the cyanidation of gold ores (68), the method being called the "manganicyanide" process. In the absence of ferrocyanides the addition of potassium permanganate merely results in the oxidation of the sodium cyanide and no increase in dissolution velocity of gold leaf is obtained, as shown by Green (67) and confirmed by us. This fact, together with the heretofore unknown deleterious effect of traces of copper, may account for the conflicting results obtained with the "manganicyanide" process and for the fact that the method has apparently been abandoned as there are no records to the contrary.

SUMMARY OF SECTION F:

The use of ferricyanide ion as an oxidizing agent in the gold cyanide process was studied. It was discovered that:

1. Soluble copper is a vigorous catalyst for the decomposition of ferricyanide-sodium cyanide solutions.
2. Traces of potassium ferricyanide overcome the deleterious effect of calcium hydroxide on the dissolution of gold leaf.
3. When sodium cyanide solutions containing an excess of potassium ferrocyanide are treated with potassium permanganate, the permanganate oxidizes the ferrocyanide to ferricyanide and the sodium cyanide is not attacked.

G. SOLUBLE SULFIDES IN THE CYANIDE PROCESS

INTRODUCTION:

That gold ores which contain compounds of arsenic and antimony are often not amenable to direct cyanide treatment is well known, and it is said that arsenic and antimony are "cyanicides" (4). It is commonly believed that part of the sulfur content of such ores is soluble in alkaline solutions to give sulfide ion, which reacts with the free cyanide content of the leaching solutions to form inert thiocyanates or with the dissolved oxygen to form sulfites or sulfates (51). It is the purpose of this section to prove that sulfide ion may behave in a manner that cannot be explained by either the oxygen-depletion or thiocyanate formation theories, and to show that concentrations of sulfide ion which cannot be detected even by colorimetric methods may retard the dissolution of gold leaf.

ACTION OF SODIUM SULFIDE:

Courtney and Maguire (27) have studied the effect of sodium sulfide on the extraction of gold from its ores by solvents containing 3.1 grams of sodium cyanide per liter. They gave no hypothesis to account for their observations, and the concentrations of sulfide ion in their solvents all exceeded 600 mg. per liter.

It seemed to us that by determining the minimum concentration of sulfide which is effective in retarding the dissolution of gold leaf, it might be possible to find out why sulfide ion is deleterious. Solvents were prepared which contained, per liter, 1.0 gram of sodium cyanide, 10.6 grams of sodium carbonate, 8.40 grams of sodium bicarbonate, and variable amounts of sodium sulfide. The pH of such a solution is about 9.9 (62). When the efficiencies of the solvents were determined by the gold leaf method (Section C), the results of Table 27 and Figure 7 were obtained.

Table 27

Effect of Sulfide Ion on the Efficiencies of Sodium Cyanide Solvents

Mg. of Sulfide Ion, as $Na_2S \cdot 9H_2O$ per Liter - - -	0.0	0.10	0.47	1.3	5.3
Minutes Time Required to Dissolve Gold Leaf - -	3.0	4.5	10.0	20.4	21.5

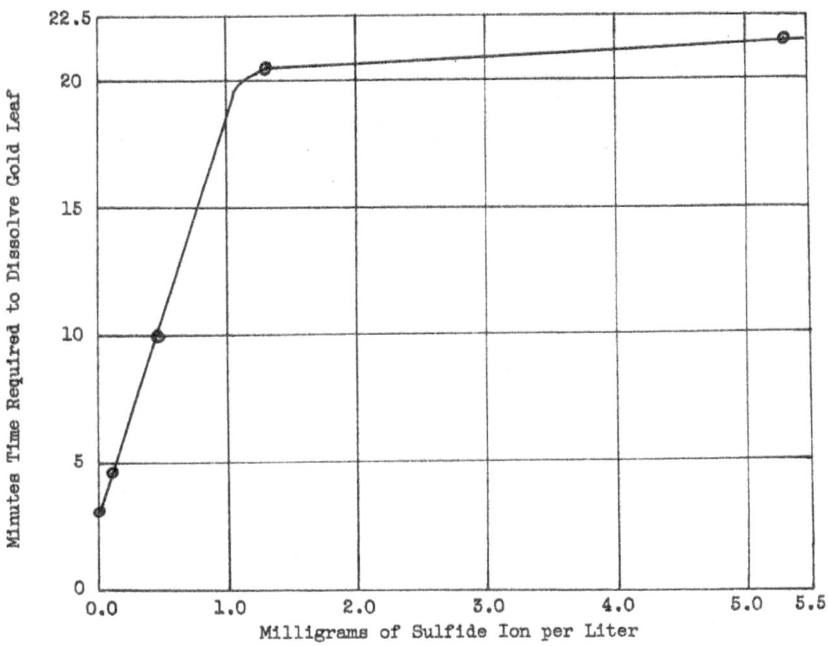

Fig. 7. Effect of Sulfide Ion on Time Required to Dissolve Gold Leaf in a Solution Containing 1.0 Gram of Sodium Cyanide per Liter

As shown by Table 27, less than 0.47 mg. of sulfide ion per liter is required to appreciably retard the dissolution of gold leaf. To oxidize this concentration of sulfide ion to sulfate, 0.94 mg. of oxygen per liter would be required; yet cyanide solutions saturated with air at 25° C contain about 8 mg. of oxygen per liter, and the oxygen solubility is not decreased by small percentages of dissolved salts (23). Moreover, the oxygen contents of the solutions would have been brought rapidly to saturation by the agitation of the test tube. With regard to thiocyanate formation, this factor could have accounted for the loss of only 0.65 mg. of sodium cyanide per liter out of 1000 mg., and we have confirmed the fact that sodium thiocyanate (25, 27) does not retard the dissolution of gold.

It must be clear, therefore, that neither the oxygen-depletion nor the thiocyanate-formation theories can explain the behavior of sulfide ion. We are of the opinion that an important factor involved is adsorption of sulfide ion by the gold at the metal-solution interface, thereby protecting the metal from corrosion by the solutions.

As sulfides are harmful at concentrations below the ranges of colorimetric methods, one of the most useful and sensitive procedures for detecting the presence of sulfide ion is the gold leaf solution test (Section C), which will, we believe, find wide application as a control test for cyanide solutions.

REMOVAL OF SULFIDE ION FROM CYANIDE SOLVENTS:

Lead and mercury salts have been used to precipitate soluble sulfides from cyanide solutions (5, 26).

We found that by the addition of lead, bismuth, thallium, and mercuric salts to sodium cyanide-sodium sulfide solutions, the efficiencies of the solvents were restored.[2] Salts of zinc, cadmium, iron, and manganese salts were not effective as agents for the removal of sulfide ion from cyanide solutions.

In order to test the ability of lead nitrate to keep cyanide solutions free from sulfur, 1.0 gram of realgar (As_2S_3) or stibnite (Sb_2S_3), ground to pass a 40-mesh screen, was added to 20 ml. of a solution containing 0.50 gram of sodium cyanide and 1.0 gram of lead nitrate per liter. PH-controlling bases or salts were also added. After standing for 30 minutes on contact with the realgar or stibnite, the solutions were filtered and the filtrates tested by the gold leaf method.

Table 28 indicates that lead nitrate is most useful in solvents having lower hydroxyl ion concentrations, and this is in accord with the investigation of Hayden (25), who studied the action of lead nitrate in preventing interference due to sulfide ion.

2. Our investigation of sulfide ion preceded our study of contact agents (Section D). We were surprised to find that when an excess of lead nitrate was added to a sodium sulfide-sodium cyanide solution, the resulting solvent had a much greater efficiency than a pure sodium cyanide solution of the same concentration (see Tables 14, 18, and 28). In this way we discovered the action of contact agents.

Table 28

Use of Lead Nitrate to Prevent Sulfide Ion from Interfering with the Dissolution of Gold Leaf

Base or Salts Added to Solvent Containing 0.50 Gram of Sodium Cyanide and 1.0 Gram of Lead Nitrate per Liter	Minutes Time Required to Dissolve Gold Leaf after Contact with	
	Realgar	Stibnite
0.50 M NaOH	80	90
Saturated $Ca(OH)_2$	70	90
0.1 M Na_2CO_3	90	90
0.1 M Na_2CO_3 + 0.1 M $NaHCO_3$	85	90
0.3 M NH_4OH + 0.25 M $(NH_4)_2SO_4$	11	10
1.0 M $NaHCO_3$	4.5	4
Nothing Added	7	8

SUMMARY OF SECTION G:

Sulfide ion, even at concentrations below 0.5 mg. per liter, may have a definite retarding effect on the dissolution of gold leaf, and this heretofore-unknown deleterious action of minute traces (less than 1 mg. per liter) of sulfide ion cannot be accounted for by either the oxygen-depletion or thiocyanate-formation theories. It is believed that the formation of an insoluble aruous sulfide film on the surface of the gold is responsible for the behavior of sulfide ion in protecting the gold from corrosion.

The gold leaf solution test, first used by Faraday, has been of great utility in obtaining insight into the chemistry of the cyanide process, and this test is apparently the most useful, rapid, and practical method for determining the efficiencies of cyanide solvents for gold.

H. GENERAL CONCLUSIONS

Experimental studies were made of various methods for accelerating the dissolution of gold, using the gold leaf solution test, and rapid methods for control of the processes were devised. The conclusions given below are new and are based upon new material.

A. STANNOUS CHLORIDE TEST FOR GOLD:

1. When the acid concentration is maintained below about 0.05 N, the shade or tint of color which is obtained with the stannous chloride test for gold is not purple but is yellow to brown.
2. The tint of color which is obtained with the stannous chloride test for gold is primarily a function of the acid concentration.
3. The gold concentrations of solutions cannot be determined from the tint of color obtained by the Purple of Cassius test.
4. When a yellow or brown solution, prepared by the stannous chloride method and containing gold in the colloidal state, is treated with sufficient hydrochloric acid to bring the acid concentration to 2 N to 6 N, the solution changes from the relatively stable yellow form to the unstable purple form. With gold concentrations of 2 to 20 mg. per liter, there is a distinct time interval during which the solution is colorless.
5. The time required for the development of the color to full intensity is less with low concentrations than with high concentrations of hydrochloric acid, and the test is therefore probably more reproducible with low concentrations of hydrochloric acid.
6. The addition of sodium chloride to stannous chloride solutions having concentrations from 0.002 N to 0.029 N, in amounts sufficient to bring the sodium chloride concentrations to 0.01 N to 0.50 N, greatly accelerates the tendency for the solutions to become turbid. For this reason it is concluded that the concentration of electrolytes should be reduced to the lowest practicable value when making the stannous chloride test.
7. Over the range of 0.0014 N to 0.14 N stannous chloride, the tint and intensity of color with the low acid-stannous chloride test are not appreciably affected by change in the stannous chloride concentration.
8. Gold may be more accurately and more precisely determined by our low acid-stannous chloride test than by the widely used and heretofore generally accepted high acid-stannous chloride test.

B. CHLORINATION IN THE PRESENCE OF CHLORIDES:

1. It is experimentally shown for the first time that the dissolution of pure gold and of gold leaf in aqueous solutions of chlorine is promoted by relatively high concentrations of sodium chloride.

2. It is experimentally shown for the first time that the dissolution of pure gold and of gold leaf in aqueous solutions of bromine is promoted by relatively high concentrations of sodium chloride.

3. Chlorinated sea water is an excellent solvent for pure gold.

4. The results on the dissolution of gold in aqueous chlorine solutions are best explained on the basis that the primary reaction product is insoluble, and that the rate of removal of this reaction product from the gold surface is the rate-controlling factor in the reaction.

C. CONTACT AGENTS:

1. Under suitable conditions salts of bismuth, thallium, and lead accelerate the dissolution of gold in cyanide solutions which are free from sulfide ion.

2. Whether or not lead salts accelerate the dissolution of gold in cyanide solutions depends upon the relationship of the cyanide ion concentration to the pH of the solution.

3. The accelerated dissolution of gold leaf in cyanide solutions containing traces of the compounds of mercury, lead, bismuth, and thallium is not due to the oxidizing power of the metal compounds. The action of the metal compounds is of a catalytic nature.

4. The data can be explained on the basis that the metals are displaced by the gold, forming a galvanic couple which accelerates the dissolution of the gold.

D. COPPER AMMONIA CYANIDE PROCESS:

1. The presence of calcium hydroxide in ammonia copper cyanide solutions decreases the stability of the gold-dissolving constituent.

2. The presence of ammonia in a cyanide solution containing gold and copper may promote the zinc dust precipitation of the gold content of the solution.

E. FERRICYANIDE PROCESS:

1. Cupric compounds catalyze the reaction between ferricyanide and cyanide ions.

2. The successful operation of the Adair-Usher manganicyanide process is dependent upon the presence of ferrocyanides and the absence of copper compounds in the leaching solvent.

3. Ozone and permanganate ion are efficient oxidizing agents for ferrocyanides in sodium cyanide-lime water solutions.

4. The presence of very small concentrations of ferricyanide ion (0.1 gram per liter) overcomes the deleterious action of lime in cyanide solutions.

F. SULFIDE ION:

1. Less than 1.0 mg. of sulfide ion per liter of cyanide solution is required to markedly retard the dissolution of gold leaf.

2. The modern theories for the action of sulfide ion, which are based upon the assumptions that sulfide ion depletes the oxygen content of the solution or forms thiocyanate ion with the cyanide ion, are inadequate. Adsorption of sulfide ion at the metal-solution interface, to form a film which excludes oxygen from the gold surface, is probably the most important factor involved.

3. It is difficult, if not impossible, to determine by chemical analysis the efficiencies of cyanide solvents for dissolving gold leaf, and this conclusion, presumably, applies also to native gold. Of all tests, the gold leaf solution is probably the most useful for determining the efficiencies of cyanide solvents.

BIBLIOGRAPHY

(1) Cited by Agricola, Georg, "De Re Metallica," (Translation by Herbert C. Hoover and L. H. Hoover), London, The Mining Magazine (1912), p. 297.

(2) Taggart, A. D., Handbook of Ore Dressing, New York, John Wiley and Sons, Inc., 1927, p. 867.

(3) Whitaker, M. C., Ind. Eng. Chem., XXXIII (1941), p. 361.

(4) Munro, N., Canadian Chemistry and Process Industries, XXIV (Sept., 1940), p. 438.

(5) Dorr, J. V. N., Cyanidation and Concentration of Gold and Silver Ores, New York, McGraw-Hill Book Co., Inc., 1936.

(6) Thomas, A. W., Colloid Chemistry, New York, McGraw-Hill Book Co., Inc., 1934, p. 116.

(7) Zsigmondy and Thiessen, Das Kolloidale Gold, Leipzig, Akademische Verlagsgesellschaft m.b.H., 1925, p. 211.

(8) Bugbee, E. E., A Textbook of Fire Assaying, 3rd ed., New York, John Wiley and Sons, Inc., 1940, pp. 245-46.

(9) Snell, F. D., and Snell, C. T., Colorimetric Methods of Analysis, 2nd ed., New York, D. Van Nostrand Co., Inc., 1936, Vol. I, pp. 402-05.

(10) Treadwell, F. P., and Hall, W. T., Analytical Chemistry, 8th ed., New York, John Wiley and Sons, Inc., 1940, Vol. I, p. 280.

(11) Yoe, J. H., Photometric Chemical Analysis, New York, John Wiley and Sons, Inc., 1928, Vol. I, pp. 191-97.

(12) Scott, W. W., and Furman, N. H., Standard Methods of Chemical Analysis, 5th ed., New York, D. Van Nostrand Co., Inc., 1939, Vol. I, p. 273.

(13) Thomas, A. W., Colloid Chemistry, New York, McGraw-Hill Book Co., Inc., 1934, p. 144.

(14) Scott, W. W., and Furman, N. H., Standard Methods of Analysis, 5th ed., New York, D. Van Nostrand Co., Inc., 1939, Vol. I, p. 439.

(15) von Bernewitz, M. W., U. S. Bureau of Mines Report of Investigations 3425 (December, 1938), p. 25.

(16) Bugbee, E. E., A Textbook of Fire Assaying, 3rd ed., New York, John Wiley and Sons, Inc., 1940, p. 74.

(17) Ibid., p. 131.

(18) Faraday, M., Trans. Phil. Magazine (4), 14 (1857), 403, London, Edinburgh, and Dublin.

(19) MacArthur, J. W., Forrest, R. W., and Forrest, W., Brit. Pat. 14,174, 1887; Brit. Pat. 10,223, 1888.

(20) Scott, W. W., and Furman, N. H., Standard Methods of Chemical Analysis, 5th ed., New York, D. Van Nostrand Co., Inc., 1939, Vol. I, p. 661.

(21) Yoe, J. H., Photometric Chemical Analysis, New York, John Wiley and Sons, Inc., 1928, Vol. I, pp. 372-73.

(22) Maclaurin, R. C., J. Chem. Soc., 65 (1895), p. 728.

(23) Barsky, G., Swainson, S. J., and Headley, N., Trans. Am. Inst. Mining and Met. Engrs., 112 (1934), pp. 660-71. See also Ref. (4).

(24) Leaver, E. S., Woolf, J. A., and Jackson, T. A., Am. Inst. Mining Met. Engrs., Contribution No. 5 (1933), 15 pp.

(25) Haden, T., Bull. Inst. Mining Met., 401 (1938), 16 pp. See also (24).

(26) Dowsett, C. W., and John, W. E., U.S. 1,952,976, March 27, 1934.

(27) Courtney, R. G., and Maguire, J. H., Thesis, University of Idaho; cited by Fahrenwald, A. W., and Newton, J., Eng. Mining J., 140 (Jan., 1939), pp. 44-47.

(28) Beyers, E., J. Chem. Met. Mining Soc. S. Africa, 37 (1936), pp. 37-39.

(29) Janin, L., Eng. Mining J., 74 (Dec. 20, 1902), p. 816.

(30) Anonymous, Ind. Eng. Chem., News Ed., 19 (March 10, 1941), p. 290.

(31) Rose, T. K., The Metallurgy of Gold, London, Charles Griffin and Co., Ltd., 1902, pp. 272-366.

(32) Elsner, L., J. prakt. Chem., (1), 37 (1846), p. 441.

(33) Greenawalt, W. E., The Hydrometallurgy of Copper, New York, McGraw-Hill Book Co., Inc., 1912, p. 374.

(34) Eissler, M., The Metallurgy of Gold, London, Crosby Lockwood and Son, 1889, pp. 237, 241.

(35) Schnabel, K., Handbook of Metallurgy, 3rd ed., London, The Macmillan Co., Ltd., 1921, Vol. 1.

(36) Aaron, C. H., Leaching Gold and Silver Ores, San Francisco, Barry and Baird, 1881.

(37) Kustel, G., Roasting of Gold and Silver Ores and the Extraction of Their Respective Metals, San Francisco, A. J. Leary, 1880.

(38) Latimer, W. M., The Oxidation States of the Elements and Their Potentials in Aqueous Solutions, New York, Prentice-Hall, Inc., 1938, p. 185.

(39) Linford, Dr. H. B., private communication.

(40) Scott, W. W., and Furman, N. H., Standard Methods of Chemical Analysis, New York, D. Van Nostrand Co., Inc., 1939, pp. 661-63.

(41) Lewis, G. N., and Randall, M., Thermodynamics and the Free Energy of Chemical Substances, New York, McGraw-Hill Book Co., Inc., 1923, p. 508.

(42) Latimer, W. M., The Oxidation States of the Elements and Their Potentials in Aqueous Solutions, New York, Prentice-Hall Co., Inc., 1938, pp. 184 and 47.

(43) Getman, F. H., and Daniels, F., Outlines of Theoretical Chemistry, 5th ed., New York, John Wiley and Sons, 1931, p. 322.

(44) Fowle, F. E., Smithsonian Physical Tables, Washington, D. C., Smithsonian Institution, 1933, p. 651.

(45) Lengfeld, F., Am. Chem. J., 26 (1901), p. 325.

(46) Cited by Euken, A., Jette, E. R., and LaMer, V. K., Fundamentals of Physical Chemistry, New York, McGraw-Hill Book Co., Inc., 1925, p. 440.

(47) Christy, S. B., Trans. Am. Inst. Mining Met. Engrs., 30 (1900), p. 896.

(48) Julian, H. F., and Smart, E., Cyaniding Gold and Silver Ores, 3rd ed., London, Charles Griffin and Co., Ltd., 1921.

(49) Gabbard, J. L., and Dole, M., Trans. Electrochem. Soc., 72 (1937), p. 129.

(50) Greenawalt, W. E., The Hydrometallurgy of Copper, New York, McGraw-Hill Book Co., Inc., 1912, p. 365.

(51) Lemmon, R. J., Chem. Eng. Mining Rev., 32 (Oct., 1939), p. 31.

(52) Rose, T. K., and Newman, W. A. C., The Metallurgy of Gold, Philadelphia, J. B. Lippincott Co., Inc., 1937, p. 413.

(53) Hunt, B., U. S. 689, 190, Dec. 17, 1901.

(54) Lawrence, H. M., Eng. Mining J., 104 (1917), p. 783.

(55) Mosher, U. S. 730,835, June 9, 1903.

(56) Edquist, V. T., U. S. 2,093,958, Sept. 21, 1937.

(57) Scott, W. W., and Furman, N. H., Standard Methods of Chemical Analysis, 56th ed., New York, D. Van Nostrand Co., Inc., 1939, Vol. 1, pp. 272-73.

(58) Jarman, A., and Brereton, E. L. G., Trans. Inst. Mining Met. Engrs. (London), 14 (1905), p. 289.

(59) Perley, G. A., Trans. Electrochem. Soc., 76 (1939), p. 85.

(60) Putnam, G. L., Jarmus, J. M., and Ochs, S. R., unpublished work.

(61) Gardiner, W. C., and Sanders, H. L., Ind. Eng. Chem., Anal. Ed., 9 (1937), p. 278.

(62) Kolthoff, I. M., and Rosenblum, C., Acid-Base Indicators, New York, The Macmillan Co., 1937.

(63) Hodgman, C. D., Handbook of Chemistry and Physics, 18th ed., Cleveland, Chemical Rubber Publishing Co., Inc., 1933, pp. 290, 772.

(64) von Bernewitz, M. W., U. S. Bur. Mines Dept. of Investigations, 3425 (Dec., 1938), p. 25.

(65) Putnam, G. L., Trans. Electrochem. Soc., 74 (1938), p. 252.

(66) Hunt, B., U. S. 689,190, Dec. 17, 1901.

(67) Green, M., J. Chem. Met. Mining Soc. S. Africa, 13 (1913), p. 355.

(68) Adair, A., J. Chem. Met. Mining Soc. S. Africa, 8 (1908), p. 331.

RECOVERING PRECIOUS METALS

FROM

WASTE LIQUID RESIDUES

A COMPLETE WORKSHOP TREATISE, CONTAINING PRACTICAL WORKING DIRECTIONS FOR THE RECOVERY OF GOLD, SILVER, AND PLATINUM FROM EVERY DESCRIPTION OF WASTE LIQUIDS IN THE JEWELLERY, PHOTOGRAPHIC, PROCESS WORKERS', AND ELECTROPLATING TRADES

BY

GEORGE E. GEE

GOLDSMITH AND SILVERSMITH

AUTHOR OF
"THE GOLDSMITH'S HANDBOOK," "THE SILVERSMITH'S HANDBOOK,"
"THE JEWELLER'S ASSISTANT," "THE HALL-MARKING OF JEWELLERY,"
ETC., ETC.

London:

E. & F. N. SPON, LIMITED, 57 HAYMARKET, S.W. 1

New York:

SPON & CHAMBERLAIN, 120 LIBERTY STREET

1920

NEW YORK

D. VAN NOSTRAND COMPANY

EIGHT WARREN STREET

PREFACE

The object in writing the following pages has been to present the reader with a thoroughly practical account of the simplest and most accurate methods of recovering the gold, silver, and platinum from the sundry waste liquid residues of goldsmiths, silversmiths, and other workers in the precious metal trades, and thereby provide a long felt want for the direct guidance of practical men engaged in the different kinds of manufacturing establishments wherein the precious metals are employed.

The text-books, as yet published, barely touch this subject, for, speaking generally, they only treat the matter from a hypothetical standpoint, and such information is altogether insufficient for service in the large precious metal manufactories wherein the manipulatory art of ornamental construction is taught and practised.

This book is written by a practical man, for practical men employed in the different branches of the precious metal industries of the nation, and the author has given full detailed descriptions of the processes and formulas, with illustrations of the apparatus to be severally employed in extracting the waste metals from all kinds of liquids, so as to make the book of special value to all concerned in the manufacture of jewellery and other kindred wares.

The author has had a large workshop training and practice, likewise an extensive commercial experience in the treatment of waste jewellery products, and has devised apparatus and methods hitherto unknown in manufacturing establishments, which are easy of execution and certain in their action; they are useful alike to the small as well as to the large manufacturer, and do not necessitate the employment of expensive chemicals or elaborate machinery to effect good results. He has treated the subject systematically, by proceeding in regular order with the different classes of wastes, so as to make the work available for all trades in which the precious metals are employed. This of necessity causes a certain amount of repetition, but for the sake of clearness it could not well be avoided.

A noted feature of the work is its simplicity of language, all technical terms and phrases, when made use of, being explained, so that the least informed operative may be able to comprehend the meaning and carry out the different operations to a most successful conclusion, and with a full appreciation of the possible results.

It now only remains to be stated that all the processes have been dealt with from an entirely practical point of view commercially, including such methods as the writer has found to give the best results in his own workshop, and it is hoped the numerous questions which for some years have been addressed to him regarding the difficulties that repeatedly arise in manufacturing establishments, will have their remedies provided for in these pages in a thoroughly sound and satisfactory manner. This is the only practical treatise on the subject.

GEORGE E. GEE.

STANDARD WORKS,
58 TENBY STREET NORTH, BIRMINGHAM,
1920.

RECOVERING PRECIOUS METALS
FROM
WASTE LIQUID RESIDUES

CHAPTER I

INTRODUCTION

THE quantity of soluble gold, silver, and platinum that exists in some of the waste liquids resulting from the manufacture of jewellery, silver plate, and personal ornament is considerable, and as these metals are of high value, it is of paramount importance that all operations in which they are used in the industrial arts should be conducted with great care, as regards unnecessary wastage resulting from the different modes of treatment, and that all the waste products be strictly preserved and correctly treated at intervals for the extraction of the precious metals.

The recovery of the precious metals from goldsmiths', jewellers', and silversmiths' waste liquid substances, such as the exhausted colouring liquids and their rinsing waters; the exhausted electro-gilding and electro-silver plating solutions and their rinsing waters; the stripping and dipping mixtures of different kinds; the pickling solutions; the waters used for washing out the polished and finished work; the wash-hands water; and all other fluids likely to contain traces of either gold, silver, or platinum is of momentous consequence to manufacturing firms employed in the making of gold and silver wares, as regards their ultimate financial success. It is also an important matter to those engaged in chemical operations in which the precious metals are concerned.

To know how to extract the gold, silver, and platinum to the best advantage, in the simplest possible way, from all the solutions and mixtures above named will prove of benefit to many, and by undertaking to make known the results of a series of experiments and researches the author has made in relation to the subject during his more than 50 years' experience in the goldsmithing and silversmithing trades, will, no doubt, be rendering a further service of some commercial value to the gold and silver working industries generally.

The recovery of silver from the hyposulphite and toning solutions, and the swilling waters of photographers, as well as the liquid residues of glass-mirror manufacturers, will receive attention, and the best methods pointed out for the extraction therefrom of the precious metal, there being little or nothing satisfactory in the text-books for the guidance of the beginner, and the more practical craftsman may not even be conversant with the most simple and successful processes. I hope, therefore, to impart some new and additional information, which will be of assistance to workers employed in the multifarious branches of the precious metal trades, and if by any means the loss occurring in waste liquids can be reduced down to the lowest possible quantity, it will be a distinct gain to science as well as to commerce.

The chemical classes in the technical schools are rendering an invaluable service to those of the rising generation who attend them, but they cannot give all that is needed, for to experiment with clean liquids is a very different thing to the treatment required for those contaminated with considerable mineral and organic matter, such as results from manufacturing processes in large establishments. This is a subject on which I have spent a great amount of time in labour and correspondence, and by laying bare some of the economies effected, it may be the means of others, more able than myself, of improving the system, and of still further reducing the working losses in those liquids to the minimum. In the old time, scant attention was paid to the recovery of gold or silver from spent liquids. In some cases with which I am acquainted there was only one tub used, as a final receptacle, for all spent solutions and swillings, which, after running through a piece of coarse bagging, were allowed to flow directly into the drain. Now this is all altered, but there is still room for improvement, for with such keen competition at the present time, and the lowering of prices, more economical methods have to be adopted for the recovery of the gold and silver therein contained. I will commence my narrative with instructions for treating colour waste.

CHAPTER II

RECOVERING GOLD FROM COLOUR WASTE

NEARLY every master goldsmith and jeweller has more or less waste of this nature to deal with, and where there is a large amount of coloured gold work made and the colouring *chemically* performed, the loss of gold is considerable. The reduction in weight of a batch of work may be estimated at anything between 2½ per cent. to 5 per cent. of the gross weight, that is to say, a loss of from 12 grs. to 24 grs. per ounce of mixed gold and alloy is incurred in the process, and this in time becomes a serious matter, unless provision for that loss is carefully provided for, and means taken from time to time to collect back the gold. The best methods of recovering the gold from the colour waste and swilling waters are not generally understood. I have frequently seen these liquids thrown away when only a part of the gold had been extracted therefrom. In some of the largest manufacturing goldsmiths' workshops it is no unusual thing for as much as 50 ozs. of work to be coloured each week, and if I take the loss of *fine gold* contained in the metal which has been dissolved from the articles and left in the colour and rinsing waters, at three-fifths of the total loss set out above, taking, of course, an average of the respective colouring standards which are usually submitted to the chemical process, it means that there will be something between 15 dwts. and 30 dwts. of *fine gold* in the colouring mixtures and rinsing waters employed for every 50 ozs. of made-up work. As gold is such a valuable metal, it will be apparent that it must pay to bestow a little extra labour and attention to these liquid residues in recovering the gold. For collecting that loss I shall give two processes, namely, the copperas process and the zinc process, which will effectually "throw down" the whole of the gold to the bottom of the receptacle in the metallic state, if undisturbed for 12 hours, in such a manner that not a trace of gold will remain in the water above it. Dissolved gold, in acid solutions which are clear, is very easily distinguished by three well-known and conclusive tests, namely, ferrous sulphate ($FeSO_4 + 7H_2O$), stannous chloride ($SnCl_2 + 2H_2O$) and oxalic acid ($C_2H_2O_4 + 2H_2O$), but in muddy and complex mixtures these salts, with the exception of the first named, are useless to the manufacturer in the treatment of the liquid wastes I am now dealing with, and nothing further need be said than that they can be, and are, used as a final test for gold, when the liquids have been cleared of all organic matter which obscures the light passing through, so as to enable the action of the testing acid upon the resultant fluids to be easily observed.

CHAPTER III

THE COPPERAS PROCESS

THE copperas process is as follows: For the collection of this kind of liquid residues, provide a large stoneware vessel like fig. 1, into which all the used colour, swilling waters, and pot-rinsings (the colour pot should be washed out each time after colouring and then dried ready for use another time) should be emptied when the colouring is finished. Ferrous sulphate, $FeSO_4$, which denotes its chemical formula, is really sulphate of iron (commonly called green vitriol and green copperas), the best precipitant that can be used for colour waste. It throws down, in strongly acid mixtures, a black precipitate of metallic gold, but with that is combined a kind of muddy sediment which it removes from the murky waters, rendering them perfectly clear, though slightly tinged with a trans-

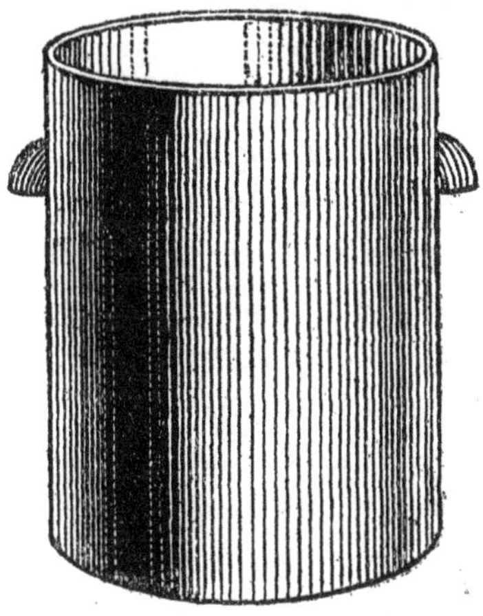

FIG. 1.—Earthenware jar for colour waste.

parent greenish shade of colour. This is the effect of the copper, which continues to remain in a dissolved state in the liquid in the form of copper chloride, $CuCl_2$. This the copperas does not precipitate.

Some manufacturers only preserve the old colour liquid in a separate receptacle, putting all the swilling waters into the general waste water tub. Now this is altogether wrong, for the reason which I will explain. The colour liquid, treated by itself only, would be much too dense for the *copperas salt* to search through the whole of that kind of liquid. It cannot therefore act to the best advantage. I have had several cases of this kind submitted to me for my advice, in which as much as 35 grs. of gold per gallon still remained in solution after treatment had been adopted for its recovery. The cause of so large a quantity of gold having been left in solution being, that insufficient copperas had been employed to decompose the strong acid liquid and liberate the gold therefrom. The remedy is simple. All the rinsing waters and pot washings from the colouring process contain gold and should be added to the colour waste that you have emptied into the stoneware jar, and in some cases, where very strong colouring mixtures are used, even that is not sufficient for most practical purposes, an equal volume of additional water being requisite in attaining the necessary state of diluteness, for by this addition the best results are always accomplished. Water is very much more preferable for dilution than that of neutralising the acidity of the mixture with alkaline salts, which not only thicken the fluid, but cause a greater abundance of sediment to appear than is beneficial in the recovery of the gold from the product resulting.

The most simple method of dealing with colour waste is to put into the stoneware vessel all the old colour, pot washings, and swilling waters which have been used at each operation during the week, and on the Saturday of every week, after the last batch of work has been coloured, to fill up the vessel with water (a vessel suitable to the extent of the business being done should be selected). After that proceed to precipitate the gold existing therein by means of a strong solution of iron sulphate, which should be made of a strength consistent with the estimated amount of gold taken from the week's work, all of which will (if this advice is followed) be contained in a dissolved state in the liquid in the stoneware jar. It is usual to preserve a portion of the old colouring salts for colouring odd jobs that will not stand the full colouring process without injury. They are kept in a stoneware jar like fig. 2 and kept covered with a lid, which preserves them in a good and clean condition, ready at any time that they may be required to be made use of. In clean neutral solutions, for 1 oz. of gold, $2\frac{1}{2}$ ozs. of iron sulphate will throw down the gold, but in the treatment of these residuums 5 ozs. at the least will be required of iron sulphate to throw down 1 oz. of gold, and in strong liquids more than that, as the acid has to be overpowered before it will give up its gold

FIG. 2. — Earthenware jar for old colour.

for the precipitation to be thoroughly effective. This is an important feature in the whole proceeding if the full measure of success is to be accomplished. For the lowest computed loss herein named, 15 dwts. of gold, dissolve 4 ozs. of copperas in 1 pint of boiling water, and add the whole of this gradually to the mixture in the jar with consistent stirring. For a loss not exceeding 8 dwts. greater than this, dissolve 6 ozs. of copperas in 1½ pints of boiling water, and proceed as before. For a loss not exceeding 8 dwts. greater than the last, or one of 31 dwts. of gold, dissolve 8 ozs. of copperas in 1 quart of boiling water, and so on in proportion to the estimated loss; gradually pour the hot copperas solution into the colour waste, stirring the liquid during the time you are pouring in the sulphate precipitating solution. The gold will at once begin to fall to the bottom of the vessel, and by the colour water resting till the morning (or if added on the Saturday till Monday) the whole of the gold will have become reduced to the metallic state and lie at the bottom of the vessel, the water above it being quite transparently clear, if sufficient copperas has been used in the operation.

The pale green water above the precipitated gold should not be thrown directly away, but transferred to the general waste water filtering tubs for a further treatment, although there should not remain a particle of gold in it if the aforesaid operation has been perfectly carried out; but to guard against accidents or imperfect management of the process it will be better to always follow this rule. The water above the precipitated gold is best drawn off with a piece of bent lead piping having a stop-cock secured on the emptying end, or with a rubber tube. The way to start the drawing off of the supernatant liquid is, to first fill the syphon tube, fig. 3, with water, then close both ends in the case of a tube having no stop-cock (when one arm of the syphon is much longer than the other it will only be necessary to close the long end), invert and place the short end of the syphon tube under the liquid, the other over an empty bucket and liberate the ends, when the water will commence to run off. Be careful and not disturb the precipitate at the bottom of the vessel or you will remove some of the gold and cause the operation to be unsatisfactory. When a syphon with a stop-cock is made use of, only one of the ends of the tube when filled with water will require to be closed before putting it into the liquid which is to be withdrawn. The advantage in using a stop-cock is that the whole of the acid water can be drawn away without having to refill the syphon tube, for by simply turning the stop of the tap, when the bucket is full, the water will cease to flow till it is turned on again. This is a very simple device to adopt. It also shortens the time of the operative engaged in doing the work. It is advisable to first withdraw by means of a pipette, fig. 4, a little of the clear green-tinted solution into a wine glass or test tube, fig. 5, and to add to it a *few drops of chloride of tin*, in order to ascertain for certain whether all the gold has been thrown down or not, and if this particular acid

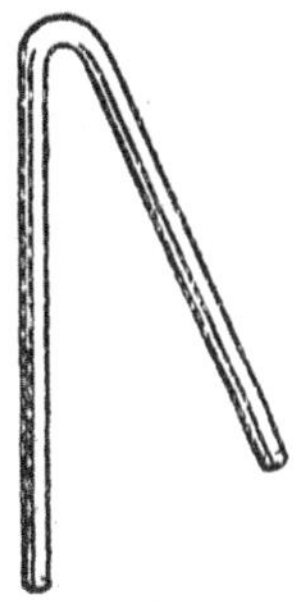

Fig. 3.—Syphon.

produces no effect in any way whatever upon the liquid withdrawn from the jar into the test tube, the supernatant water (this is the fluid standing above the sediment) may be safely drawn off without fear of any loss of gold. Chloride of tin ($SnCl_2$) is the most delicate test known to science for detecting the presence of the smallest possible quantity of gold in auriferous liquids. It has been stated that it is capable of detecting 1 gr. of gold in 1 million grs. of solution when transparently clear so as to transmit the light through it; that means it will detect 1 gr. of gold in about 13 gallons of liquid. Subsequently I will describe the way to make the chloride of tin and also the method of using it, pointing out its remarkable features in the detection of gold. The *tin salt* sold under that name in the crystallised form is of uncertain composition, and will not do, as it turns the green liquor being tested with it a cloudy grey colour, and this lessens its value as a test by taking away its best distinguishing quality, for when there is no gold in the liquid no change of colour should with the chloride of tin take place. *Protochloride of tin, stannous chloride, and chloride of tin*, by whichever name the "testing acid" is called, are all

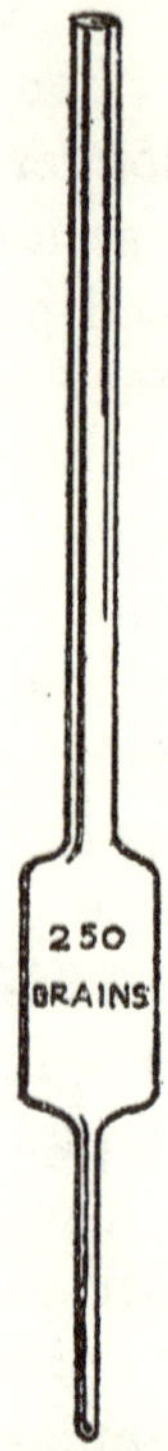

FIG. 4.—Pipette.

FIG. 5.—Test tube.

the same thing. The water standing above the muddy sediment lying at the bottom of the vessel should always be sampled and tested for gold with this liquid, before being drawn off. The acid water above the dark coloured mud (this contains the gold) will, when it is in a fit state to be drawn off, appear quite clear, with the exception of a tinge of green, a tint given by the copper which remains in solution in the form of dissolved copper chloride (cupric chloride $CuCl_2$). The copper under the copperas process is not precipitated along with the gold by the iron sulphate (ferrous sulphate, protosulphate of iron, green vitriol, and green copperas salts are all identical in substance, although differently named), but continues to remain unaffected in the watery waste, and is no detriment to the recovery of the gold. If there is an *atom of gold* remaining in the liquid when testing with the tin chloride, that acid will immediately detect it by turning the sample water from its greenish tinge to a reddish brown, or if more to a black colour, the shade varying in intensity in proportion to its diluteness and to the amount of gold existing therein; when this is the case it shows some gold to be still in the liquid state, and more copperas solution must be added to the bulk of liquid until no portion of gold remains; and, no matter how infinitesimal it may be, the "tin test" will distinguish it at once by producing a change in the colour. The copperas should not be added to the waste liquid in the crystallised state, but in solution. Repeat the testing operation after the further addition of copperas. When pouring in the latter solution to the bulk stir well, and allow to settle again before repeating the test, and if, on the second occasion of testing, no gold is detected in the sample taken from the stoneware jar, the water standing above the sediment may safely be removed, and conveyed to the waste water filtering tubs to avoid the loss of any fragment of metallic gold that may happen to swim on the top of this kind of liquid waste, the surface of which is often found covered with a thin lustrous film of pure gold.

The iron in the copperas salt is prevented from precipitating along with the gold by the large volume of hydrochloric acid (HCl) usually existing in liquids such as those I am now dealing with, for that acid has a greater affinity for the iron than the gold, and displacement between the two metals takes place, the gold going to the bottom of the vessel in the metallic state as a brown powder, while the iron remains in solution as muriate of iron, called muriate of iron when in a liquid and chloride of iron when in crystals, as a scientific distinction. The small amount of nitric acid (HNO_3) existing in the liquid oxidises some of the iron in the copperas salt, a portion of which then becomes ferric sulphate, $Fe_2(SO_4)_3$, or $Fe_2S_3O_{12}$, or Fe_23SO_4, and that is the reason for a larger quantity of copperas being required to throw down the whole of the gold from these mixtures, as the former takes little part in the precipitation of the gold through losing most of its active properties as a reagent for that metal.

The silver dissolved from the work during the process of colouring is precipitated as *chloride of silver* (AgCl) at the time of colouring by the hydrochloric acid and common salt used in the colouring mixtures, although some small portion of the silver may remain in a soluble condition in strong HCl and common salt; this is, however, when the colour waste is diluted with water reacted upon, and becoming liberated from the fluid settles down with the other as AgCl to the bottom of the vessel which is used as a receptacle for receiving these kinds of products. The truth of this conviction is easily ascertained by pouring off the water above, collecting and dissolving some of the sediment (AgCl) in a solution of potassium cyanide and immersing therein a piece of clean copper wire, when it will almost immediately become coated with metallic silver. The silver chloride may be separately collected and reduced to the metallic state by adding to it about two or three times its quantity of water acidulated with a little sulphuric acid (H_2SO_4) in the proportion of 1 part acid to 10 parts water, and placing in contact with it a plate of metallic zinc to extract the chlorine; the silver will then take the form of a blackish powder of metallic silver, which may be dried, after pouring off the liquid, and melted, with one-fifth its weight of soda ash as a flux; or it may be washed free of organic matter and dissolved in nitric acid and water and used for making an electro-plating solution; or the chloride of silver as it is taken out of the stoneware jar may be well washed and dissolved straight away with cyanide of potassium into a silvering solution, without reducing the silver (in the manner before stated) to the metallic state. Chloride of silver is insoluble in water. I give these processes for what they may be worth, but cannot recommend goldsmiths and jewellers to put them into practice commercially, as some of the metal is always lost in these washing operations unless very carefully performed. The far better plan is to collect all the sediment in one operation, boil it dry in a small boiler furnace, stirring the mass from time to time with an iron rod, and then either melt it into a bar yourself, have two or three assays taken by as many refiners, and sell to the highest bidder, or let the refiners take a sample of the burnt powder and make their offers for it in that condition, for in washing these sediments some of the metal is frequently carried away with the wash waters. To perform some of these operations commercially, although they may appear exceedingly simple, is quite another thing to the treatment indulged in scientifically in chemical laboratories for experimental purposes only, as difficulties will be encountered with the first kind, when being carried out in factories, that do not present themselves to those of the second kind performed in laboratories.

Pure iron sulphate is of a green colour in the crystallised state, and white in solution, but the crystallised green vitrol or green copperas of commerce is of a pale bluish green owing to its being contaminated with copper sulphate, but when in solution it is of a green colour, and has an inky taste. This salt turns *concentrated* colour liquid containing gold a very dark brown to black, but this shade ultimately changes to one of reddish brown, which remains fixed although every vestige of gold may have been separated from the solution —the undecomposed nitric acid being the cause—by oxidising some of the iron sulphate; the difficulty then is when to cease adding the copperas solution, but if sufficient of the green vitrol solution is poured into the colour waste when *considerably diluted with water*, it will quickly reduce the dark shade of the residue colour liquid to one of a pale green, which readily admits the light through it to such an extent that it appears to leave the water well-nigh colourless, a result which proves almost to a certainty that all the gold has been thrown down. It cannot, however, be safely used for concentrated fluids on account of its low density compared with that of the colour liquids, without considerable stirring to force it through those accumulations, for without much stirring the copperas solution, being lighter in weight, will not pass through such dense residuary fluid substances from end to end, but remain as a layer on the top; therefore, to be successful, you will see the paramount importance of well thinning the concentrated waste colour with water to enable the precipitant to search right through it, and to act upon it in every part. In dilute colour preservings the copperas, on its first addition, changes the colour of the fluids to a dark brown or browny black, but afterwards, as the gold subsides, the colour changes to one of very clear pale green. A good plan is to put the copperas solution into the colour waste during the day, stir frequently, and then let it remain all night; by the morning all the gold will have gone down and the operation completed. Sulphate of iron rapidly precipitates gold in these mixtures, but the particles being so very fine, time is required for their complete subsidence. The precipitated gold is removed from the jar at stated periods only, according to the fixed rule of each factory, and not each time the acid water is withdrawn.

The chloride of tin has the advantage of being the most delicate of all "tests"—that is to say, it will indicate the presence of a smaller quantity of gold when in the liquid state than either sulphate of iron or oxalic acid.

For drawing off a sample of the liquid for testing purposes a *pipette* is a very useful little instrument. It consists of a glass tube blown with a bulb in the centre, and drawn fine at one end (see fig. 4). The fine end of the instrument is put into the liquid, the other end into the mouth of the operative, and slight suction applied until the bulb is filled with fluid; then remove from the mouth and stop the top end with the thumb, and the liquid may thus be safely transferred to the glass test tube, into which the liquid will run from the pointed end when the thumb is removed from the top. Next pour into it a few drops of the chloride of tin, and very closely watch if any change of colour takes place. If no discoloration is observed, the gold has all been separated from the liquid, and the latter may then be decanted or syphoned off the sediment, leaving the vessel ready for use again as circumstances permit. The pipette should hold about 250 grs. of fluid. The test tube should not only be held between the eye and the light for examination, but also so held that the light is reflected from it instead of passing through it. This will show if the liquid is in any way changed from a greenish tint to a faint reddish brown hue, and if it does so change some gold must still be present, for a *very minute* proportion of gold gives a manifest reaction when this test is employed, its colour being so characteristic that once having seen it you will not in future be likely to be mistaken. Liquids looked at through *transmitted* light always appear lighter tinted than from *reflected* light.

It sometimes happens, after complete precipitation of the gold has been effected by the copperas solution, that minute particles of the gold swim upon the surface of the liquid, and it is often difficult to make them sink. This can, however, generally be brought about by beating the liquid with a glass or porcelain rod. After the supernatant water has been removed from the solid matter as far as is expedient, it is sometimes advisable to add a little coarse sawdust to the latter in order to dry up the remaining fluid. This greatly assists in the drying and burning down of the bulk, and prevents the resulting product forming into so refractory a substance, or by another method the remaining water may be got rid of by filtration of the sediment.

The precipitated gold, by a further method, may be separated from the liquid acid water by *filtering* the bulk through flannel or unglazed calico, but the syphoning is by far the best and most useful method, if conducted with moderate care and attention. It is, however, for individual manufacturers to adopt whichever method is most conducive to their interest.

The colour liquid—unlike the general waste waters, which, in every case, are considerably dilute—requires diluting with water besides the rinsing waters used in the colouring process, otherwise the liquid will be too thick and dense for the copperas to act to the best advantage. The colour liquid should be sufficiently dilute that when the copperas is added it should clear the solution sufficiently for the chloride of tin to pass freely through the liquid put into the test tube for sampling, so that by its action it may be at once perceived if any gold is left in the solution, no matter how infinitesimal the portion may be. When in the right condition, the chloride of tin can be seen travelling with sparking-like action to the bottom of the test tube, without the least change in the colour of the fluid, when it contains no gold.

Ferrous sulphate or copperas quickly oxidises in the air, losing some of its active properties and giving turbid solutions. It should therefore be kept in air-tight vessels when not required for immediate use. It will assume a rusty brown colour on the surface when it has taken up oxygen from the air, in which case that portion causes very little, if any, precipitate of gold, and this may not be generally known as being the reason of some imperfect results in gold precipitation.

It is frequently better economy to prepare the precipitant for the gold in the factory, and this may easily be done in the following manner: Into a stoneware acid-proof vessel put some iron filings or scraps of iron, then some water, and to this is added a certain proportion of sulphuric acid (oil of vitriol). This mixture must be prepared a few days before it is actually required for use, to enable the whole of the iron to become dissolved. The under-mentioned is a useful formula to employ :—

Iron filings	1 oz.	or	20 ozs.
Oil of vitriol	1 ,,	,,	1 pint
Water	8 ozs.	,,	1 gallon

More iron than the above will dissolve in this

mixture, but then it is apt to crystallise in the vessel, and will require to be redissolved in water before it can be used, otherwise it will be no disadvantage. One half pint of this solution will precipitate about 1 oz. of gold, and a teacupful of the crystals (when redissolved in water) a like amount. This knowledge, when possessed by the workman, will enable him to judge pretty nearly how much to take, in either form, for precipitating the estimated amount of gold in the waste liquid. After this is added, if it does not remove the dark colour from the liquid, as much more of the precipitant as is required to clear the solution is imperative, for the fluid must appear perfectly clear before the water can be drawn off, and I emphasise this point, in order that complete success may be the result of the operation.

The chloride of tin testing liquid should *always* be prepared on the premises, and not purchased from a chemist; then you will be absolutely on the safe side as regards its capacity to act properly. It is made as follows:—

Pure grain tin . . .	1 oz.	or	5 ozs.
Pure hydrochloric acid . .	4 ozs.	,,	20 ,,

Both the metal and the acid must be *quite pure*, and the hydrochloric acid is best saturated with the tin—that is to say, as much tin should be dissolved therein as it is capable of taking up when put under gentle heat. Put the tin into a glass dissolving flask (fig. 6), then the hydrochloric acid, and heat gently by means of a sand bath (fig. 7), placed on a small gas stove (fig. 8), until the whole of the tin has gone into solution, and when cool, after having removed the flask from the stove, put the contents into a stoppered glass bottle and preserve in a dark place. This liquid is acted upon by light, therefore do not omit this advice. A single drop of this liquid

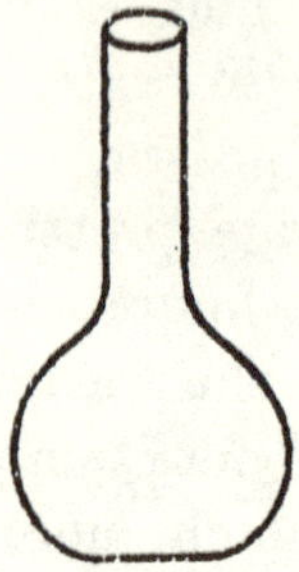

FIG. 6.—Glass dissolving flask.

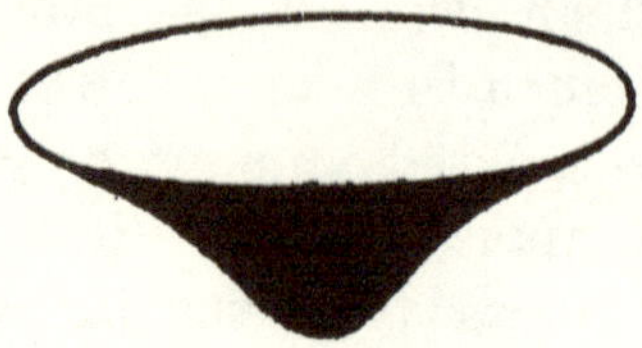

FIG. 7.—Sand bath.

on being allowed to fall into the clear pale green sample of 250 grs. selected for testing, if there is more than the minutest quantity of gold in it,

FIG. 8.—Gas stove.

will turn the liquid a black colour, even through *transmitted* light, and looking at the same through *reflected* light it will appear jet black. Two-thousandth part of a grain of gold, if existing in the quantity of liquid put into the test tube (equal to 1 gr. in 26 quarts—6½ gallons—will, with a few drops of *chloride of tin*, change the pale green colour of the liquid to a browny red colour, and if it contains more gold, from that to a brownish black colour, and still further to an inky black colour, due in proportion to the amount of gold there has been left in the solution after precipitation with the copperas mixture. The more gold there is the darker will be the colour. This reagent freely penetrates through the whole of the liquid sample, and will show you with accuracy whether the treatment with *copperas* has been properly performed.

No other acid or chemical salt is required to be added to the gold liquid resulting from "colouring" to prepare it for the reception of the ferrous sulphate (copperas). For this salt will, particularly when the colouring waste liquids are weakened with water, remove all the chlorine from its combination with the gold to unite with the iron of the ferrous sulphate for which it has greater affinity, thus converting the latter substance into chloride of iron (ferric chloride Fe_2Cl_6). I have already stated the small amount of nitric acid (HNO_3) existing in the liquid, and resulting from its combination with the saltpetre, one of the ingredients used in the colouring mixture, oxidises and converts some portion of the ferrous sulphate into ferric sulphate $Fe_2(SO_4)3$ (a higher state of oxidation than ferrous sulphate) without separating the gold from the liquid. The *ferrous compound* takes up the chlorine which holds the gold in solution to form *chloride of iron*, and by this reaction the complete precipitation of the gold is effected, for when the chlorine (the actual solvent of the gold) is removed, the gold gradually settles down to the bottom of the vessel as a dark brown metallic powder. After allowing time for the gold to settle and the solution to become cleared of its turbidness, the clear pale green liquid is ready to be drawn off from above the gold deposit in the manner already described. When this has been done, the vessel is ready for fresh quantities of colour exhaust, rinsings and water to be put in as occasion requires, the fresh portions being subsequently tested in the same way as the first, and this method is repeated again and again, until sufficient gold has accumulated upon the bottom. The gold recovered in this manner is usually very pure, because other metals should not be precipitated by the ferrous sulphate (copperas). The silver chloride, when not previously removed from the vessel, will of course cause the purity of the gold to be reduced in fineness. The copperas solution does not change the silver from its form of chloride to the metallic state.

To recover the gold almost pure from colour liquid waste, dilute the bulk with water, and this will precipitate the whole of the silver. Allow time for it to settle, then draw off the liquid above the precipitate of silver chloride. The water drawn off will contain the gold and copper in their liquid state, which, after being treated with the copperas solution, the former will become precipitated as nearly pure metallic gold, the copper being still left in the water above the gold, and with this will be most of the iron released from the copperas salt, a little oxide of iron being mixed with the gold.

I will now give a brief description of the reactions evolved in both strong and weak colouring waste products, and also in the rinsing waters from the same when separately treated, to determine whether the whole of the gold has been removed from the various liquids before the testing device is brought into use. Either chloride of tin or sulphate of iron may be employed as the final testing agent, but if the latter is made use of it must be the *pure salt*, and when dissolved only the clear white liquid taken for the purpose of testing, for if the solution is at all turbid no positive opinion can be arrived at, as it will impart its own rusty-looking brown colour to the liquid sample under examination, and will then not prove sufficiently distinguishing.

Sulphate of iron, if freshly made from the pure salt and a few drops of the liquid are added to a very dilute acid solution containing the merest trace of gold, gives a beautiful light blue colour to the liquid; but in a stronger acid solution, with the smallest increase in the proportion of gold and a corresponding increase in the proportion of the testing drops, a dark blue will be imparted to the liquid, and when there is a fraction more gold in the solution the depth of colour will increase to a brownish black, and to a higher degree the richer is the solution. After a few hours' rest for the gold to settle down, the liquid above it will change back to a pale green colour of transparent clearness which readily admits of the rays of light passing through it.

Chloride of tin, a few drops of this liquid on being added to a very dilute acid solution containing the minutest atom of gold, gives a brownish red colour to the liquid, but in a stronger acid solution with the slightest increase in the proportion of gold, a brownish black colour will at once be given to the liquid, and when there is more gold in the solution the depth of colour will increase to a *coal black*. But unlike the sulphate of iron, the chloride of tin does not clear the solution so as to admit the light passing through it, but remains the same colour into which it is changed for a considerable time afterwards. It will be observed that the more gold there happens to be in the sample liquids selected for trial the darker will be the resulting colours from the action of both testing mixtures.

Sulphate of iron is the best and quickest precipitant for bulky solutions, seldom requiring more than a few hours to throw down the whole of the gold and clear the solution sufficiently for testing purposes.

Chloride of tin is the most delicate test, and by far the best one to be employed in ascertaining if the sulphate of iron has precipitated all the gold from the waste liquid colour and its belongings.

The colour rinsings contain a considerable portion of gold, as the following examinations will show. Frequently after colouring a batch of work I have, for experimental purposes, swilled the pot out with water and added this to the rinsings through which the work had passed, then put a portion of this water into a test tube and added a few drops of tin chloride; in each case it immediately assumed the colour of port wine. As to further experiments, after first well washing the inside of the test tube with weak diluted hydrochloric acid, and following this with clean water, I proceeded to treat varying quantities of the rinsing water by putting a few grains to begin with into the test tube, and afterwards filling it up with water—the proportions of rinsings being increased each time and less water added—with the view of ascertaining the distinguishing effects of the *tin test* on the different samples. The colours produced varied from a browny red to a dark black, with only a few drops of the testing fluid, in proportion to the amount of additional water that had been used in the experiments. Next, a hot solution of copperas was stirred into the rinsing waters and allowed to rest for an hour or so, when the liquid became as clear as water. A drop or two of chloride of tin, on being added to a portion of the clear liquor, after transference to the test tube, produced then no change of colour, thus showing most conclusively that no gold after this treatment was left in the rinsings. On the water being poured carefully out of the basin and the small black sediment dried, it was, after rubbing with a burnisher, proved to be metallic gold of almost its natural colour. In nitric acid it was insoluble. These results settle firmly two facts, perhaps hitherto not very generally known, namely: (1) that a considerable quantity of gold is left in the rinsing waters used in the colouring process; and (2) that the tin chloride "test" is a fine and eminently satisfactory one in its search for gold when in the dissolved state, for when gold is present it affords an infallible proof of its existence as evidenced by the results of these investigations.

Oxalic acid ($C_2H_2O_4 + 2H_2O$), the acid of sorrel, is another test for liquid gold, although not so fine a one as the tin test. This salt when dissolved in water (1 oz. of the salt to 5 ozs. of water) and some of it added to a solution of gold, causes it to be precipitated as a brown powder, in clear solutions, in the same manner as the sulphate of iron; but the precipitation does not occur so rapidly, for it requires not less than forty-eight hours for the whole of the gold to be precipitated by oxalic acid solution. This test liquid when added to a solution of gold acts very slowly in the cold, but on applying heat, which is a chief point, it acts more quickly and liberates carbonic acid gas from the mixtures. It, however, precipitates gold to the metallic state without any other metal that may be present going down with it. It is a powerful poison. But apart from this, it is not a suitable precipitant for goldsmiths' and jewellers' residue solutions, nor for a "test" to finally determine the presence of gold after the adoption of the copperas treatment. The chloride of tin is much the better one, not only on account of its atomic weight, but also for the other reasons I have stated. And the sulphate of iron will precipitate the gold from all acid solutions so effectively that this particular testing fluid will show no action in any part of the clear liquids resulting from its effects, thus determining at once, when such is the case, that no gold is present in the solutions subjected to its influence.

Stoneware acid-proof vessels are best for preserving these residue liquids, for being smooth inside they are better for cleaning out the gold sediment, as it will not stick to the sides like it does to wood, which becomes soft and porous, unless thickly coated inside with asphaltum cement.

The action of sulphate of iron on the colour rinsings and pot washings produces, without further dilution, a light blue colour through *transmitted* light, and one of a darker hue through *reflected* light, taking the quantity of rinsings commonly made use of (about two quarts) for a single colouring operation.

The action of the chloride of tin on colour rinsings and pot washing produces, even when they are further diluted with an equal bulk of water, a browny red colour through *transmitted* light, and one of a dark brown through *reflected* light. The larger is the batch of work being coloured and the stronger is the mixture employed, the more gold is of course left in the rinsing waters, and this causes the colour tinges brought about by either of the testing liquids to be any shade between a light blue and a dark brown with the iron test, or between a red brown and a dark black with the tin test. The exact shade of colours obtained depends, in all cases, upon the quantity of water used in the operations.

The gold sediment which accumulates in the stoneware jar is usually collected every six months (or oftener, according to the amount of work that has been coloured), and put into a small boiler furnace made of cast-iron (fig. 9), to be gradually dried of its moisture, and then heated hot enough to completely destroy all organic matter. It must be stirred repeatedly during this operation to prevent its agglutinating. The burnt dry powder is next well pounded to break up hard lumps, then passed through a fine sieve, and this operation of pounding and sieving is repeated until all the coarse parts are reduced to fine powder to enable the whole of the product to freely pass through the sieve. The fine powder is then mixed with a reducing flux in the following proportions :—

Waste colour residue . .	10 ozs.	or 100 ozs.
Soda-ash (Na_2CO_3) . .	3 ,,	,, 30 ,,
Fluor spar (CaF_2) . .	1 oz.	,, 10 ,,

This is an excellent flux, which the writer has devised for the purpose of melting this kind of residue. The fluxes used must be reduced to quite fine powder, and then thoroughly well mixed with

FIG. 9.—Boiler furnace for drying precipitates.

the gold residue. Put the preparation into a clay crucible (the sort called a London round is the best, as the action can be more easily watched, and it stands the fire and fluxes better than most of the other kinds) and transfer it to the melting furnace. A good heat will be required for its complete fusion; continue the heat until the mass becomes quite liquid, which is facilitated by an occasional stirring with a thin iron rod. This helps to break up any undissolved portions and bring them within the action of the flux; thus a more perfect fusion is obtained by inducing uniform liquidness, and allowing of all the light particles of gold to pass downwards through the thin flux to the bottom of the crucible, so as to become collected together into one united lump of gold.

If any obstacle in the mixed product prevents the flux from taking the form of a thin liquid, such as iron oxide (a small portion of which may have gone down with the gold during precipitation), a few crystals (about 2 per cent.) of nitrate of soda ($NaNO_3$) added to the fused mass in the crucible will assist the fusion and further oxidise any portion of the iron that may be present in the residue, and this is then dissolved by the fluor spar and withdrawn from the gold into the slag; or a little more of the mixed flux may be used instead if preferred, for although an oxidising flux oxidises the base metals first, when it is used in too large quantities it causes some loss of gold as well by volatilisation, particularly when the latter is in a very fine state of division. The writer has found this colour, waste flux very satisfactory in his operations, and the proportions given ample for the collection of the gold by giving the product a good heat and seeing that it becomes thoroughly liquid before withdrawing the crucible from the fire, for if this is not done the whole of the gold will not be recovered from the residue. The whole contents of the crucible should be reduced to a thin liquid consistency to cause the small globules of melted metal to pass through the mass of foreign matter, by which in the first stage of fusion they are surrounded, to the bottom of the crucible, and by keeping up a good circulation in the fused contents of the crucible by an occasional stirring, the result will be found very satisfactory and easy of accomplishment.

Well-burnt precipitate of this description takes two-fifths of its weight of flux to melt it properly. The carbonate of soda or soda-ash is sold in a dry fine powder; it is a good reducing flux on account of the carbon it contains, and of its melting into a thin fluid; it also acts as an oxidising agent to iron: while the calcium fluoride or fluor spar possesses oxide dissolving properties for iron and other metals; and a small amount of it is the equal to a larger quantity of most other fluxes. Both these fluxes are cheap, which is also greatly in their favour. Fluor spar melts to a very limpid fluid, and when used along with the soda-ash forms an excellent compound flux for wet-colour precipitates. The flux floats on the top of the melted metal, dissolving and withdrawing into the slag any particles of iron oxide that may have fallen down with the gold during precipitation or formed in burning the residue, and it is readily parted from the metal underneath when the crucible is broken to recover the button of gold.

Ferrous sulphate ($FeSO_4 + 7H_2O$ being its molecular formula) in the crystallised state contains 56 parts of iron, 32 parts of sulphur, 64 parts of oxygen, and 126 parts of water = 278, which represents its molecular weight and combining proportions. One ounce of the salt contains 4 dwts. of iron as its percentage of the compound.

Stannous chloride ($SnCl_2 + 2H_2O$ being its molecular formula) in the crystallised state contains 118 parts of tin, 71 parts of chlorine, and 36 parts of water = 225, which represents its molecular weight and combining proportions. One ounce of the salt contains 10½ dwts. of tin as its percentage of the compound.

Oxalic acid ($C_2H_2O_4 + 2H_2O$ being its molecular formula) in the crystallised state contains 24 parts of carbon, 2 parts of hydrogen, 64 parts of oxygen, and 36 parts of water = 126, which represents its molecular weight and combining proportions. This is explained by saying that it takes 24 parts by weight of carbon, 2 parts by weight of hydrogen, 64 parts by weight of oxygen, and 36 parts by weight of water to unite chemically to produce 126 parts by weight of crystallised oxalic acid. This salt being an organic substance composed of *non-metallic* elements, it forms an excellent precipitant for gold when it is absolutely necessary for the gold to be precipitated free from any other metal. The action of oxalic acid on liquids holding gold in solution being that the hydrogen takes up the chlorine with which the gold is associated to form hydrochloric acid, and this frees the gold, causing it to fall down from the liquid as a metallic powder, whilst the carbon, also a reducing element, takes up oxygen from the liquid, thus forming during the process hydrogen chloride (HCl) and carbonic acid (CO_2). The latter mostly escapes as a gas as the work of precipitation gradually proceeds. Chlorine (Cl) is the actual solvent of the gold in waste colour liquids, of which it takes three atoms to form chloride of gold ($AuCl_3$) and any excess will be rejected after combination has taken place, the excess either being chemically united with the hydrogen of the water to form hydrochloric acid, or mechanically mixed with the watery fluid.

Chemically, each elementary substance has an abbreviation of its full name, called its symbol, which also represents its atomic weight—Fe, for instance, the symbol for iron, does not only mean iron, but 56 parts by weight of iron. Sn, the symbol for tin, does not only mean tin, but 118 parts by weight of tin.

When the small figure follows after the symbol it represents the number of atoms present in the molecule of that particular element—thus O_4 and Cl_2 in the above formulas mean that the molecule of the element oxygen, as shown by the first named symbol, contains *four atoms* of oxygen, or a multiple of four times its atomic weight of 16 by taking the latter as one atom; and that the molecule of the element chlorine, as shown by the second named symbol, contains *two atoms* of chlorine, or a multiple of twice its atomic weight of 35·5 by taking the latter as one atom.

When the large figure is in front of the symbol it represents, of an element, the number of atoms, and of a compound, the number of molecules there are in the whole compound substance; thus the compounds $7H_2O$ and $2H_2O$ represent the weight of *seven molecules* of water with regard to the first named, and of *two molecules* of water with regard to the second named. The molecular formula for water being H_2O, a combination of 2 parts hydrogen and 16 parts oxygen = 18, which represents its molecular weight and combining proportions, and that is taken to represent one molecule of water consisting of two atoms of hydrogen and one atom of oxygen.

The parts of the different elements as expressed in the paragraphs above may be taken as the number of grains of each substance which will chemically combine with each other to form ferrous sulphate, stannous chloride, and oxalic acid.

The drying and calcining furnace as shown in the illustration (fig. 9), is a very economical and much approved apparatus for the drying of precipitates of all kinds to be met with in manufacturing jewellery establishments, and it is also applicable for the burning of all the organic residues arising during the course of the multifarious working operations. It consists of a shallow cast-iron pan resting upon a frame of brickwork. It is heated by a coal fire, but any other kind of fuel may, if desired, be used for the purpose of heating. The boiler pan is covered with an arch (or a movable fume hood may be used instead, if preferred) having on the top an iron pipe for carrying off the fumes evolved during the operation of drying and burning the residuary products to powder. Two cast-iron doors are fitted in front of the brickwork, one for receiving the fuel, and the other to allow for the burnt ashes to be removed. The degree of heat is regulated by means of these doors, for by closing the top or fire door and opening the lower or ash door, the heat can be increased considerably when required, and by closing the bottom door and opening the top one, the heat can be slowed down in demand to the exigencies of the operation. Coating the inside of the boiler with a paste of lime and water, or with whiting and water (or with linseed oil instead of water), well rubbed on the iron surface, and then dried, tends to prevent the liquid substances attacking the iron in case they contain strong mineral acids.

Sometimes it is advisable with waste colour liquids, after you have got off what water you can with the syphon, to add two or three separate portions of boiling water and well stirring up the sediment each time, allowing the precipitate to settle down again before drawing off the water for any subsequent addition to be made; this is to free the sediment of any acid and destroy any corrosive action that it may be possessed of before putting it into the drying pan. When rendered sufficiently neutral, which may easily be known by dipping the finger into the last rinsings and tasting the liquid; it will then have a sour taste. Well diluted colour waste liquids do not require these preventive measures being adopted. After putting the gold sediment into the cast-iron boiler, dry gradually at first, then heat it until all is burnt to a powder; this is greatly facilitated by a repetition of careful stirrings with a thin iron rod. The residue will in most cases turn red in burning. This is owing, probably, to some oxide of iron having gone down with the gold, through portions of the ferrous sulphate salt becoming oxidised to ferric sulphate ($Fe_2(SO_4)_3$), which corresponds with ferric oxide (Fe_2O_3), a red powder commonly known as colcothar, crocus, and red oxide of iron, a material used in the polishing of gold articles. This oxide when once formed is insoluble in water and not readily dissolved by moist chlorine.

Precipitates of the precious metals, reduced to

dry powder in a boiler furnace of the description here represented, are effected without any loss of gold or silver, as is frequently the case when they are washed.

Gold dissolved in colour menstruums cannot be recovered therefrom by simple filtering operations alone, notwithstanding what is said to the contrary, for the gold will pass the filtering materials, such as paper, calico, baize, woollen cloth, and other substitutes, when in a state of solution, as readily as the clear water itself does; only the organic substances will be left behind on the filters, the whole of the gold being entirely lost, if the filtered liquid is allowed to pass straight into the drains without chemical treatment.

Ferrous sulphate (copperas) causes no effervescence or frothing to take place in waste colour liquids holding gold in solution. Neither does it cause fumes to be evolved. In aqua-regia containing gold the same characteristics are of identically a like kind.

Stannous chloride (tin salt) causes no effervescence, fumes, or frothing to take place in testing the samples selected from the waste colour liquid, nor does it with those containing gold in solution in aqua-regia.

Oxalic acid added to either of the above liquids produces effervescence and bubbles of gas. For the complete precipitation of the gold by this salt, sufficient oxalic acid solution must be added until it does not produce effervescence, and then it is best to slightly heat the solution, when, if there be any gold present, it will be precipitated as a dark brown powder. Carbonic acid gas escapes in bubbles soon after the oxalic acid is added to the gold liquid, which, in a cold solution, is rather slow in action, thus giving time for the bubbles of gas taking place throughout the solution to be observed; and to permit of the numerous small particles of metallic gold, forming in the body of the liquid by the reactions evolved, to more slowly fall down in the form of a brown powder.

Stannous chloride causes no discoloration of the fluid to take place in aqua-regia when it contains no gold, but with aqua-regia containing gold it causes an immediate dark brownish black colouring, the effect of the change being similar in its results to that produced on the waste colour liquids containing gold in solution.

Ferrous sulphate in aqua-regia, free of gold, turns the liquid a black colour at first, but this quickly changes to a mahogany colour which remains stationary, the nitric acid in the mixture converting a part of the iron salt into ferric sulphate, which gives the red brown tinge to the liquid.

Gold precipitated with ferrous sulphate causes very little organic sediment to fall down with the gold in waste colour residues, but if a larger quantity of the precipitant has been used than is necessary, and a portion of the iron in the copperas salt has become oxidised and fallen down as an oxide of that metal, due to the chemical action of the nitric acid or other causes, it will intermingle with the gold and prove an obstacle in the melting of the product. The iron can, however, be removed by a method which I will give, without interfering with either the gold or silver. The copperas solution does not reduce the chloride of silver to the metallic state.

Neither will the oil of vitriol pickle employed to dissolve out the iron from the gold. But if a few pieces of clean sheet zinc are used along with the pickle in contact with the silver chloride it will do so by the hydrogen evolved, the chloride of silver being then decomposed and converted into metallic silver, and sulphate of zinc ($ZnSO_4$) is produced in the liquid state. To remove the oxide of iron, the precipitated gold is subjected to the action of the following pickle, after all the liquid standing above it has been carefully syphoned away without disturbing any of the sediment.

Sulphuric acid H_2SO_4) . .	1 oz. or 1 pint
Boiling water (H_2O) . .	8 ozs. or 1 gallon

This mixture should be poured over the precipitate in sufficient quantity to entirely cover it. The water must be boiling, and is first poured into the jar holding the gold sediment, and then the acid is added to the water in small portions at a time, and great care will have to be taken to prevent its flying about and injuring the person in charge of the operation, as its scald or burn will destroy almost everything it touches. It is safest to add the oil of vitriol by means of the copper boiling-out pan, taking hold of the mouthpiece with a long pair of tongs, and extending them to their full length. Do not pour the acid into the hot water all at once, but very gradually, for on the acid first coming in contact with the hot water, detonations will occur, but as you proceed with the pouring in the explosions will be less forcible until only a hissing sound is made, at which stage sufficient acid has been added to the water, and the dissolving of the iron will proceed at a rapid rate until it has all become dissolved into the water. When the bubbles of gas cease to be evolved, and the liquid has cooled, it may be drawn off with the syphon, and another addition of the mixture of acid and water put on the sediment, if the iron has not all been dissolved by the first quantity. The iron oxide by this method is converted into ferric sulphate, leaving the solution a clear-tinted liquid of varying degrees of brown, according to the amount of iron that has been extracted from the gold. The sediment left after this operation will be the precious metals, and these are scraped out of the stoneware jar and put into the iron boiler and dried, ready for melting into a solid lump by means of the reducing flux before mentioned. The contents of the vessel, after the sulphuric acid mixture has been added, must be well stirred to bring every part within the reach of the acid mixture. Sulphuric acid is the strongest of the three mineral acids, and has a much greater affinity for metallic oxides than either nitric acid or hydrochloric acid has, and after treating waste colouring precipitates with the sulphuric acid pickle and drying the products, they may be melted down with one-fifth their weight of *soda-ash alone* for the flux, as there will then be no iron oxide to retard the fusion, and this flux soon liquefies to a thin fluid, which then permits the small particles of gold to easily pass through it down to the bottom of the crucible, from which they are recovered in a solid mass without any further trouble or difficulty of any kind.

In manufacturing establishments it will not be necessary to adopt all of these methods to recover the gold and silver in their metallic form from waste colour products, for if the former directions with regard to the diluting of the acid mixtures, the precipitation of the gold, the drying of the sediment, and the melting of the latter by means of a good reducing flux are carefully complied with, all the gold and silver will have become separated from other materials and found in a single button of solid metal upon the bottom of the crucible after cooling.

Briefly speaking, the various steps in the copperas process may be summed up as consisting of the following operations:—

(1) The preserving of the colour exhaust, rinsings and colour-pot washings in a separate vessel.

(2) The diluting of the liquid with an equal volume of water.

(3) The precipitating of the gold with a strong, hot solution of copperas and stirring of the mixture.

(4) The testing of the clear green acid water resulting, by means of a special test liquid of tin chloride.

(5) The syphoning of the acid water as completely as possible, when freed from all the liquid gold, and carefully safeguarding the sediment which settles at the bottom of the vessel.

(6) The drying and burning of the sediment in a small boiler furnace of cast-iron.

(7) The pounding and sieving of the product to a uniformly fine powder.

(8) The melting of the resultant powder by means of a suitable flux into a button of gold.

(9) The remelting of the button of gold, and the pouring of it into an ingot mould, ready for trial by assay with the view of sale.

The acid water withdrawn from the gold sediment (if these directions are faithfully performed) contains nothing of value. The copper and iron, and traces of other metals, should they be present in the liquid, may be thrown away, or if preferred, the liquid may be put into the general waste water filtering tube for further treatment.

CHAPTER IV

THE ZINC PROCESS

THE principle of this method, which is, I believe, new to the jewellery trade, consists of the fact that metallic zinc is an excellent agent for the detection and precipitation of gold from its liquid state in acid mixtures. Zinc can therefore be used with advantage for the recovery of the gold from the waste liquids resulting from the process of colouring gold. If a plate of zinc is immersed in these products, under the conditions about to be described, it will completely precipitate all the gold existing therein, and leave the liquid quite clear without any colour tint, for not only will the gold be "thrown down" but the copper as well.

The liquid colouring waste containing the gold, silver, and copper is put into a stone jar, the same as described under the "copperas process," but in addition it has a cross-bar, to which is attached a zinc plate (fig. 10). The solution containing the gold should be diluted with at least twice its volume of water, after the old colour, rinsings and pot washings have been emptied into the collecting vessel. Large or small quantities of liquid may be treated with equally good results if sufficiently diluted, for diluted colour and its auxiliaries will not then act on the metallic zinc with more than the necessary energy, as it is not owing merely to the dissolving of the zinc that the gold is precipitated, but to the evolving of "nascent hydrogen." The object is to expose the greatest amount of surface to the liquid as is required to reduce the gold without adding too much to the size and weight of the zinc plate employed, for if too much zinc were made use of in the operation, and the liquid consisted of too much free acid, it would act to disadvantage, for the zinc would be dissolved out of all proportion to its needs, causing a great and unnecessary fluffy sediment to be thrown down with the gold, with which would be ultimately formed a quantity of zinc oxide (ZnO), to be afterwards dealt with in reclaiming the gold. The metallic zinc gives the best results in the form of a sheet, and the gauge size may be about that used for making thick gutter spouting, which will expose the necessary surface to the action of the liquid for the development of the greatest amount of *nascent hydrogen.* Hydrogen which has left the generating vessel is not nearly so effective in the precipitation of metals as that which is produced in the same vessel in which it is to act. This is explained by stating that "nascent" means, *beginning to exist, or newly born,* in which state it is much

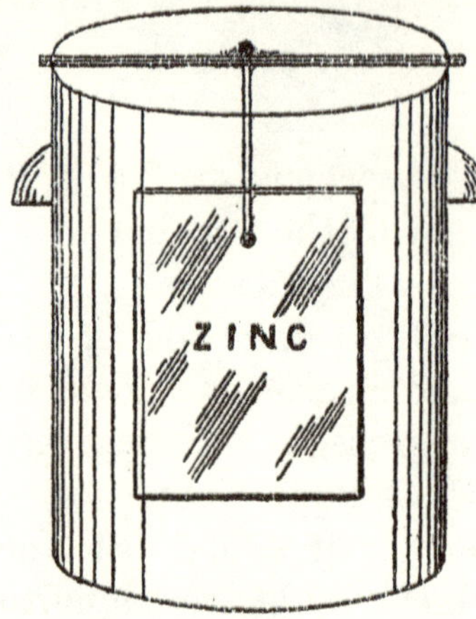

FIG. 10.—Stoneware precipitating vessel.

more vigorously active. By bringing metallic zinc into waste colour liquids, properly diluted, the free acid immediately begins to act upon the zinc, and though ever so slightly, it causes nascent hydrogen gas to be given off, one equivalent volume combining with one equivalent volume of chlorine to form hydrogen chloride, the excess of hydrogen rising to the surface of the liquid and escapes, while the remaining equivalent volumes of chlorine which assisted in holding the gold and copper in solution are extracted by the zinc as it is being slowly dissolved (the chloride of gold and copper are then decomposed and reduced to the metallic state), and chloride of zinc ($ZnCl_2$) is produced; the latter, which is a very soluble salt, remains in solution in the water above the gold, silver, and copper sediment.

In using the zinc, it is not necessary that the surfaces should be clean, as tarnished surfaces prevent the gold depositing upon them, and the action of the acid liquid does the work of presenting fresh surfaces to the influence of the mixture. The larger is the volume of free acid in the liquid residues resulting from colouring gold the more capable is it of dissolving the zinc, and the larger must be the quantity of added water to save the unnecessary waste of zinc. The proportion of zinc used need not be very exact so long as it is in excess of the computed amount of gold in the solution. It depends more on the amount of surface exposed, and to the very gentle development of hydrogen that takes place over the whole surface to bring about the more successful precipitation of the gold. About 1 oz. of metallic zinc to 1 gallon of liquid, after being diluted with water, will give good results, but the quantity of zinc required will depend upon the amount of gold present in the liquid. The best and most simple plan will be to suspend in the centre of the vessel from a cross-bar a large piece of sheet zinc reaching nearly to the bottom of the liquid, and let it remain for several hours, stirring occasionally during that time so as to bring fresh portions of the liquid in contact with the zinc sheet. The reactions thus set up within the liquid will, in a satisfactory manner, completely throw down all the gold and copper into metallic powders in a few hours, without assistance of any other kind, providing the liquid is not too strongly acid to corrode the zinc unnecessarily and cause it to be dissolved too quickly. The zinc when placed in the colour liquid first turns browny-black, caused by the bubbles of hydrogen gas which escape from its surface. This reaction prevents the gold from depositing upon it, and these effects are continued until the whole of the gold has become liberated from the liquid, and afterwards until the zinc is all dissolved, if not removed from the vessel. It will not be advisable to let the zinc remain too long in the solution if acted upon too strongly, as too much zinc chloride would be formed, and nothing is to be gained by continuing the operation longer than is required to effect its purpose; usually, in weak solutions, a day or a night is ample for the purpose. The occasional stirrings should not be neglected, for this greatly facilitates the displacement of the metals from their chemical union with chlorine in bringing fresh portions of the liquid to react on the zinc sheet, and more hydrogen gas is thereby given off to combine with its equivalent of chlorine and liberate it from the gold and copper, to form in the watery liquid dilute hydrochloric acid, which then being unable to hold the gold and copper in solution, they become precipitated as metal. This acid readily dissolves the zinc into the solution, and holds it in the form of *liquid zinc chloride*, and as the action progresses the hydrogen which is being set free (1 part only combines with 35½ parts chlorine) escapes in vaporous fumes in an invisible form, the chloride of zinc taking the place of the gold and copper, which will then have become reduced from their liquid condition to metal. The silver is already at the bottom of the receptacle in which this operation is being performed, in the form of silver chloride, it having been reduced to that condition by the chemical reactions set up during the process of colouring the gold. The cause of this is explained in treating of the copperas process. The chloride of silver, by the zinc process, may readily be reduced to metallic silver at the same time that the precipitation of the gold and copper is taking place, by putting a few cuttings or scraps of zinc on the bottom of the vessel in direct contact with the silver chloride, when the act of conversion will commence at once, and nascent hydrogen gas will be evolved, which, uniting with the chlorine, withdraws it from the silver, leaving the latter in the form of a dark grey metallic powder, which will remain in a mass at the bottom of the vessel in association with the other precipitating metals which are at the same time leaving the liquid. This operation of treating the silver chloride does not require any further attention than frequently stirring to bring the scraps of zinc in contact with fresh portions of the silver chloride (which is very small in quantity in these waste products) to cause the development of sufficient hydrogen to extract all the chlorine from the silver, for it is owing to the nascent hydrogen which is being set free by the agency of the metallic zinc that possesses the property of reducing the silver chloride to the metallic condition, as it is well known that hydrochloric acid consists of a solution of *hydrogen and chlorine* gases in water, and when zinc is brought into the presence of these two elements in solution they become decomposed from their chemical combination; the chlorine, which has a stronger affinity for zinc than for hydrogen, leaves the latter, which then escapes, while the chlorine unites with the zinc, and chloride of zinc is produced, which remains soluble in the liquid residue.

The precipitation of gold and other metals by means of zinc shavings and zinc dust is not new to commercial enterprises, as it has been used in gold mining, in connection with the process known as the "cyanide process," for quite a long time. But the principle of this method has not, it is believed, been applied to the recovery of gold and silver from the residuary liquid products of any manufacturing branch of the precious metal trades. The writer claims the right as to the inceptive of rendering the system completely applicable to almost all the gold and silver manufacturing establishments, for the recovery of the gold and silver from the liquid residues, accumulating from, and as a result of, the different working operations. To many it is undoubtedly known that zinc will precipitate the precious metals, but they have been unable to provide workers in those metals with a simple method whereby it can be rendered commercially successful, therefore such knowledge is valueless to the commercial business man. The great secret of success lies in the fact of being able, not only to reclaim the precious metals, but to do so completely and commercially in the most simple manner possible. It is the knowledge of these things that makes the operations of some practical value, less troublesome, and less expensive to use, by being reduced to simple working form, and, on the details of the "zinc process" being carefully attended to at the commencement, very little attention is afterwards required, until the gold has all been precipitated from the dark, cloudy, acid water with which it is in chemical combination. By properly regulating the strength of the gold solution so as to act on the zinc with moderate energy, and by using sufficient zinc in proportion to the quantity of gold in solution, every particle of the latter metal, and also the copper, will be thrown down into the metallic state. The resulting liquid will then have (after standing awhile) the appearance of clear water without any tinge of colour in it.

By this method the acid water standing above the precipitated metals should be tested with the chloride of tin before syphoning or pouring off, to make sure that all traces of gold have been "thrown down." This is done in the following manner: After the zinc plate is withdrawn from the mixture, and when the whole of the precipitate appears to have settled to the bottom of the vessel, after standing unmolested for a time, a small portion is drawn off with the pipette (fig. 4) and put into the test tube (fig. 5) for examination under the action of a few drops of the tin chloride. If no change of colour takes place, it will be sufficient proof that the whole of the gold has been thrown down ; but if any change of colour does take place, not excepting the faintest brown colouring, yet no subsidence be produced, it will indicate the presence of gold, and the zinc plate must be replaced to effect the complete precipitation of the gold still remaining in the solution. From time to time samples should be taken from the solution and tested for gold, and when the solution appears perfectly clear, and remains so after the tin chloride is added, the supernatant water can with safety be decanted or syphoned away from the sediment, for it will be proof positive that no liquid gold is then remaining in the solution.

Metallic zinc immersed in old colour liquids precipitates the gold and also the copper in so perfect a manner as to leave a clear white solution for a final trial to be taken by means of the chloride of tin test. No gold becomes deposited upon the zinc plate in liquids impregnated with hydrochloric acid, and no other substance is required to prepare such solutions to receive the zinc plate. The only attention that they require is being diluted with sufficient water at the commencement, to avoid the zinc plate being too forcibly attacked by the acid. The chloride of zinc, formed in the solution by the dissolving of the zinc plate, is not precipitated by protochloride of tin, which being more dense, readily searches the selected sample throughout the length of the test tube, without any chemical reactions being set up,

4

and without any change taking place in the colour of the fluid, when it is entirely freed from gold.

Chloride of zinc, if used in the form of a saturated solution of its salt will not precipitate gold from waste colour products, because there is then no hydrogen liberated, no chemical change takes place, the zinc would be distributed throughout the whole liquid in a more finely divided state, thus causing the solution of zinc chloride to be the less saturated only without decomposing the gold.

In testing with protochloride of tin a clear sample of the zinc-treated colour liquid, when the quantity of gold is very minute no precipitate will be perceptible for some hours, and you will have to be entirely guided by the change in the colour of the liquid, for when it contains gold it will acquire a red-brown to black colour immediately on becoming mixed with the protochloride of tin; the lighter is the colour produced the less gold there will be in the solution, and the darker is the colour produced, the more gold there will be left undecomposed in the liquid. The bulk of the solution will, when a change of colour takes place in the selected sample undergoing examination, require further treatment with zinc, otherwise a loss of gold will result, through the imperfect precipitation of the whole of the gold from the liquid in the stoneware jar by means of the aforesaid zinc plate, and the latter will require to be hung in again for some time longer to complete the process.

In making the protochloride of tin or "tin salts," the tin should be granulated by melting it and pouring it into cold water to break it up into small fragments, in which form they dissolve more readily, the water being stirred in a circular direction during this operation, and the pouring done at some height above the surface of the water; or the tin may be rolled thin, and cut into small pieces to expose a larger surface to the action of the acid. The tin should be dissolved in concentrated hot hydrochloric acid, as tin does not readily dissolve in dilute hydrochloric acid, and in concentrated cold hydrochloric acid the action is slow. It is of importance that this liquid should be of the right composition for gold-testing purposes. It is not absolutely necessary that the tin should be in excess in order to neutralise the acid exactly, although it is much the better for it, and if a small amount of tin is left undissolved in the bottom of the glass flask (fig. 6), the clear solution should be poured from it, and the liquid will then be ready for use if it presents a clear white solution. In testing selected samples of acid liquids with this fluid, when they are found to contain gold, it proves, without doubt, the whole of the chlorine has not been extracted from the gold and the liquid changed into a different compound mixture to the original.

The theory of the action of stannous chloride $SnCl_2$ (protochloride of tin) is that it has the power to combine with excess of chlorine to form stannic chloride $SnCl_4$ (perchloride of tin), and thus rid the gold of the remaining chlorine, is evidence of its delicacy as a test for gold in solution. Stannous chloride has also a great affinity for oxygen as well as chlorine, and owing to the great readiness with which it combines with these elements (both of which exist in the colour liquid) it forms a powerful reducing agent for gold in the soluble condition, but through this affinity for oxygen some oxychloride of tin is formed, which in variable quantities becomes mingled with the gold precipitate, forming purple of cassius, which consists essentially of metallic gold and oxide of tin, and the change in the colour of the liquids holding gold in solution is due to these reactions. A mixture of protochloride of tin and of perchloride of tin, gives a purple precipitate of stannic oxide mixed with finely divided metallic gold, and that is one reason, the expense being another, why stannous chloride cannot be employed with advantage in manufacturing establishments as a first means to remove the gold from liquid residues in which the gold is soluble, as too much tin oxide would be thrown down with the gold; but as a testing agent for gold it is superior to any other known substance in the detection of the minutest quantity. Both stannous chloride and stannic chloride in solution are heavy colourless liquids readily passing through lighter solutions. Gold being a feeble base, and, unlike most common metals, is easily separated from its various states of combination; its compounds being generally unstable, and are speedily decomposed without the assistance of complicated and high-toned scientific operations.

Sometimes a small amount of the gold precipitated by means of zinc swims, and a thin film of fine gold is observed to float upon the surface of the solution in the metallic state; when this is the case, the bulk of the acid water may be drawn away without its being lost, or even disturbed for the matter of that, for by carefully placing the short end of the syphon, or the end of the rubber tube, some distance beneath the top layer, the water below can be withdrawn without interfering with the layer of gold at the top; then as the water is lowered by being taken out of the vessel, the film of gold on its surface will sink down with it, and when nearing the sediment at the bottom of the vessel the whole of the remaining liquid, together with the sediment, may be thrown upon a filter of calico stretched on a wooden frame, or placed over the meshes of a cane-bottomed riddle; the gold and the organic substances remain upon the calico, the liquid only passing through it, or a rapid agitation of the liquid at the surface only, with a glass or porcelain rod, will cause the gold film to break up and fall to the bottom of the receptacle in which it is being treated. Either of these devices can be successfully brought into use when dealing with float-gold.

In precipitating gold by means of metallic zinc the precipitate will appear more voluminous than that resulting from the action of ferrous sulphate, as all the metallic substances in the liquid will be thrown down with the gold, and the operation will require to be closely watched to avoid too much zinc getting into the solution, which will result in a light spongy substance of oxychloride being formed, and this settling down above the gold powder, becomes when dried in the boiler furnace oxide of zinc, and a product is thus produced which, besides adulterating the gold, increases the mass and the oxide of zinc making it more difficult to melt, and a larger quantity of flux is therefore required to effect a clear fusion. It is true the light fluffy substance could all be washed away from the gold, but some gold always goes with it and a loss is incurred by the process. The washing of jewellers' waste products is bad practice, and it is not to be recommended, especially to those unacquainted with chemical operations. But if this method is preferred, after you have drawn off what liquid you can with the syphon without disturbing the sediment at the bottom of the vessel, for this will contain the gold, add three or four saucepansful of boiling water, each saucepanful being added separately, and allowed to cool each time and the water carefully syphoned off; this will clear most of the light fluffy substance away from the more solid sediment, and also clear it of any free acid and chloride of zinc crystals that may have formed and separated out from the liquid and become mixed with the gold, if the point of saturation had inadvertently been reached in the process of precipitation.

By another method both the oxides of copper and zinc may be removed so as to leave only the gold and silver behind in the operating vessel. For instance, by bringing these oxides into a strong mixture of hydrochloric acid and water both chloride of copper and chloride of zinc is formed, each combination being soluble in water may, when solution is effected, be drawn off the gold and silver sediment with the syphon and thrown away.

The residue left after precipitating with metallic zinc, contains the gold, silver, and copper, and probably some oxide of zinc. The two last named substances can be dissolved away from the gold and silver, after the supernatant liquid has been drawn away from the sediment, by means of the following acid mixture:—

Hydrochloric acid (HCl)	.	1 oz. or 1 pint.
Water (H_2O)	. . .	4 ozs. or 2 quarts.

If some of this solution (sufficient for the work in hand) is poured over the precipitate and stirred, it will dissolve out the copper, and also other base metals, should they be present in the residue, without affecting the gold or silver. The pickle

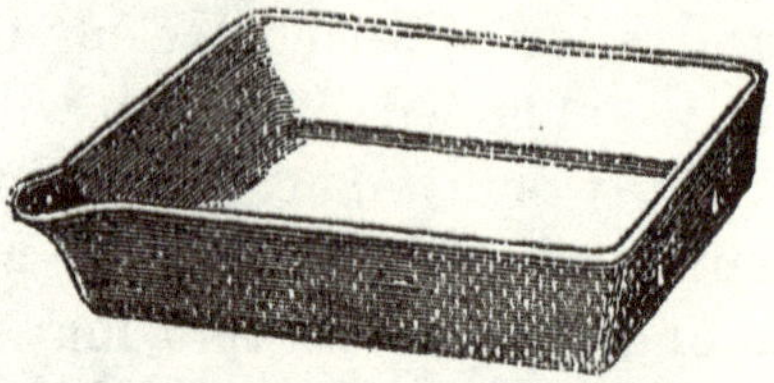

FIG. 11.—Dissolving vessel for base metals.

must be used boiling hot to be thoroughly effective. The precipitate may be treated more conveniently if it is removed from the precipitation jar and put into an enamelled iron, or into a porcelain or earthenware vessel of a shape similar to illustration, fig. 11, which has a large interior surface of shallow depth, so as to enable the residue to be spread out well over the bottom for the acid mixture to act to the best advantage, and a sufficient quantity of the pickle must be poured over the precipitate to entirely cover it. There will be no danger of explosive detonations occurring in using this mixture, like those previously mentioned in describing the sulphuric acid pickle, and it can be mixed (the acid with the water) before adding it to the residue. In due proportion to the impurities in the sediment will the effervescing action appear visible and bubbles of hydrogen gas arise. (When the base metals consist of metallic oxides no hydrogen is liberated in dissolving in HCl—the hydrogen combines with the oxygen of the oxides to water, and the chlorine with the metals to chlorides.) When the bubbles of gas cease to be evolved, and perfect quietness exists, the liquid is poured away from the sediment remaining through the lip of the operating vessel. If it is suspected that any base metal is left undissolved, a fresh portion of the acid mixture must be added, and the sediment gently stirred with a glass or porcelain stirrer, when, if no further bubbles are perceived after this addition, the operation of removing the base metals is completed, and the sediment may be put into the boiler furnace and dried of all moisture, at which stage it is ready for melting into a solid lump by means of a good reducing flux. The best flux for this purpose (when the base metals have been dissolved out) is soda-ash, in the proportion of 2 ozs. of the salt to each 10 ozs. of the powdery residue. But when the precipitate is dried and burnt direct from the precipitating jar, with all its organic and metallic impurities, this is not the best flux to use, particularly by itself. The burnt powder will then require two-fifths to one-half its weight of flux to effect a thorough fusion, and the following formula for gold reduced by the zinc process will prove excellent :—

Waste colour residue . . .	10 ozs.	or	100 ozs.
Bisulphate of potassium ($KHSO_4$)	3 ,,	,,	30 ,,
Common salt (NaCl) . . .	1 oz.	,,	10 ,,

There are other fluxes which may be used for

reducing and recovering the gold from "wet colour" precipitates, but I have found none so effective as those I have named, in dealing with the products resulting from the two processes under consideration. The bisulphate of potash, commonly called salenixon, is a flux of great power capable of searching through the most refractory substances, and will melt almost anything; it has solvent properties for metallic oxides, such as iron, zinc, and all the base metals, and it forms with the common salt, when melted together, a thin fluid through which the gold readily passes to the bottom of the crucible, no shots being left behind in the flux; the common salt is the best flux for oxide of zinc, as it converts it into chloride of zinc, and it is then retained in the liquid state; and both fluxes are inexpensive to use, a property adding further to the value of the mixture. The action of this flux is very energetic, all the common metals being attacked by it, but the crucible should not be more than three-fourths filled with the product at first, because the sulphurous acid which is given off might cause the fusing mass to overflow the sides of the crucible if filled to the top, although both salts possess counteracting properties to the boiling up action.

The zinc process, which is purely one of hydrogenation, is based upon the readiness with which hydrogen in the "nascent" state, that is, at the moment of its commencing to be produced, attacks the chlorine, in which state it has a far greater combining capacity for that element than when generated by the ordinary chemical methods. Hydrogen, even more than oxygen, is practically insoluble in water, which does not appear capable of dissolving more than about 2 per cent. of its volume of the gas.

As the colour liquid has usually an excess of acid, and as no precipitate of gold takes place until all the free acid is neutralised, which causes a considerable loss to the weight of zinc, there will always be found in the liquid a quantity of zinc chloride; but by previously diluting the colour liquid with twice its volume of water this compound is not so largely formed, for the solution is then made much weaker, and the free acid being in a measure sufficiently neutralised to present the condition which better serves for the evolution of nascent hydrogen gas (H_2) and the elimination of the chlorine (Cl) from the gold by the zinc, for which it has a greater combining capacity, and, by these two elements thus uniting, the gold leaves its associate and returns to the metallic state, a higher attractive force for the chlorine than that of the gold which is in union with it is being set up, which causes the chlorine to leave the gold it was originally combined with to combine with the zinc, as it is being gradually dissolved without unnecessary waste into the watery mixture, if the fundamental principles are complied with.

The zinc process for the reduction of the gold and copper is based upon the fact of the electrolytic action which takes place between the acids and the metals in solution and the one introduced in the metallic form, whereby a chemical change or simple displacement of the metals is effected. If you take into account the silver, which is always to be found in the form of chloride at the bottom of the vessel in liquid colour wastes, the substances used and produced may be thus expressed:

Substances Employed.

$AuCl_3$ +	$AgCl$ +	$CuCl_2$ +	$3Zn$
Gold chloride	Silver chloride	Cupric chloride	Zinc metallic

Substances Produced.

$3ZnCl_2$ +	Cu +	Ag +	Au
Zinc chloride	Copper metallic	Silver metallic	Gold metallic

Gold is precipitated from colour water by other substances than the two I have named, but none of them are the equal to the "copperas process" which has held its own and foremost place for some considerable time, and is likely to do so in the future, as it is simple to operate, clears the solutions of all cloudiness, is economical, and thoroughly effective in recovering the last atom of gold. While claiming the second place for the zinc or "hydrogen process" which, with a little practical experience in its management, will be found to be nearly the equal of the former in every way, so far as regards the principles of simplicity, the low cost of recovering the gold, the freedom from objectionable features, and the easy method of working, should commend the process to the notice of precious metal workers in reclaiming the gold and silver from their waste liquids, for it will be found to consist of all the virtues claimed for it, namely, that of a powerful operating agent for extracting gold and silver from solution.

I will now enumerate in a summary form the various operations appertaining to the zinc process:

(1) The pouring of the exhausted colouring mix-

tures, rinsing waters, and pot swillings into a stoneware jar kept specially for the purpose.

(2) The dilution of the liquid with twice its volume of water.

(3) The precipitation of the gold and copper by means of a sheet of metallic zinc (about 1 oz. to the gallon) with an occasional stirring of the solution.

(4) The testing of the clear white liquid resulting with protochloride of tin solution.

(5) The removal of the clear acid water as near to the precipitated metals as possible.

(6) The drying and burning of the precipitate to a uniformly fine powder.

(7) The mixing of the powder resulting with the reducing flux, also in a powder, and the melting of the mixture into a metallic button.

(8) The remelting and casting of the button into the form of a bar for sale to the highest bidder.

Having now described the general principles of these two processes, which my practical experience and researches have enabled me to put to a series of tests with the view of establishing easy working methods for the complete recovery of the gold and silver from muddy acid liquids, that may be successfully used by anybody with a modicum of ability engaged in manufacturing articles that are required to undergo the chemical process known in the trade as "gold colouring." I have entered at considerable length into all the details connected with this subject commercially, without bringing into the narrative the numerous theories that have been advanced from time to time, as to the best methods for recovering the gold from the dark murky acid liquids resulting from manufacturing processes in the precious metal trades.

A brief reference to the more commonly advised methods, however, ought to be given alongside those I have introduced, and I will next proceed to give one or two distinct projects of other authorities with my criticisms thereon, which will be sufficient, I feel, for the purpose of describing the different modes of recovering the gold from the waste liquids resulting from wet-colouring.

In giving an outline of the several methods introduced by scientific men for the extraction of the precious metals from these liquid wastes of goldsmiths and other workers in those metals, it is only fair that the views of such persons should receive due respect and consideration; but notwithstanding the superior training received from a thorough course of chemical and scientific study in the laboratory, such advice has not always given satisfactory results in the workshop, through the omission of some simple, though most important, detail necessary in practical operations; and this is not to be wondered at when unacquainted with a knowledge of workshop practice in large manufacturing establishments, for possession of this knowledge is quite as essential as that of the arts and sciences, if not more so, to achieve commercial successes in a speedy and satisfactory manner in business enterprises. The waste problem has hitherto proved very difficult of solution, even to the practised eyes of the most experienced and intelligent workman, but I trust when these articles are read and well digested, something really serviceable will have been established to minimise the enormous losses resulting through the ineffective treatment of waste liquids appertaining to the manufacture of jewellery. The first of these methods, which has been much extolled by some, may be mentioned by name as the old iron binding-wire process.

CHAPTER V

THE OLD IRON BINDING-WIRE PROCESS

It consists in hanging in the waste colour liquid a matte of used-up old pieces of iron binding-wire. The liquid is to be undiluted, and it is stated the reactions evolved within the vessel will precipitate all the gold from the liquor. Then, to test if all the gold has been thrown down, withdraw a portion of the fluid into two jugs, and add some sulphate of iron. If this causes a cloudiness to appear, traces of gold are still left in the liquid, and a small handful—about $\frac{1}{4}$ of a lb.—of sulphate of iron is to be added to each gallon of the old colour to precipitate the remaining gold. Then we are told that the gold in the liquid exists in the *mechanical state* as well as in the chemical state.

Comments and Criticism of this Method

In order to obtain a clear idea of the abovenamed process for the reduction of chloride of gold from old colour liquid by waste iron binding-wire to metallic gold, several important facts will have to be taken into consideration, which, when thoroughly inquired into, will tend to show that the process is not only slow, but far from good practice. Firstly, it is very inferior to the method of employing freshly prepared sheet zinc instead of a matte of partly oxidised iron, for solid metallic iron is slow to dissolve, partly through the acid liquid being nearly rendered neutral to it by the process of colouring the gold, which causes the iron to be less attacked, and insufficient hydrogen gas is therefore given off to combine with the chlorine to dissolve the iron and liberate the gold. Secondly, the old iron binding-wire is oxidised to ferrous oxide (FeO) on the surface, through the numerous heatings to a red-heat it has received by being used in tying together sundry parts of articles which have to undergo the hard-soldering process, and in dissolving this oxide in hydrochloric acid and free chlorine (the chief constituents of colouring mixtures) no hydrogen is liberated from it, because the hydrogen of the acid firstly combines with the oxygen of the iron oxide to form water in the solution, and this, when dissolved, does not precipitate any gold. No hydrogen is being liberated from the iron matte until the whole of the oxide is dissolved away. And if the nitric acid in the mixture is not neutralised by the other ingredient in the solution, more iron is oxidised to the condition corresponding to ferric oxide (Fe_2O_3), and this does not precipitate any gold either; it will only contaminate the precipitate formed with oxide of iron. It will be made apparent that the iron binding-wire process is too slow and uncertain in its action to be eminently satisfactory in its workings to meet the requirements of jewellers, for there is more information coming. The iron matte system makes the liquid in concentrated colour waste turn to an inky black colour, from which it is very reluctant to change, and you have to add a quantity of hydrochloric acid to the bulk of solution to attack the iron more forcibly if you require the operation to be finished in anything like a reasonable time, as the iron is only feebly attacked by the old colour liquid. The solution appears, when samples are withdrawn for testing, too dense and cloudy for a test precipitant to be used, such as sulphate of iron even, which has been recommended, for the latter will not traverse through the liquid, but remains as a layer at the top of the sample to be tested when it is added to it. *In testing sample liquids stirring should not be done.*

In this formula it will be noticed that a portion of the iron matte treated liquid is recommended to be withdrawn into two jugs (why two jugs?), and some sulphate of iron added, about ¼ of a lb. per gallon, if a cloudiness (it will be already cloudy) is produced on first testing with a little of the sulphate of iron liquid. Now, this quantity of copperas salt should be more than sufficient to clear the solution and precipitate all the gold out of a gallon of the strongest mixtures of old colour waste by itself, and that alone makes it appear that the iron matte was in a degree ineffectual in extracting the gold, and that view is in accordance with my experience and experimental researches with the subject of the old binding-wire method.

Iron binding-wire will not satisfactorily precipitate gold from colour waste to meet the requirements of manufacturers engaged in industrial concerns for the purposes of profit, for the iron turns the liquid a very dark colour, and ultimately to black. The iron wire does not, of itself, clear the solution sufficiently for any distinguishing test for gold to be taken, the liquid remaining permanently a cloudy colour throughout.

If, on the other hand, too much iron oxide is formed in the mixture you cannot get past a reddish-brown tinge of colour which that substance imparts to waste colour liquids.

Protochloride of tin will, if a large quantity is used, ultimately clear the liquid sample of its black colour, by changing it to a bright olive-brown colour, but it takes considerable time (some days) for the clearing to be effected, and no distinguishing characteristics peculiar to gold can be observed during the change, and the resulting sediment will be supercharged with oxide of tin by the addition of the tin salt.

Protosulphate of iron will clear the liquid, resulting from the iron matte method, in a satisfactory way, if a sufficient quantity is employed for the purpose, but if this salt is required to complete the work of reducing the gold, why make use of the iron matte at all? The final results of my investigations into the method of the iron binding-wire reduction of gold may be summed up in a few words, namely, that the process is unsatisfactory and troublesome to an eminent degree, involving as it does extra time, expense, and labour, with indifferent results, and is one more of detriment than of success to the recovery of the gold, and especially is it so if undertaken by inexperienced persons, as no distinctive indication is shown when the gold has all been thrown down.

The second method is one prescribed for a large firm of manufacturing jewellers, with whom I am intimately acquainted, by a metallurgical chemist and public analyst. To give it a distinguishing name I will call it the complex method.

CHAPTER VI

THE COMPLEX PROCESS

THE firm to which reference is made were losing a very large quantity of gold through the bad treatment of their waste colour liquid, when it was, after duly considering the matter, decided to seek the advice of a scientist, and the first instructions which were received not proving satisfactory, another application was made for further advice to enable the firm to overcome the difficulty. The following information was then given:—

"I now confirm my telephone message *re* the waste colour water. My assay of the partly treated sample you sent me a few days ago shows that it still contains 34 grs. of gold per gallon. Certain indications made apparent since you have added the lead salt I advised, lead me to conclude that the best method for you to adopt now is, to add scrap zinc to the clear solution of your wash water until it loses all its yellowish colour. You can test a portion after treatment with zinc as follows for gold: Take some of the liquid that appears by you to have lost its yellow colour and place in a clear glass vessel. Add some sodium sulphite and a little sulphuric acid, and warm gently. Should a precipitate appear it means that the gold is not all removed and requires further treatment with zinc."

This formula, after having some time spent on it in patient endeavours to effect a solution of the difficulty by a member of the firm (quite capable of performing the operation), without any satisfactory results being obtained, was discontinued. Then I was consulted in the matter. The gentleman brought with him a pint of the black muddy liquid, and also the original letter of instructions, which I asked permission to retain. I then requested my client to call again in an hour, in which time I would have the gold separated from the liquid, and at the end of that time I was quite prepared with a solution of the difficulty, having separated the gold in less time. For when he again called I showed him the gold in a sediment at the bottom of the vessel, with the liquid quite clear standing above it, and it is needless to say he was quite surprised. I then, in his presence, tested the supernatant water standing above the precipitated gold with chloride of tin, to show that there was no gold left in the liquid. The tin chloride produced not the least change in the colour of the fluid after I had separated from it the gold.

This is what I did with the black liquid brought to me; after having poured it into a large basin, and diluted it with an equal volume of water, I next dissolved 8 dwts. of copperas in 2 ozs. of boiling water and added this to the bulk of solution with gentle stirring, and then let it stand for a time for the gold to fall down to the bottom of the vessel. Every particle of gold was in that simple manner extracted from the black liquid. This jeweller has since informed me that in the four following months after adopting my method he recovered £75 worth of gold from his waste colour liquid alone, whereas in any previous four months it had not exceeded £25, and I may mention that I had the privilege of seeing the button of gold that realised the larger amount.

Comments and Criticism of this Method

The reduction of gold from murky acid solutions by means of lead salts in manufacturing establishments is impossible of satisfactory results, and as lead has but few soluble salts, of which the nitrate, the acetate, and the cyanide are the principal, it must be taken as a practical fact that neither of these salts are eligible as precipitating agents for the reduction of gold to the metallic state from its soluble condition in liquid colour waste, and I need not go over the same ground again in explaining the reasons why this is so, particularly after having so fully described the chemical reactions required to effect the complete reduction of the gold from acid liquids, and the most useful substances to employ for that purpose. Suffice it to say, that all the likely salts of lead thicken the solution to be treated after being added thereto, and render it so dark and cloudy that it would be difficult, even for the most expert, to decide whether the liquid has been purged of its gold or not, and a great quantity of lead oxychloride is caused to accumulate in the vessel, and this is objectionable when the gold has to be melted, although it takes less flux in proportion to its bulk to cause it to melt as the lead assists the fusion, but it produces an impurer gold when lead salts are used, as the precipitated gold is more or less contaminated by them in the form of oxides when the sediment is dried, as that operation liberates the chlorine, and to remove the lead, if so desired before melting the gold, the whole sediment will require digestion in dilute nitric acid (HNO_3), which will dissolve any excess of these oxides, leaving the gold unchanged. Solutions of either of the lead salts above named have their lead precipitated in coming into strong "colour liquid," as heavy powders of lead oxychlorides ($PbCl_2$). On, however, heating the solution—an operation unfitted to the workshop—the oxychlorides dissolve, but again precipitate as the liquid subsequently cools. If lead peroxide (PbO_2) is dissolved in hot diluted hydrochloric acid, lead chloride is produced, which, on adding it to old colour liquid, causes a portion of the chlorine to be liberated from the gold and some of the latter to fall down, though not all of it. Peroxide of lead cannot be dissolved in water unless a large excess of hydrochloric acid is added. The supernatant liquid standing above any precipitates which the lead salts may have produced will still contain gold in solution, as evidenced by the assay report of the scientist appearing on page 68, and also of my own experience as the results of experimenting with several of these substances. It is advised in the report, after the lead salt has been added, when the result is not as expected, to add scrap zinc to the clear solution (the solution will not be clear) of the waste water until it loses all its yellowish colour (there never is any yellowish colour in colour waste). Then the advice as to *testing the solution* from which the gold is supposed to have been thrown down, with sodium sulphite and sulphuric acid, is equally absurd, for sodium sulphite would cause effervescence and discoloration of the liquid selected for sampling, and the sulphuric acid would add further fuel to the fire, and provide no safe means whatever to guide the operative workman in his task of satisfactorily accounting to his employer for the uttermost amount possible of gold that can be reclaimed from those liquid wastes—in short, the lead method is thoroughly useless, and impracticable in the workshop.

The tests for the reduction of gold most easily applied by persons in factories, and unaccustomed to chemical manipulations, will be by those of sulphate of iron and metallic zinc.

The chief merits of the two processes I have advised, and described at some length, are their simplicity, their absolute freedom from any difficulties, their certainty of effecting the recovery of the whole of the gold and silver, and the easy manner by which the processes can be put into action; no expensive jars, chemicals, and other utensils being at all required in carrying out the operations to a thoroughly successful issue.

Of all the precipitating substances the solution of sulphate of iron is the most immediate in its results, it is cheap, and is generally readily procured. Its employment, however, requires a certain amount of precaution; otherwise portions of iron may be thrown down from the sulphate of iron solution, and becoming mingled with the deposited gold may introduce difficulties into the operation and confuse the final result. But in old colour waste by means of the hydrochloric acid existing therein, and with careful manipulation, every portion of the iron should be retained in solution, and the gold permitted to fall down alone.

Another formula which I have found very effectual in melting the substance precipitated by metallic zinc, when it contains much oxide of zinc—oxide of zinc is very infusible—is the following :—

Waste colour residue . .	10 ozs.	or	100 ozs.
Soda-ash (Na_2CO_3) . .	3 ,,	,,	30 ,,
Common salt (NaCl) . .	1 oz.	,,	10 ,,
Fluor spar (CaF_2). . . .	1 ,,	,,	10 ,,

Melt the gold residue with this flux in a clay crucible of the kind before mentioned, the fusion being continued until it ceases to froth. The whole mass is then stirred with a thin iron rod, and after the crucible is withdrawn from the fire and allowed to cool, and, when broken, the gold will be found underneath the slag in a metallic button. This quantity of flux is usually sufficient to give a clear fusion of melted metal from unwashed precipitates.

If in carrying out the instructions—recapitulated in the synopsis of each process—complete success does not always follow the efforts of the operative engaged in the work, do not at once put the fault to the cause of the formula, but consider as to whether it may not be due to imperfect manipulation of some part or other of the process, for I can assure the reader that each of the two methods I have recommended are absolutely reliable in every detail, and have been performed by myself with perfect success, in the extraction of the last atom of gold from liquid colour wastes.

CHAPTER VII

RECOVERING GOLD FROM ELECTRO-GILDING WASTE

Every electro-gilder, and goldsmiths and jewellers who do their own gilding, have more or less of this waste to deal with, and the best and most simple methods for the recovery of the gold do not appear to be well known. The methods usually adopted for the recovery of gold from exhausted or spoilt cyanide solutions are either to evaporate the solution to dryness and then melt the residue, or to precipitate the gold by means of an acid. The first method is pretty generally practised by refiners of the precious metals, and the second method by those employed in manufacturing and electro-gilding establishments. There is also a third method sometimes employed, and that is to extract the gold from spoilt or exhausted solutions by trying to deposit the gold on to a piece of copper or other metal by means of the electric current; but this method is very slow and does not then extract the whole of the gold, the cyanide of potassium in the solution is not entirely decomposed, for some of it continues to be chemically united with a portion of the gold, and this prevents the latter from being completely deposited on the copper cathode. The anode used is an insoluble substance, such as platinum or carbon.

The *evaporation method*, particularly with large quantities of solution, is a tedious and expensive process, and requires special utensils for the performance of the operations. With small quantities of solution it is much more readily accomplished. This method, in any case, is, however, unsatisfactory, for the reason that for bright-gilding the solutions have to be renewed at short intervals; in some establishments, where a large quantity of bright-gilding has to be done daily, I am informed that a new solution has to be made every fortnight, for they become so loaded with foreign matter that their working properties for the bright finish become impaired, and a new solution is necessary to produce the right finish. There is also the rinsing waters to be dealt with, for there are traces of gold in the liquid clinging to the articles as they are removed from the solution to the rinsing waters, and these waters will have to be treated for the extraction of the gold, as well as from the solution, for they increase to large bulk, and the gold they contain, though small in quantity, is of high value, so that it pays to consider the best means available to effect its recovery; and to evaporate all this liquid would prove, on account of the time required to effect it, an expensive operation, and one that cannot be undertaken by every electro-gilder or manufacturing jeweller, as the cost would prove greater, in many cases, than the value of the gold obtained from the bulky liquid. Smelting establishments that are accustomed to recover the gold from cyanide solutions, evaporate the liquid down in iron pans until it becomes dry.

The residue is then scraped out and the whole

melted in a crucible, and the cyanide of potassium and the carbonate of potassium that is present in the residue acts as the flux for the fusion of the product, and this is generally found sufficient, as old electro-gilding solutions usually contain very little gold, so that, as a rule, when the dry method is adopted no other reducing salt is needed to be added to cause the mass to become thinly fluid, to allow of the small particles of gold passing through it and gathering together into a button at the bottom of the crucible, as both cyanide of potassium and carbonate of potassium are excellent reducing salts, and readily melt into a thin liquid.

This is called the "Dry Method" for the extraction of gold from old cyanide of gold and potassium solutions; and most of the treatises on electro-metallurgy recommend, after the solution has been evaporated to dryness, and the residue finely powdered, to wash it with water after it has become cold, to dissolve out the cyanide of potassium and other contaminations. It is not, however, a good plan to wash the residue in commercial undertakings, for there will always be found some gold in the wash waters upon testing them with suitable reagents, as you will be unable to heat the residue to a full red-heat so as to decompose the whole of the gold and reduce it to the metallic state on account of the salts of potassium in combination with it fusing before the gold can be brought to a red-heat. Another device is to intimately mix the nearly dried residue with an equal weight of litharge (oxide of lead PbO) and melt it into a mixed button of gold and lead. The lead being afterwards extracted from the resulting alloy with a mixture of hot nitric acid and water as follows:—Mixed alloy 1 oz., nitric acid 2 ozs., water 2 ozs., when the lead will go into solution and the gold will remain as a loose spongy powder—sponge gold. Both these methods are superfluous in commercial operations, for they take up extra time and labour, and can well be dispensed with, particularly as they bring forth less possible results in the end. In evaporating and burning to dryness a solution of gold in cyanide of potassium, when the water has been eliminated in the form of steam vapour the resulting product will rise and ferment, and the drying must be completed at a very low heat to avoid the fusing of the alkaline salts in association with the gold, to prevent which it is a good plan to add a little coarse sawdust to the sediment and stir the mixture repeatedly until sufficiently burnt to a dry powder. In the evaporation method, if carried out in the boiler furnace (fig. 9) the poisonous fumes are not so developed as they are in some other processes, notably the one of extracting the gold with acids; this is called the "Wet Method." In the former operation of dry reduction, the boiler has a fume hood with chimney attached which carries off all the dangerous fumes into the open air, and thus prevents them entering into the workshop in which the operation is being conducted. The iron of which the boiler is composed is not attacked by the cyanide, and almost any iron pan may be used for small operations, the iron having a tendency to reduce the gold from its solution in cyanide of potassium. The sediment, after the water has been removed, is transferred to a clay crucible (not a plumbago) and melted, the mixture being self-fluxing, owing probably to the large contents of cyanide, and that substance is an excellent reducing flux for gold and other metals usually associated with it in electro-gilding solutions which have been used for gilding articles of silver, copper, and brass. Usually, in melting waste products of the above kind, the contents of the crucible rise and foam somewhat, and in order to avoid boiling over, the crucible had better be not more than three parts filled with the saline mass. The frothing, which takes place when the crucible has arrived at a red-heat, soon afterwards subsides, and then the whole melts down to a thin, still liquid. When, after stirring the mass with a thin iron rod heated to redness before immersing it in the mass, and continuing the heat for a short time longer for the gold to settle down to the bottom, the crucible may then be removed from the fire and allowed to cool, when it is broken and the gold recovered in a lump.

The *precipitation method* is one that is frequently followed: it consists of endeavouring to precipitate the gold by means of an acid. It is called the "Wet Method" for the extraction of gold from old solutions of the double cyanide of gold and potassium. These solutions, when either sulphuric, hydrochloric, or nitric acid is added to them slowly, cause a deposit of gold cyanide to take place. Hydrocyanic acid and carbonic acid gases are given off in each instance, and as these fumes are deadly poisonous, the most perfect ventilation is necessary to avoid danger to the health of the operative engaged in the work. It is advisable to conduct the operations in the open air, or under a properly arranged flue, carefully avoiding to breathe the vapours of prussic acid which are evolved in the process.

Old cyanide liquids should, when they are to be treated with an acid, be put into extra large acid-proof stoneware vessels, because a great amount of froth is produced when the acid is added, and the latter should only be added in small portions at a time to avoid frothing over the sides of the vessel in which the liquid is contained, and the solution should be stirred while the acid is being added. This method is not completely satisfactory, for the double reason, that it fails to extract the whole of the gold from the liquid, and gives off dangerous prussic acid and other fumes.

In solutions of the double cyanide of gold and potassium, the gold is precipitated by nitric acid, muriatic acid, and oil of vitriol, but not with sulphate of iron, nor with chloride of tin. The muriatic acid is generally preferred as the precipitant, but the fact that it leaves some chloride of gold in the liquid appears to be overlooked. Nitric acid leaves behind a light precipitate, some of which is easily lost if the washing of the sediment is adopted. Oil of vitriol is to be preferred for gold, as it reduces the cyanide to a single salt, and precipitates the gold as *aurous cyanide*, which is insoluble in water; and whilst muriatic acid also precipitates *aurous cyanide*, it is not so fixed a compound, for when it comes to be washed a little chloride of gold dissolves out in the operation. Oil of vitriol, owing to its dense nature and to its being chemically stronger than the other two acids, is to be preferred for precipitating the gold from spoilt or exhausted cyanide solutions, for then there will not be any chlorine in the liquid to combine with the gold.

It is better to dilute the gold cyanide solutions with an equal volume of water, and to add the acid carefully and in small portions at a time, so that the action will not be too violent; allow the effervescence to subside somewhat before adding another portion; repeat this operation several times until no further effervescence is produced, and then give the liquid a rest for the froth to evaporate away and the gold to fall down. The liquid can then be tested (to see if it has been freed from all gold) with iron sulphate, or tin chloride, but if the latter is made use of it is advisable to add a little hydrochloric acid to the sample in the test tube before pouring in the drops of tin chloride, to convert into chlorides any possible soluble sulphates that may be contained in the liquid. With the iron sulphate this precaution is not necessary when the gold has been precipitated with oil of vitriol. Neither is it with the tin chloride when the gold has been reduced from the solution by means of muriatic acid, as the alkaline substances will then be in the form of soluble chlorides, and no reactions are then developed.

Gold cyanide solutions, when the gold is precipitated by mineral acids, must have sufficient acid added to them to render the liquid quite clear before the sulphate of iron or the chloride of tin can be applied as testing agents.

Chloride of tin added to a sample of the actual gilding solution will produce an immediate milky colouring with effervescence and frothing, and a white cloudy precipitate will begin to fall at once in flakes as a curdy deposit, and the action of this reagent may cause an overflow of the operating vessel if not sufficiently capacious to prevent it. It is caused probably through the evolving of carbonic and cyanogen gas, and to the conversion of some of the cyanide of potassium into chloride of potassium, through the action of the hydrochloric acid contained in the testing mixture. But when the cyanide solution has been completely decomposed and made clear by mineral acid, and if all the gold is then precipitated, the chloride of tin will show no action on, or change in, the colour of the liquid under examination.

Rinsing waters, from cyanide gilding solutions, will present a grey colouring, followed by a cloudy deposit if a few drops of chloride of tin are added to such waters, and if the liquid in which the white cloudy colouring is produced is allowed to rest for a time it will become quite clear, the gold (if any is contained therein) will fall down to the bottom of the test tube, along with the foreign substances of the liquid, the chief being chloride of potassium, the cyanogen gas of the cyanide being quickly dispersed with effervescence through the reactions set up by mixing the two solutions together. Chloride of tin, in both of these solutions (before being treated with acid for the reduction of the gold) produces abundant organic precipitates; these are the more copious and dense the larger is the proportion of gold and cyanide which they contain. Sulphate of iron added to a sample of the actual gilding solution will produce an immediate reddish-brown cloudy substance which agglomerates together and remains suspended in the liquid. It does not fall down to the bottom of the test tube, but if the liquid is shaken up by closing the mouth of the test tube with the thumb the concrete mass passes down, and, mixing with the liquid, turns the whole an intense red colour, which afterwards changes to a pale reddish-brown shade and remains so for a long time, no clearance being effected under some days, when a copious precipitate will be found half filling the test tube, and on making another addition of copperas to the apparently clear liquid at the top a red cloud is produced as before, showing that gold is still retained in the liquid.

Rinsing waters from cyanide gilding solutions will present a variety of colours, such as brown, blue, greenish, and reddish, in proportion as they are stronger in gold and cyanide, by adding a little sulphate of iron solution to them, which precipitates the gold with abundant organic sediment. Sulphate of iron (copperas) will ultimately precipitate all the gold from gilders' wash waters, and will reduce the red colour to clear liquid, and when the gold is all thrown down the copperas solution will show no change in the colour of the liquid. The operation is considerably hastened by pouring into the waste waters a little oil of vitriol. The copperas must be freshly made to effect the above results. Stale made copperas solution added to either gilding solutions, or to their rinsing waters, turns them into a black cloudy substance, but otherwise the reactions are exactly the same as those produced by adding freshly prepared copperas solution.

Gold cyanide solutions when the gold is decomposed by mineral acids leave a residue strongly impregnated with acid, and to destroy the corrosive action likely to be set up by the acidified precipitate when the drying operation is put in hand, it is a common practice to wash the precipitate with water so as to remove from it its acid properties; and to avoid the loss of any gold, when washing is adopted, the following method is recommended by the writer as the safest to pursue. After all the gold has been completely thrown down, and the clear liquid drawn off the sediment, remove the latter from the precipitating jar and put it into a glass vessel like the one shown in fig. 12—these may be obtained in all sizes up to several gallons capacity. Such vessels make good and useful utensils for washing acid precipitates, for, being conical in shape the smallest trace of sediment is prevented from being washed away—portions of the precipitate are very light and sink with difficulty—and the wash waters can be easily examined for particles of floating gold precipitate when they have become clear. The water having to come upwards, as it were, to the mouth of the vessel, it is easily poured away through the lip before any of the precipitate can reach the outlet, and in operating in that manner the syphon can be dispensed with. The water should be stirred while adding it to the precipitate, and

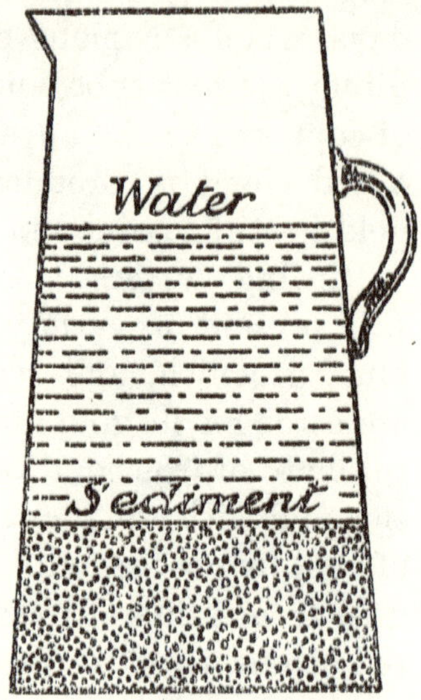

FIG. 12.—Vessel for washing precipitates.

the latter allowed to settle until the water above is clear, when it is poured off; then fill the vessel again with clean water, stirring it as before, allow to settle and pour off; repeat this four or five times, or until the liquid does not turn blue litmus paper red. This will indicate that all the sulphuric acid has been washed out; and the water can then be all drained away, the sediment dried and fused into a lump of metal.

Some operatives in dealing with liquid wastes of this description, after precipitating the gold with acid, pour the whole upon a filter of calico or other suitable cloth stretched on a wooden frame. The sulphuric acid and excess of prussic acid is supposed to pass through; the sediment left on the filter is washed by pouring water on to it two or three times and allowing it to run through, it is then dried and melted. I do not, however, consider this method so good as the one I have recommended, for the simple reason, that if it had not been ascertained that all the gold has been thrown down before doing this, that portion which is still soluble in the liquid will go through the filter and be lost, and sometimes when a precipitate is in a very fine state of division, some of it also is liable to pass through the pores of the filter.

There is always a certain amount of organic sediment that cannot be washed out of these precipitates without endangering the safety of the gold, and for that reason it is better not to attempt the impossible. It is far better, when the acid water has been removed, to proceed with the operation of drying, and to employ a little extra flux in the melting of the product. The following are suitable proportions to mix together to effect good results when the precipitate is so treated :—

Gilding waste residue . .	10 ozs. or	100 ozs.
Soda-ash	3 „ „	30 „

Double cyanide of gold and potassium precipitated by sulphuric acid and dried in the boiler-furnace gives off hydrocyanic acid and carbonic acid gases, leaving as the residue a mixture of gold and sulphate of potassium; and the latter (when not washed out) acts as an assistant in fluxing the gold, when the product is being melted, and less flux, about three-tenths only, of its weight, is required to be added to effect its complete fusion and recover the whole of the gold.

In melting waste residues of this description, after the drying has been accomplished and they have been intimately mixed with the dried soda-ash, the resulting product should always be put into a clay crucible (plumbago crucibles are acted upon by the fluxes, portions of the materials of which they are composed become detached, and, mixing with the fluxes retard the fusion by causing the mass to assume a semi-fluid condition only) and the heat given at first should be moderate, or the contents may rise up and overflow; should this be likely to happen, a few pinches of *dried common salt* thrown into the crucible will at once cause the fusing mass to subside. It is necessary to melt the whole product to as liquid a state as possible, to enable the fine particles of gold to separate from the flux and collect together in a liquid mass at the bottom of the crucible. When the melted mass has become thinly fluid and tranquil, then, and not before, is the time to remove the pot from the fire and place it in a safe place to cool. The pot is afterwards broken by a blow from a hammer on its lower part, when the button of gold will fall out. This is subsequently assayed and sold to the refiner who makes the best offer.

The *Battery Method* is sometimes resorted to in recovering the gold from old or spoilt cyanide solutions. The galvanic battery or the electric dynamo may be employed for the purpose. The plan adopted consists in attaching a piece of sheet copper to the negative wire, that is, the wire on which the articles when being gilt are suspended, and a piece of platinum, or in the absence of this, a piece of carbon or even iron is attached to the positive wire as an *anode*, for neither of these substances are attacked by the solution. The battery or dynamo is then put in action, when, after continuing the operation for some time, nearly the whole of the gold will be deposited on the copper sheet; and when the solution ceases to deposit any more gold on to the copper sheet, which is easily ascertained by weighing it at intervals, the operation is stopped, and the gold is recovered by dissolving it in aqua-regia, diluting the solution with water, and precipitating therefrom the gold with a solution of iron sulphate, when it may be redissolved in aqua-regia, converted into crystallised chloride of gold, and used again for making a new gold solution. With regard to the quality of the sulphate of iron crystals to be used in making the copperas solution, for delicate operations it must be understood that they should be perfectly free from red or brown powder (peroxide of iron Fe_2O_3), as this powder is insoluble in water, and will readily deposit to the bottom of the vessel when the crystals have been dissolved. But when the iron sulphate salt cannot be procured in a perfect condition, it is a good plan after dissolving the precipitating agent in water, to allow the insoluble powder to settle, and then pour off the clear liquid for use. By adopting these means any iron oxide that may be present in the salt will be prevented from falling down and mixing with the gold precipitates produced through its employment, as it would otherwise do if some such precautionary measures were not taken to remove it beforehand.

Having now given a descriptive account of three of the old methods of recovering the gold from cyanide gilding solutions, which may be briefly recapitulated as follows :—

1. The *evaporation* of the solutions to dryness and the melting of the residue into a button of metallic gold in a clay crucible without any flux. The alkaline substances combined with the product acting as the flux.

2. The *precipitation* of the gold in the solutions by a mineral acid as aurous cyanide (AuCy), the washing out of the mineral acid, and the melting of the remaining sediment by means of a good reducing flux, and approximately weighing three-tenths, or 30 per cent. of the gross weight of the precipitated substance.

3. The *electrolytic* extraction of the gold from the solutions by depositing it in the metallic state on to a sheet of copper which acts as the cathode, while a piece of carbon or of iron does duty as the anode, the removal of the gold from the copper

sheet being afterwards effected either by mechanical or chemical means.

The first of these methods is a slow process, and somewhat dangerous when large quantities of solution have to be evaporated, as prussic acid fumes are given off during the operation. The second method is even more dangerous than the first, on account of the excessive prussic acid vapours that are evolved by the action of the acid on the cyanide of potassium, and of its absolute necessity to be almost always performed out in the open air to avoid injury to the health of the workers, for which reason it is unsatisfactory, and with large quantities of solution it becomes an expensive operation. The third method is less injurious than either of the other two, but it does not extract the whole of the gold, some small portion being always left in the solution when so treated, as it does not entirely decompose the whole of the cyanide. The operation, however, requires little attention when once set in action, and that is in its favour. It is also economical in working, but slow, even under the most favourable conditions, and cannot, therefore, be recommended for universal adoption commercially.

The process I shall next describe for recovering gold from cyanide gilding solutions will, without a doubt, be a great improvement on old methods, as it is so simple to perform, is self-acting, *will throw down every atom of gold*, and requires no special apparatus—in fact, the solutions may, if necessary, be treated without the use of separate vessels, and the method is almost entirely free from danger to the workpeople employed in the factory. Large or small volumes of solution can equally be treated by my process with excellent commercial results, the only apparatus needed being a vessel large enough for holding the whole of the solution, after it has been diluted with water, which has to have its gold extracted therefrom.

Copper, iron, aluminium, and zinc precipitate metallic gold from solutions of its cyanide by a simple displacement of the metals, but in each case the cyanide has to be in excess to cause the different metals to dissolve in it, when "nascent hydrogen" is freely evolved and cyanide of the metals used is formed, otherwise the operations prove unsuccessful in throwing down the whole of the gold. How the latter difficulties may be easily overcome in all solutions, and the process rendered thoroughly workable, I shall point out to the reader as this discourse proceeds. Zinc is the most useful metal to employ to effect the reduction of the gold, for the reason that it is not only the most electro-positive metal to the metal in solution—gold, but that it is readily attacked by the liquid, and is cheap. The extraction of the precious metal from cyanide solutions may therefore be readily and completely effected by the action of metallic zinc.

CHAPTER VIII

THE ZINC-ACID PROCESS

THE principle of this method consists of the fact, that if a thin sheet of metallic zinc is immersed in a cyanide gold solution in a vertical position, and a very small quantity of mineral acid is carefully added to the auriferous liquid, the gold will become precipitated and fall to the bottom of the vessel in which the operation is being conducted as a metallic powder. The gold will not, as stated by some authorities, be deposited in a coherent state on to the surface of the zinc sheet. For if this were to take place no direct advantage would result from the method. If the gold adhered firmly to the zinc sheet, the moment it becomes covered over with gold all action would cease, as no deposit takes place when two electro-negative metals of the same kind, such, for instance, as these would then be—gold in a solution of gold. The secret of success lies in the addition of a very little mineral acid to the gold cyanide solution when the zinc plate is first immersed, and to add a little more after a short time. The mineral acid will prevent the gold depositing itself on the zinc sheet, and also provide a free surface for the action of the liquid to display its powers thereon to the best advantage in the generation and discharge of "nascent hydrogen" to the end of the process, when, after about 12 hours of such action, the whole of the gold will be completely thrown down from the solution under treatment.

In treating gold cyanide solution with metallic zinc without the mineral acid, and when the cyanide of potassium is not present in the solution in excess, the action soon stops altogether, owing to the formation of a layer of hydrogen over the surface of the zinc sheet; but when the cyanide of potassium is present in large quantity the zinc dissolves in it, and hydrogen continues to be evolved, until the cyanide loses its active properties, and cyanide of zinc is thus so far formed in the solution, which takes the place of the gold cyanide, as the gold becomes precipitated therefrom; but in old, or spoilt, cyanide of gold solutions, the cyanide of potassium is either partly or wholly exhausted before they are placed aside as of no further use. This causes unsatisfactory results, and my object in introducing a little mineral acid to stimulate the action of the zinc and thereby perfect the operation by making it continuous, was to completely overcome all the besetting disadvantages which prevailed without its employment.

The proportions of metallic zinc and mineral acid need not be large, but the more surface of zinc that is exposed to the action of the liquid, and with just sufficient acid to prevent a layer of hydrogen forming on its surface, is all that is required to cause a rapid precipitation of the gold. No gold will then be deposited on the zinc, it will all fall to the bottom of the vessel in the metallic state, leaving the solution quite clear and colourless. The solution can be treated in its concentrated form, or it may be diluted with an equal volume of water; either will give satisfactory results. The following proportions of substances make up a good formula to work to, when diluted, as stated above :—

Gold cyanide solution .	1 gallon	or 10 gallons.
Metallic zinc . . .	½ oz.	„ 5 ozs.
Oil of vitriol . . .	1 „	„ 10 „

The gold solution from which the precious metal is to be recovered may be treated in its own vessel, or placed in a stoneware jar when it is preferred to be used diluted (see fig. 10), or any other suitable vessel may be used when the solutions are large.

The quantity of zinc that is required will depend upon the amount of gold and cyanide of potassium existing in the solution to be treated. The proportionate parts I have given will be satisfactory up to 10 dwts. or more of gold per gallon of solution; more gold than this, with its equivalent of cyanide of potassium, may require at the beginning to have the zinc, but not the acid, proportionately increased. From ½ to 1 oz. of zinc to 1 gallon of solution will be sufficient to throw down all the gold from any solution when diluted with an equal volume of water. The vessel holding the solution should not be filled to the brim, as a little frothing takes place all over the surface of the liquid, through the hydrogen bubbles which arise, and in starting the operation it is much better to immerse the zinc plate in conjunction with a proportionate part of the oil of vitriol, by which means the agitation is greatly reduced and soon passes off. It is also an advantage to add a part of the remaining acid at a time, and repeat the operation, rather than add all the acid at once, particularly if a little extra acid is required to make the zinc more active, for in that manner the action of the zinc is regulated to the needs of the solution; all that is necessary is for the zinc to be feebly attacked by the acid, and no more acid should be added than will just bring about this result—a good test is when the liquid barely turns blue litmus paper red.

When the gold is all reduced the solution will become quite clear like water, and then a sample of the liquid can be withdrawn and placed in a clear glass test tube, and tested for gold with a few drops of a clear white solution of copperas, and if all the gold has been thrown down no chemical action or change of colour will appear visible to the eye.

If it is intended to test the sample liquid with the protochloride of tin test, it is necessary to first add a little hydrochloric acid to the selected sample before the tin chloride is employed, for it is only reasonable to assume that a little *sulphate of zinc* may be present in the liquid, and the hydrochloric acid prepares the liquor in a suitable condition to prevent chemical action being set up between two opposite salts like chloride of tin and sulphate of zinc when they come in contact. The chloride of tin if added direct may cause a white cloudy colouring to appear (protochloride of tin is chemically acted upon and a white cloudy coloration is given to the liquid by sulphate of zinc), but when a little hydrochloric acid is added previously no alteration of colour will arise (this acid does not precipitate sulphate of zinc, but converts it into *chloride of zinc*, a very soluble salt, without the chemical change being observed), and when the copperas test is employed, the addition of hydrochloric acid is not required, for the simple reason that there will be a sulphate salt coming in contact with another sulphate salt, and no reactions then occur.

Gilding solutions, which have been used for some time in gilding articles of alloyed gold, silver, copper, brass, and other common metals, contain portions of those metals in the solution, and these are all precipitated by the zinc sheet, so that when the sediment is dried and melted a metallic button of mixed metals is the result.

It is advisable to occasionally stir the solution during the period that precipitation is taking place, in order to bring different portions of the solution into the immediate vicinity of the zinc sheet, as this will greatly assist the operation of extracting all the gold from these solutions in a perfectly exact manner. No gold will remain in the liquids standing above the precipitates, after being carefully treated by this method, and the poisonous gases are not so developed as they are in the processes for extracting the precious metals by strong mineral acids.

Metallic zinc in the form of sheets is more convenient to use than either turnings, clippings, granulated into flakes, filings, or any other form of finely divided zinc, as the sheets can easily be hung in the solutions, and readily withdrawn when all the gold has been thrown down, or at any other stage of the operation. It is not necessary—when a little acid is used—that the zinc should have clean surfaces, for if coated with the ordinary oxide it will work quite satisfactorily. For when the zinc is immersed in the solution its oxide is attacked and dissolved, and the zinc turns a blackish colour, and if any gold attaches itself to the surface at first, which is unlikely, it is spongy and non-adherent, and is quickly removed by the uninterrupted action of the sulphuric acid on the zinc. If, however, the surface of the zinc was scraped quite clean and placed in the gold cyanide solution without the acid the gold would deposit on the surface in a thin layer by chemical or electrolytic action, but only in a thin layer, and the moment it became covered all over, all action would immediately cease and the operation then prove a failure. The oil of vitriol provides a simple remedy for this innovation by giving to the zinc a free surface, for the continual production and escape of the hydrogen gas which arises from the reactions caused within the solution, which is a distinct advantage. Again, it must be noted that without the acid, a white zinc salt (hydrate of zinc) is sometimes formed on the zinc, and this interferes with the action of the liquid also by putting a stop to the operation. Too much oil of vitriol must not, however, be added to the gold solution to cause too rapid an action on the zinc sheet; a very little, and added occasionally as required, is much the best form of treatment.

When all the gold has been precipitated with accuracy — indicated by either of the before-mentioned test liquids—the clear solution may be syphoned off and thrown away, as it will then contain nothing of any value. The whole product is then ready to be transferred to the drying furnace and there heated until all moisture has evaporated. No washing is required to be done by this process, and it will be unnecessary to filter the remaining liquid away from the sediment, as that will soon vanish under the heat of the furnace. The dried sediment is next reduced to a fine powder, mixed with a good reducing flux, and melted in a clay crucible, when a button of mixed metals will be obtained, without any loss occurring, such as is often experienced in washing precipitates in manufacturing establishments.

Note.—The crucibles in which these products are melted should be of the kind known as the London round white fluxing pots. They are specially manufactured for the purpose by Messrs Morgan, at Battersea Works. Ordinary blacklead and plumbago pots will not answer, owing to the action of the fluxes on the material of which they are composed. These special crucibles, when being used, are sometimes placed within plumbago pots as a precaution against loss should the former crack, which, however, seldom happens when the necessary care is used in performing the work.

In precipitating gold from cyanide gilding solutions by means of zinc sheets and a little oil of vitriol, in order to avoid excess of the latter getting into the mixture it will be found much the better course to pursue to use as little as possible, that is to say, add less of the acid in the beginning than the maximum quantity required, and, when no chemical action is perceived to be going on, to add a further small portion, and so on, until the end of the operation is reached. This will prevent too much oil of vitriol (H_2SO_4) getting into the solution.

The chemical reactions which occur in the cyanide

of gold solutions by the "Zinc-acid process" are rather complicated, but the principal change is probably one of exchange of metals in the solution. The gold is precipitated as metal by the chemical action on the zinc, and the latter must then, as a liquid, take the place of the gold in the solution. The zinc being dissolved by the combined effects of the cyanide and the acid, which causes hydrogen to be evolved, and this acts as a reducing agent to the gold, which falls to the bottom of the vessel in a metallic powder. The cyanogen of the gold cyanide unites with the zinc as it becomes dissolved, forming a fluid substance of cyanide of zinc; sulphate of potassium being also formed by the action of the oil of vitriol on the potassium salts contained in the gilding solution, portions of which (as hydrate KOH) fall to the bottom of the liquid along with the gold, as does probably some small particles of hydrate of zinc; the residue thus obtained, when dried and burnt, contains the whole of the gold, oxide of potassium, and a little oxide of zinc. The zinc, therefore, in cyanide solutions to which a small portion of H_2SO_4 is added, precipitates as residue metallic gold and hydrate of potassium, while there remains in the liquid both cyanide and sulphate of zinc.

CHAPTER IX

RECOVERING GOLD FROM RINSING WATERS AND WASH WATERS

The rinsing waters made use of during gilding operations, as also those resulting from the making of gold salts, and from other gold-washing processes, should never be thrown away without previously testing them. They should be set aside and preserved in order that the gold may be recovered therefrom at some future time, since nearly all the washing waters from gold precipitates, as usually performed in factories, contain a little gold. They may all be put into one and the same vessel, where they can be treated altogether for the recovery of the gold by the means I am now about to make known. Testing in the proper manner will prove the truth of the assertion, that all these wash waters contain a little gold in solution, as it is usually of high caratage, the loss is greater in value than it may appear to the mind to be at the moment, and without due consideration of the facts. The greater the quantity of alkaline salts, or of free acid, in the wash waters, the more capable are they of holding the precious metals in solution. The gold can, however, all be thrown down from such liquids in a very simple manner, for by immersing in them sheets of zinc, from 1 to 2 ozs. of zinc per gallon of solution will be quite sufficient, provided the rinsings are made slightly acid, if not already so, and sufficient time is given to effect the purpose. The precipitation is so completely successful that all the gold contained in solution in such products is thrown down in one night to the metallic state, leaving no trace of it in the fluid in the morning,

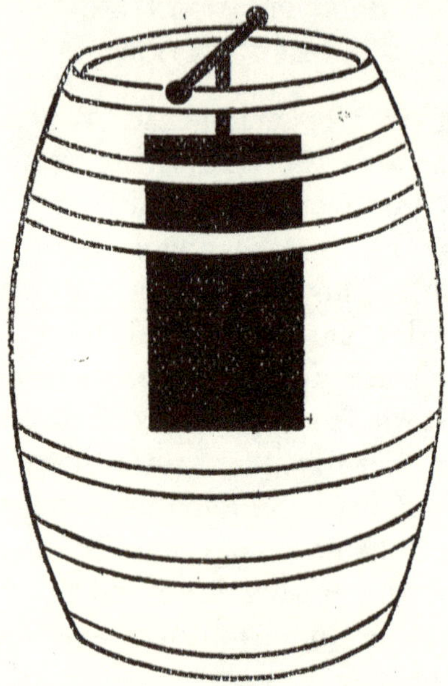

FIG. 13.—Wooden precipitating vessel.

after carefully carrying out this method, which is very simple and economical in its application to these kinds of waste liquids.

A receptacle for holding the rinsings and wash waters resulting from electro-gilding should be procured of a size well adapted to the extent at least of a day's work being done in the factory. A tank or old paraffin barrel (fig. 13) is a suitable vessel in which such liquids may be collected together and preserved.

The liquid may be drawn off the sediment every morning, after it has been ascertained that the gold has all been precipitated from it, by means of a syphon—one with a suction tube is preferable (fig. 14), and is a useful little instrument for syphoning solutions which cannot with safety be touched with the fingers. The shorter leg is placed in the liquid and the longer one closed with the finger or an indiarubber pad placed against it, then with the mouth suction is carefully applied at the top end of the side tube until the liquid fills the longer leg of the syphon. When there is danger of inhaling poisonous vapours, the suction of the mouth may be replaced by an indiarubber ball (see sketch) being secured to the suction tube. The longer leg of the syphon is closed after putting the short leg into the liquid, as before, and the ball is squeezed in order to remove the air, when, by its elasticity, the ball resumes its former state, and thus, by continuing this action a short time, a suction is produced which fills the long leg with liquid and sets the syphon in action.

FIG. 14.—Syphon with suction tube.

The precipitated gold need not be removed from the vessel after the first lot of clear solution is withdrawn, but each time this is done the vessel will be ready again for fresh quantities of rinsings to be put in as often as is required during the next day, this device being repeated time after time until sufficient sediment has accumulated upon the bottom of the vessel, when it is removed, and may be mixed with the gilding solution precipitates and treated as one product, or, if preferred, melted separately.

The wooden tub which is used to hold the rinsings should have two coats of asphaltum cement applied both inside and outside, and the recess at the bottom of the inside should be filled level with melted pitch, and smoothed quite even, as neither alkaline nor acid liquids have scarcely any action on this substance.

It is no earthly use in alone simply filtering these solutions, as the gold will pass through the filtering material as easily as does the water.

When a zinc plate is employed to precipitate the gold from these kind of rinsings, which are always very dilute, a little acid is usually required to be added to them to generate *nascent hydrogen* as the reducing agent; and as a guide to the right quantity, I may remark that when blue litmus paper is faintly turned red upon immersing it therein, sufficient H_2SO_4 has been added to cause the zinc to quickly perform the necessary requirements.

No washing of the sediment resulting from this method of recovering the gold is at all needful, nor in any of the operations dealing with jewellers' waste products, for there will always be found some gold in the washing waters on testing them with suitable reagents.

The sediment in the operating vessel is scraped out at stated intervals, dried and burnt to powder in the boiler-furnace, and then melted direct, with two-fifths of its weight of soda-ash, when a lump of metallic gold—usually alloyed with other metals—will be obtained.

The following formula for adoption will be found excellent in melting this product :—

Electro-gliding waste residue .	10 ozs. or 100 ozs.
Soda-ash	4 „ „ 40 „

It is my intention to give full and complete directions for dealing with each kind of residue separately, then any of my readers who decide to follow the information given may continue the operations to the end, without having to refer to back pages for instructions to complete the different stages of the work to be undertaken.

The residue after drying, which should be done slowly at first, the heat being continued until all organic matter is destroyed, is intimately mixed with dry soda-ash in the proportions given above. This acts as the flux in melting, and with the residue of potash contained in the burnt powder acts as a powerful flux in reducing the metal from dry wastes of this description.

Soda-ash is not a pure carbonate of sodium commercially; it contains about 50 per cent. of carbonate of soda; the remainder usually consists of common salt and sulphate of soda. This combination of substances forms a readily fusible triple compound salt which combines with the non-metallic matter of the residue, forming a very liquid slag which rises above and separates from the metal by the act of gravity, on its becoming melted.

When this description of waste has been well burnt, and allowed to become cold, it is reduced to fine powder in a mortar; next spread out on a large sheet of strong smooth paper, and well mixed with the aforesaid flux, and when this is done, the mixture is put into a clay crucible which has previously been warmed to prevent cracking, as it would be liable to do if put into a red-hot furnace in a cold, moist condition. The crucible should be of sufficient capacity to contain the mixture, but it must not be filled too full, for this kind of residue has a tendency to rise as the salts begin to melt. If the crucible furnace is not of the proper size to take a crucible large enough to hold all the mixture, a portion can be reserved and put in the crucible afterwards, when the first lot has sunk down. Continue the heat of the furnace until the mass within the crucible forms into a thin, steady liquid, stirring occasionally with a thin iron rod so as to render the fused mass uniform in composition, and to enable the small shots of metal, as they become melted, to fall through the flux to the bottom of the crucible. When this has been accomplished, withdraw the crucible from the fire, and allow it to cool; then break the pot at its lower part by a blow from a hammer, and a metallic button-shaped substance will fall out, consisting of an alloy of gold, silver, and copper in small proportions. This is remelted, this time in a plumbago crucible, under a layer of fine charcoal powder to protect it from the action of the air, and cast into an ingot mould to form a bar of metal which may then have a parting assay taken from it in order to ascertain its value, when it can be sold to the refiner who makes the best offer. The slags resulting from melted fluxes are thrown into the sweep.

The following are the operations, reduced to a brief account, for the extraction of the gold from cyanide solutions, their rinsing waters, and the making of gold salts by the zinc process :—

(1) The dilution of cyanide of gold solutions with an equal volume of water. The rinsing waters will be already sufficiently diluted.

(2) The precipitation of the gold by the immersion of a sheet of zinc, in the proportion of from ½ oz. to 1 oz. to the gallon of gold solution, and the addition of a small quantity of oil of vitriol occasionally, until the solution becomes quite clear.

(3) The testing of a sample of the clear liquid with a colourless solution of iron sulphate to certify that all the gold has been thrown down.

(4) The syphoning off of the clear acid water as near to the sediment as possible.

(5) The removal of the sediment, and the drying and burning of it in the boiler-furnace to powder.

(6) The mixing of the powder with two-fifths of its weight of soda-ash, and then melting the mass in a clay crucible to recover the metal in a solid lump.

(7) The remelting of the lump of metal obtained, under a layer of fine charcoal powder in a plumbago crucible, and casting into ingot form, so that a *parting assay* may be taken from it with the view of sale to the refiner of precious metals who makes the best offer, for it cannot be worked up again.

Another method which is capable of extracting all of the gold from cyanide gilding solutions I will next bring under the notice of my readers, although, like the previous one, I have never heard of its being adopted in the jewellery trade for the purpose of extracting the gold from waste cyanide liquids, nevertheless it possesses features of merit, and may be utilised with advantage and profit in the recovery of the gold from the waste liquids under consideration. The method may be appropriately classified as the hydrogen sulphide process.

CHAPTER X

THE HYDROGEN SULPHIDE PROCESS

THE reduction of the gold from cyanide of gold solutions is effected by this method by passing hydrogen sulphide gas (H_2S) (also called sulphuretted hydrogen, and hydrosulphuric acid) into the solution until the liquid reddens blue litmus paper, and the gas freely escapes, indicating that the operation is completed, when it is allowed to rest for several hours for all the gold to fall down. The gold solution should be diluted with twice its bulk of water. For the production of the hydrogen sulphide, otherwise sulphuretted hydrogen, which is a highly poisonous gas, and must not therefore be inhaled, a good sized wide-mouth jar (similar to a pickle jar) is required, a cork pierced with two holes—a thistle funnel with a long tube being fitted in one, and a gas delivery tube in the other. See fig. 15 for the arrangement of the apparatus.

The tube of the thistle funnel should reach almost to the bottom of the jar. Place in the jar some iron sulphide (FeS), replace the cork with the bent delivery tube and thistle funnel, and pour diluted sulphuric acid (H_2SO_4), in the proportion of 1 part acid and 8 parts water, down the thistle funnel until the jar is about half filled. The development of the gas begins immediately, and passes through the bent tube into the gold solution, when sulphide of gold (Au_2S) begins to precipitate to the bottom of the vessel holding the gold solution, and in a very short time the whole of the gold becomes reduced to the condition of sulphide of gold, a black precipitate readily decomposed to metallic gold on applying a dull red burning heat with exposure to air, such as is given in the iron

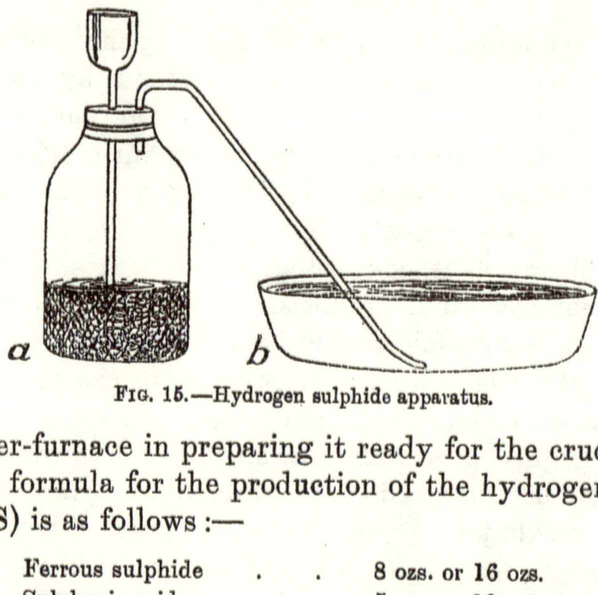

FIG. 15.—Hydrogen sulphide apparatus.

boiler-furnace in preparing it ready for the crucible. The formula for the production of the hydrogen gas (H_2S) is as follows :—

Ferrous sulphide . .	8 ozs. or 16 ozs.
Sulphuric acid . . .	5 „ „ 10 „
Water	1 quart or 2 quarts.

The first quantity will require a jar with a holding capacity of 2 quarts, and the second quantity one of 1 gallon capacity. The ferrous sulphide should be broken up into lumps about as big as hazel nuts. When the gold cyanide solution has been successfully treated, by the removal of all the gold from the liquid, the acid is poured off the iron sulphide in the gas-generating jar and clean water poured in, so as not to cause any further development of gas, and when the apparatus is required to be again used, the water is poured away, and a fresh mixture of acid and water is put into the vessel.

Sulphide of gold is insoluble in water, and so are the sulphides of nearly all the metals—sulphides of potassium, sodium, calcium, barium, strontium, and magnesium being the exceptions.

The chemical reactions in the hydrogen sulphide generating jar proceed readily, when once started, without the application of heat; the sulphide of iron decomposes the sulphuric acid, its hydrogen combining with the sulphur of the iron sulphide forming "sulphuretted hydrogen gas," which by means of the delivery tube provided for its conveyance, enters the vessel containing the gold solution, throwing down the gold as sulphide of gold (Au_2S), whilst the undecomposed or free acid in the generating jar dissolves and unites with the iron, as the sulphur becomes liberated, forming sulphate of iron, which is retained in the liquid. Hydrogen sulphide gas is possessed of acid properties in solution with water, and reddens blue litmus paper. Its acid properties decompose the cyanide and acts, therefore, in these solutions as an aid to the complete reduction of the gold by causing it to remain insoluble in the liquid. The changes which take place in the gas generating vessel are shown in the following equation:—

$$\underset{\text{Ferrous sulphide.}}{FeS} + \underset{\text{Sulphuric acid.}}{H_2SO_4} = \underset{\text{Ferrous sulphate.}}{FeSO_4} + \underset{\text{Hydrogen sulphide.}}{H_2S.}$$

Hydrogen sulphide gas produces no precipitate in zinc sulphate, iron sulphate, or nickel sulphate solutions when acid is present. To reduce these metals from acid liquids by means of sulphuretted hydrogen gas, the acid will require to be first neutralised or destroyed by an alkali like ammonia, as the combinations of these metals with sulphur are soluble in dilute acid solutions. In alkaline solutions they are, however, reduced by sulphuretted hydrogen to metallic sulphides of the respective metals and remain insoluble, while sulphide of gold precipitated from acid liquid becomes redissolved by alkaline sulphides—that is to say, if sulphuretted hydrogen gas is passed through an acid solution containing platinum or gold till no further precipitation takes place, the metallic precipitates formed will be insoluble in acids, with the exception of aqua-regia, but soluble in excess of alkaline sulphides, notably those of potassium, sodium, and ammonium, with which they form double salts.

Hydrogen sulphide gas may be produced by dilute oil of vitriol acting on sulphide of potassium (K_2S), commonly called liver of sulphur. The same arrangement is made use of in its preparation as the one just described and illustrated (fig. 15), (*a*) being the generating vessel, and (*b*) the precipitating vessel. The liver of sulphur is employed in small pieces, and is put into the jar along with the water before the cork containing the thistle funnel and delivery tube is fixed in, and then the oil of vitriol is added, by degrees, by means of the thistle funnel. In this process of making the gas the reactions occurring in the generating vessel are partly the same as those in the previous process, the evolution of the gas depending upon the following change of substances:—The liver of sulphur decomposes the water and oil of vitriol, the sulphur of the former uniting with the hydrogen of the two latter and thus forming the required gas; the potassium combining with the liberated oxygen and free acid to form sulphate of potassium (K_2SO_4), which remains behind in the solution, the sulphuretted hydrogen gas alone passing through the delivery tube into the solution from which the gold is to be precipitated. It may be expressed thus:—

K_2S	+	H_2SO_4	=	K_2SO_4	+	H_2S
Potassium sulphide.		Sulphuric acid.		Sulphate potassium.		Hydrogen sulphide.

The proportions of the different substances employed in the preparation of the gas from potassium sulphide may be as follows:—

Potassium sulphide . .	8 ozs.	or 16 ozs.
Sulphuric acid . . .	2 ,,	,, 4 ,,
Water	1 quart	,, 2 quarts.

When the gold has all been precipitated, which requires some time before it is all precipitated to the bottom of the vessel, and the liquid must appear perfectly clear before it can with safety be drawn off to avoid loss of gold. From time to time samples of the liquid should be taken and put into a clean glass test tube or tumbler, and a few drops of a clear solution of ferrous sulphate ($FeSO_4$) added, when, if no discoloration takes place, there will not be any gold left in the solution, and it may be drawn off and transferred to the general waste water receptacles (if you are not satisfied as to the effectiveness of the process), and there undergo a further process of treating and filtering before being allowed to run away into the drains.

The gold obtained after precipitation by hydrogen sulphide gas is scraped out of the vessel (which should have a very smooth surface, else it is difficult to collect together all the finely precipitated gold) and introduced into the boiler-furnace and subsequently dried to powder, then slowly heated red-hot with access of atmospheric air; this treatment causes the sulphur to unite with the oxygen of the air, and to be evolved in the condition of *sulphurous acid gas* when the product becomes reduced to metal for melting purposes. The same flux (soda-ash) being in that case employed, and in the same proportions (two-fifths of its weight) as recommended for melting the precipitate resulting from the "zinc-acid process," when, if the operations are well performed, there will be found less percentage of loss from wastage than is frequently the result from some of the methods employed. Gold sulphide is completely reduced by melting, being separated into metallic gold and sulphur.

When sulphuretted hydrogen gas is employed in the "nascent" state to precipitate the gold from diluted cyanide solutions and their rinsing waters, only a little acid is required to be added to assist the action of the gas, as this gas itself possesses the nature of an acid at the time it is being first generated, and sufficient of the gas in its nascent state must be allowed to pass in until its action on blue litmus paper shows that property by turning it red when immersed in the solution into which the gas is being conveyed; but if it should not show this reaction some gold may remain in the liquid, and it will be advisable to add to the gold solution a further small quantity of oil of vitriol by degrees, until it does give this change of colour to the litmus paper. A solution of cyanide is alkaline and the sulphuretted hydrogen gas is acid.

Sulphide of potassium (K_2S), commonly called liver of sulphur, will precipitate gold to the insoluble condition from some auriferous liquids, but they must be acid. The chief of which is the reduction of gold from weak aqua-regia and chlorine solutions by this well-known alkaline sulphide salt, and if it will throw down the gold from weak acid liquids containing it in the form of chloride, why will it not also precipitate the gold to the insoluble condition from the double cyanide of gold and potassium solutions it may be required to know, as it generates the same kind of gas (H_2S) as that produced by liver of sulphur in the generating vessel already described and illustrated (fig. 15)? The reason is, because the liver of sulphur when added as a liquid to the gold cyanide does not, having no acid properties, effectively decompose the cyanide of potassium to cause the precipitate of sulphide of gold produced to remain insoluble, for almost immediately as fast as the gold is precipitated it becomes redissolved again, through the action of the alkaline sulphide salt of potassium forming a double salt in which the gold is soluble. The acid properties of the hydrogen sulphide which is thus generated, being so feeble that the alkalinity of the solution is not sufficiently neutralised to cause the gold to be precipitated in a form so as to remain in the insoluble state, the precipitate of gold sulphide being redissolved by the alkaline sulphides in the solution nearly as fast as it is formed. Sulphide of gold thus produced being soluble in excess of sulphide of potassium.

Potassium sulphide will not, therefore, precipitate the gold from solutions of the double cyanide of gold and potassium in a commercially practical manner, although it apparently renders them clear looking, but if a very little solution of copperas is added to gold cyanide solutions after treatment with sulphide of potassium, a red-black cloudy colouration will be imparted to the liquid, and this continues to be produced to the end of the operation, indicating that gold still exists in the solution after every endeavour to extract it has been exhausted by means of sulphide of potassium, thus showing that gold is soluble in sulphocyanide of potassium. But when the solution is acid the gold thrown down is insoluble. The immersion of a small sheet of zinc, and the addition of a very little oil of vitriol to the resulting double sulphocyanide solution, will completely precipitate the whole of the gold to the metallic state, and render the liquid standing above the sediment which has fallen down quite clear and transparent; and if then to this liquid the copperas testing fluid be again made use of, no chemical action will take place, and no colouration of the liquid will be observed to have been imparted to it, thus indicating that every atom of gold has been removed from the solution, and fully demonstrating that the "Zinc-acid method" of recovering gold from double alkaline solutions has properties equal to those it possesses for recovering gold from acid solutions. This knowledge is, I believe, now given to the community for the first time.

A very useful and efficient portable iron boiler-furnace for drying and burning precipitated substances, or any other waste products of precious metal workers, is that manufactured by Messrs R. Cruickshank, Limited, Birmingham, and herewith illustrated, fig. 16, and one with a holding capacity

FIG. 16.—Portable iron boiler-furnace.

of over 6 gallons of liquid can be purchased for 27s. 6d.[1]; larger sizes being made, with a corresponding increase in the price. It is thus described by the manufacturer:—"They are constructed on the slow combustion stove principle, the fire box, collar for supporting piping and outside jacket being cast in one piece. The furnace boiler which fits into this is also cast with a broad flange, which fits firmly on the outer rim and forms a flue space between the inner boiler and outside jacket. The

[1] Pre-war price.

boiler is fitted with a dome cover, in the centre of which is another disc 6 to 9 inches diameter, which also lifts out. All the separate parts are fitted with handles for convenience in removing; it stands on four legs, and requires no foundation. It is made of plain cast iron, and may be used as a boiler for liquids, as also to burn any waste material."

A small fire is lighted under the boiler, or a gas burner may be used, and this is often found sufficient for many purposes, as the iron casing prevents the escape of heat, an outlet being provided to allow the fumes to escape.

Gold sulphide sediments have most of their sulphur removed by drying and burning for some time with access of air in iron boiler-furnaces, as the iron acts as a reducing agent to the gold; the sulphur, on being liberated, passes through the piping in sulphurous vapours, and metallic gold is left, not pure, but mixed with some sulphur and other non-metallic extraneous substances (all of which will leave the gold on the residue being melted) that are precipitated by the gas from the solution, which subside to the bottom of the vessel along with the gold, and form a part of the residue to be heated in the boiler furnace.

If you have not an iron boiler-furnace of either of the illustrated descriptions I have spoken of, an iron pan may be used, if desired, in which to perform the drying and burning. The burnt dry powder is then mixed with two-fifths of its weight of soda-ash (soda-ash takes up sulphur from the sulphides and makes the slag more liquid), next transferred to a clay crucible, and melted until it is in clear fusion and does not rise in the crucible.

The whole contents may then be poured out into an open casting mould, or may be left in the melting pot, which is broken after the gold has become solid, when it will be found in the form of a button in the bottom of the crucible underneath the slag, and from which it may readily be separated. This method causes the loss of the crucible, but clay crucibles are cheap, and it is not safe to use one a second time, unless it is set inside a plumbago one, in case it breaks. Acidified products act on the crucibles, as also do most of the fluxes used for the reduction of the residues of gold to the solid state, and it is a good plan to coat the inside of clay crucibles (when melting residues of the above kind) with a paste of whiting and water, well rubbing it into the pores of the clay, and afterwards—before using—give the pot a sufficient warming to thoroughly dry the whiting. This will safeguard the crucible, by offering considerable resistance to the action of the fluxes. A plumbago crucible, when employed to protect one of clay, in the manner above stated, will stand from 30 to 40 different times of heating. In order to avoid too much frothing when melting dried cyanide and other alkaline waste products they should be well dried and burnt, as far as it is possible to do it, without causing the residue to agglomerate and form into a solid refractory substance.

If the chloride of tin test gives a white cloudy colouring to liquids from which the gold has been precipitated by means of a zinc plate and a little oil of vitrol, and a white spongy sediment is produced, it is caused through the presence of sulphate of zinc in the solution. Chloride of tin is precipitated by sulphate of zinc, and clean metallic zinc becomes tinned in a solution of chloride of tin. On the other hand, chloride of tin is not precipitated by chloride of zinc, nor is there any change of colour given to the liquid by the admixture. The remedy, when the above difficulty presents itself, is to add a little hydrochloric acid to a sample of the liquid holding sulphate of zinc in solution, before testing it with chloride of tin. Hydrochloric acid does not precipitate sulphate of zinc, but without even changing the colour of the fluid converts it into chloride of zinc. This acid, therefore, shows no outward chemical action, but simply prepares the liquid in a clear condition so that the "tin-testing fluid" may be employed without causing any undesirable chemical reactions to be set up when a few drops of the latter mixture are added to the sample liquid about to be tested; and when there is no gold remaining in the solution from which the sample has been taken, not the least colouration will then be imparted to the liquid which is being tested by the tin chloride. Thus, when a little hydrochloric acid is present in the solution, the tin chloride may be used, with perfect accuracy, for testing liquids from which the reduction of the gold has been effected by any of the known methods, and so delicate is this test, that it will indicate the presence of one-thirteenth of a grain of gold in a little more than a gallon of solution, by imparting at once a faint reddish-brown colouration to the quantity of fluid (250 grs.) selected for sampling, if placed in an ordinary "test tube." The above fraction of a grain of gold is equivalent to the detection by stannous chloride of 1 part of gold in 1,000,000 parts of solution, thus showing that the presence of so small a portion as 6 grs. of gold is easily distinguishable in 78 gallons of the waste liquids of jewellers.

CHAPTER XI

RECOVERING GOLD FROM DIP-GILDING WASTE

GILDING by dipping is a method performed without the use of the electric current. It is an inexpensive process, for there is no danger of depositing more than a mere film of gold upon the work, as a given quantity of gold may be dissolved and mixed with various salts of potash or soda, and then used until the gold dip is exhausted, or very nearly so. Solutions which hold gold in the state of a double salt cannot be treated like those which hold the gold in the form of a single salt by an acid. Neither sulphate of iron (copperas), chloride of tin (tin salt), or oxalic acid (acid of sorrel) will precipitate gold from the former liquids, unless they are first treated by a mineral acid to restore the gold to a simple acid salt. This method cannot be recommended commercially on account of the unhealthy vapours which arise when a strong mineral acid is added to liquids holding double alkaline salts in solution, to the cost of the necessary acid, and also for the reason that a great amount of foaming takes place, which requires extra personal care to be bestowed to prevent the overflowing of the sides of the operating vessel. The reactions in solutions employed for gilding by dipping on adding mineral acids to them, being similar in effect to those of the double cyanide of potassium and gold solutions which are used for electro-gilding.

A good dip-gilding solution will allow of the mixture being replenished with gold, time after time, before being entirely discarded as of no further service, so that the rinsing waters resulting from this process will constitute the bulk of the waste liquid requiring to have the gold extracted therefrom. I have designed a very simple contrivance for the reception of the swilling waters, to which is added the dip-gilding mixture when it has to be thrown away. This contrivance is automatic or self-acting, and is thoroughly reliable in extracting the gold from all waste liquids of the above kind, as well as from others of a similar description.

The principle of the method is confined to a two-tub precipitating and filtering arrangement, so constructed that it may be employed for the recovery of every particle of liquid gold from the exhausted solutions of gold salts, and the rinsing waters resulting from such processes as are generally known as "water-gilding or dip-gilding," galvanic-gilding or contact-gilding, and likewise those of "electro-gilding" by means of the electric current. This method may be termed:—

The Self-acting Process.

The precipitating vessels may consist of disused paraffin barrels (fig. 17), or constructed of waterproofed wood, and made to design. The wooden vessels, of whatever kind, should be rendered impervious with pitch to the liquid inside, or with a mixture of pitch and tar, or one of pitch and paraffin, and have a perforated false bottom covered over with some good filtering material, and above this a layer of coarse deal sawdust is placed to a depth of 4 inches to act as a filter, and as suction for holding down the gold. There should be a space of about 1 inch between the two bottoms for the reception of the filtered liquid. The filtered liquid, owing to the volume of water

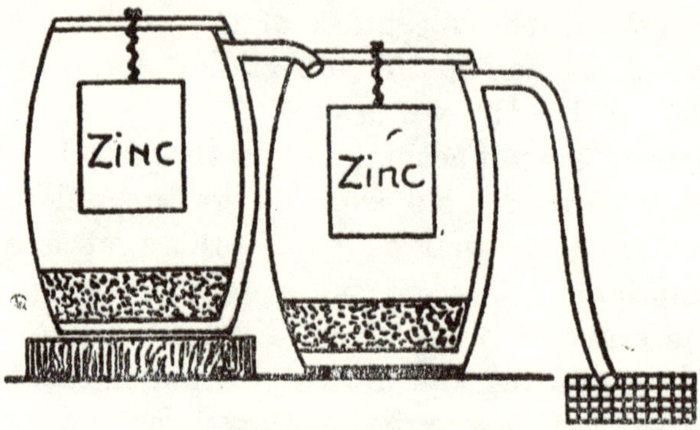

FIG. 17.—Precipitating and filtering vessels.

above, rises up through a narrow waterproof side-passage to the outlet near the top of the tub, and then flows through the spout into a second tub constructed on the same principle. From the latter tub the liquid rises in the same way as from the first, but from the second tub the water passes out free from gold, and is allowed to drain away. A $\frac{3}{4}$-inch or a 1-inch piping may be utilised for conveying the water from tub to tub, providing the fixing is properly secured. There must be sufficient distance between the outlet (through which the liquid passes from vessel to vessel) and the top of the first vessel, to prevent overflowing when the rinse waters are being poured in—about 3 inches of space will be deep enough.

A zinc sheet is suspended in each tub in a vertical position, and a little sulphuric acid (oil of vitriol) is added gradually to react on the zinc, or a very small quantity of hydrochloric acid (spirits of salts) may be used if preferred, or even some of the exhausted oil of vitriol pickling solutions. The acid is introduced to increase the activity of the zinc and prevent the gold depositing thereon.

Zinc in an acid solution containing gold decomposes the acid and the water. It displaces the hydrogen from the two latter and converts the former substance, when oil of vitriol is used, into sulphate of zinc ($ZnSO_4$), and when spirits of salts is used, into chloride of zinc ($ZnCl_2$), both very soluble salts which remain in the liquid. The zinc while being dissolved gives off "nascent hydrogen gas," and this displaces and precipitates the gold in the solid form as it passes through the solution in its endeavours to rise to the surface and escape. A small percentage of sulphuric acid is sufficient to incite the action of the zinc, and as the hydrogen becomes separated from the water, the oxygen goes to the zinc plate and oxidises it, which in turn is dissolved by the free acid into what is now called the sulphate of the oxide of zinc ($ZnSO_4$), and this passes into solution while the gold passes out and assumes the metallic state.

Owing to the solvent action of the oil of vitriol on the zinc, the zinc will dissolve in quantities out of all proportion to the weight of gold which is being separated from the liquid waste. Very great care should therefore be taken in adding the right proportion of acid to the waste liquid, or larger plates of zinc will be required, for it is not owing to the quantity of zinc which is dissolved that the gold is separated from solution, but to the volume of *nascent hydrogen* that is generated in the waste liquid collecting receptacles upon which the full measure of success depends. Sulphate of zinc does not precipitate gold. The following formula will be found suitable for the reduction of the gold from all classes of the liquid wastes now under consideration :—

Dip-gilding solution . .	1 gallon	or 10 gallons
Sheet zinc . . .	½ oz.	„ 5 ozs.
Sulphuric acid . . .	1 „	„ 10 „

The gold (as will also the other metals which may be in solution) falls down out of the liquid on to the filter beds as a metallic powder, and when it is desired to remove the gold from the vessels, the supernatant water above the sediments is decanted off, or syphoned away from it, when the whole sedimentary product, gold, sawdust, and filtering cloth is removed from each vessel and dried in the boiler furnace, at a slow heat at first, to oxidise any hydrate of zinc that may be mechanically mixed with it, the heating being continued until the sawdust and the cloth is burned to powder. The sawdust will prevent the mass forming into a hard clinker-like substance as the burning nears completion, particularly if an occasional stirring is introduced into the operation it will do much to prevent caking.

There will be no necessity to dilute these waste liquids with water after they are put into the collecting vessels, for they will be already sufficiently diluted to permit of the precipitation of the gold, as the volume of the rinsing waters will be much larger in proportion to those of the gold-dipping mixtures, and this prepares the solution in quite a suitable condition to receive the zinc sheet and its excitant. But when the dip-gilding mixtures are treated separately they will require to be diluted with two volumes of water.

It is advisable to test the water running off from the second tub to waste occasionally with one of the "testing fluids" to certify that it is quite free from gold.

The sediment is taken out of the tubs at regular periods according to the rule of the establishment, when it will be found that a quantity of organic matter and foreign substances have fallen down with the gold, which (with the sawdust) adds largely to the bulk of the residue, but during the operation of drying and burning the foreign substances become destroyed or oxidised, and the bulk is reduced by about one-half of its former volume. If the precipitation is imperfectly performed, and much oxide of zinc is allowed to get into the residue, it will add to the difficulty of melting, and a much larger quantity of flux will be required—as also with sawdust-mixed products—to effect a complete fusion. The following proportions of residue and of fluxing salts will, however, speedily overcome any little difficulty of that kind :—

Dip-gilding waste residue .	10 ozs.	or	100 ozs.
Soda-ash (Na_2CO_3) . . .	$4\frac{1}{2}$,,	,,	45 ,,
Fluor-spar (CaF_2)[1] . . .	$1\frac{1}{2}$,,	,,	15 ,,

[1] Common salt may be used in place of the fluor-spar if preferred.

The residue, which is precipitated by means of zinc, consists of gold, silver, copper, and probably more or less zinc; this, after drying and burning, is mixed with the compound flux and the whole melted in a clay crucible. The above is an excellent flux for the purpose of melting such residues, and originated as the result of some of my experiments in melting refractory substances. The advantage of the fluor-spar is the increased fluidity which it gives to the slag produced from residues mostly non-metallic, from which the metal does not readily separate, and a series of shots may, in a thick fluid, remain in the slag. The fluor-spar assists in rendering the slag so fluid that all the metal the residue contains passes through it, after it has become melted, and settles to the bottom of the crucible. The resulting button of mixed metals, after being removed from the clay crucible, is re-melted under a thin layer of charcoal powder, as before directed, and cast into a bar. This is then sent to the refiners as a rough bar with the view of sale, or in exchange for new metal.

CHAPTER XII

RECOVERING GOLD FROM OLD STRIPPING SOLUTIONS

SILVER and base metal articles that have been electro-gilt, also old rolled gold-plated work, now and then require to have the gold stripped off the base metal. There are various methods of doing this, and when the base metal consists of either silver, copper, gilding metal, brass or german silver, it is customary to use an acid stripping solution. The operation of removing the gold is technically spoken of as *stripping off the gold*, and this is effected by immersing the articles in a compound acid solution, which has the property of dissolving the gold without attacking the baser metals to any remarkable extent, if kept free from water. The acid stripping solution used for gold is practically a kind of weak aqua-regia dipping mixture, which, after it has been used for dipping a lot of work, becomes saturated with gold in a liquid state, and this requires recovering at stated intervals, which is not a very difficult operation if you go about it in the right way. The gold may easily be recovered from old or exhausted acid stripping solutions by treating them with a strong solution of *copperas*, which will throw down the metal to the metallic state. The following directions constitute

The Copperas Process

Procure a large stoneware vessel similar to fig. 18, into which put five times the quantity of water that you have of exhausted stripping solution—that is to say, if you have 1 gallon of stripping solution from which it is desired to recover the gold with absolute certainty, 5 gallons of water must be mixed with it

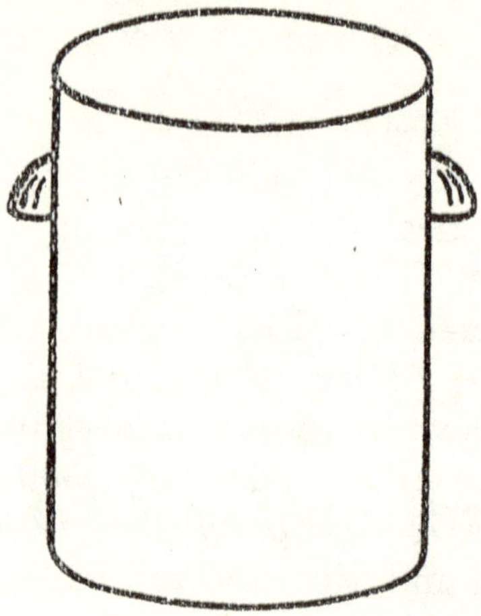

Fig. 18.—Stoneware precipitating vessel.

to destroy the power which the nitric acid has of attacking the iron sulphate and converting some of it into ferric nitrate $Fe(NO_3)_3$, or (FeN_3O_9), an ineffective ingredient which obstructs the precipitation of the gold when excess of nitric acid is present, by altering the nature of the real gold precipitant; and also to restrain the corrosive powers of the mixed acids, by distributing them throughout a larger area of the diluted solution, which is being prepared to receive the copperas liquid. Nitric acid causes waste of precipitating salt, and has to be destroyed or neutralised before the copperas mixture can exercise the proper functions of effecting the complete reduction of the gold.

Pour the water into the large stoneware vessel and then pour the stripping solution into it, a little at a time, stirring with a piece of cane or with a rod of smooth hard wood all the time the pouring is being done. The water should not be added to the stripping solution, or a violent reaction may take place when it is being poured into very strong vitriolised acids. When all the acid has been added to the water, the diluted mixture is then ready for the precipitation of the gold, to be effected by a strong solution of iron sulphate (copperas), which will remove all the gold from the liquid and deposit it at the bottom of the vessel as a dark sediment, consisting of metallic gold mixed with other non-metallic substances which have separated out from the murky solution.

The copperas salt should not be added to the acid solution either in the crystallised form or in dry powder, but in the form of a liquid brought about by dissolving the crystals in hot water, for the precipitation of the gold is then more successful. It will soon be ascertained, by a little experience, how much copperas solution is required to precipitate the gold from a given volume of the diluted stripping solution, when the necessary quantity can be taken each time to effect the intended purpose. A strong copperas mixture is prepared on the following scale :

1	Crystallised copperas, 1 oz.	Water, 5 ozs. or ¼ pint.
2	„ „ 2 ozs.	„ 10 „ „ ½ „
3	„ „ 4 „	„ 20 „ „ 1 „
4	„ „ 6 „	„ 30 „ „ 1½ pints.
5	„ „ 8 „	„ 40 „ „ 2 „

and so on, in proportion to the volume of liquid to be treated, to its acidity, and to the amount of gold it is supposed to contain.

Add the copperas solution to the dilute acid solution gradually, and with stirring. The gold will be "thrown down" as a finely divided powder, into the metallic state. After sufficient time has been given to allow the gold to settle, which will only be a few hours, a little of the solution should be withdrawn from the vessel into a test tube or clean white glass tumbler, and tested whether the whole of the gold has been separated from the liquid (the existence of gold is indicated by a change in the colour of the liquid), and if not, more of the solution of copperas must be added until the liquid standing above the precipitate is rendered perfectly clear; and on sampling this a second time, if a few drops of a clear solution of copperas produces not even the faintest change of colour, it will be proof positive that every trace of gold has been removed from the solution, for a single atom of gold would cause a faint blue tinge to be imparted to the sample put into the test tube. The chloride of tin "test," which is far more delicate, and will detect gold in more dilute solutions than the copperas "test," may be used as the *testing agent*, if preferred. It can be added direct to gold stripping solution samples without any preparation to receive it, as muriatic acid is always one of the ingredients of the mixture, and if no action is perceived on the addition of a few drops of this liquid, not excepting even the faintest brown tinge of colouration, it can be taken as a certainty that no gold is left remaining in the solution.

When the gold has settled to the bottom of the precipitating vessel, the clear solution above it is drawn off by means of the syphon, or with a rubber tube, and fresh water poured into the vessel and the contents stirred up well and allowed to settle. The clear liquid being syphoned off as before, and another addition of water is then made, again stirred and allowed to settle. This is repeated in all three times (the washing is not imperative) so that the acid is nearly all removed by the operation, and the sediment is ready to be scooped out of the vessel and transferred to the boiler-furnace for drying and burning, and this reduces the preparing mass into a fit state for the last and final operation, namely, that of mixing with the necessary flux and melting the dry gold powder in a clay crucible to the solid condition.

Stripping solutions usually contain a large quantity of liquid gold, particularly when they have been used for some time for gold-stripping purposes, and extra care is required to be bestowed in the operation of precipitation when they are strongly acid to avoid the loss of some of the gold, as a portion may easily be thrown away with the supernatant waters through imperfect precipitation, and it is advisable to pass them on to the general waste-water tubs, instead of letting them flow direct into the drains. If, however, the samples taken from the bulk of the solutions are correctly tested, and the results appear as above indicated, there will be little danger of any of the gold being wasted, for the copperas salt is an excellent economist in the gold-working trades.

If the substance resulting from stripping was

attempted to be melted alone without any flux being added, a clean melt could not be obtained, the mass in the crucible would agglomerate and form into a half-fused matte of metal. In order, therefore, to melt such kinds of metallic products and obtain a good clear fusion and free the metal of non-metallic matter it is necessary to use suitable fluxes. The following flux is an excellent one for the purpose of melting this production :—

Gold-stripping waste residue .	10 ozs.	or 100 ozs.
Soda-ash	3 „	„ 30 „
Fluor-spar	1 oz.	„ 10 „

This flux will dissolve all extraneous substances that are mixed with the metallic mass, leaving the particles of metal free to fall through the thin fluid and unite together on the bottom of the crucible. The precipitate resulting from gold stripping will chiefly consist of gold, when the precipitation is brought about by the copperas method, as that salt does not precipitate the other metals from acid solutions. The stripping residue, when dried, is thoroughly mixed with the flux and the whole is melted in a clay crucible. The crucible should not be more than three-fourth parts filled with the mixture, as it rises a little during the earlier part of fusion. The amount of flux needed to melt this residue into clean metal is given in the formula, and as the residue is chiefly metallic it will be found amply sufficient for rendering the slag so thin that all the metal will easily run through it and settle at the bottom of the pot, leaving no shots remaining behind in the slag above the melted metal.

The following is a brief summary of the operations for recovering gold from acid stripping solutions by the copperas method :—

(1) The dilution of gold acid stripping solutions with five times their volume of water, and stirring of the mixture while the acid is being added to the water.

(2) The precipitation of the gold with a strong solution of copperas, sufficient being added to render the liquid clear.

(3) The testing of the clear liquid with a colourless solution of copperas (sulphate of iron), or of stannous chloride (chloride of tin).

(4) The removal of the clear acid water as near to the precipitate as possible without disturbing it.

(5) The removal of the precipitate from the precipitating vessel, and the drying and burning of it to powder.

(6) The mixing of the powder, after reducing it to fineness, with two-fifths of its weight of the aforesaid special flux, or the same weight of soda-ash alone may be used to advantage if it is otherwise preferred.

(7) The melting of the prepared mass in a clay crucible to recover the metal in a solid form, and then remelting this into a rough bar ready for assaying.

This formula is more particularly applicable for the recovery of gold from stripping solutions in which oil of vitriol constitutes the basis mineral acid. The additions of acid employed in making up the mixture, consisting of 20 per cent. of hydrochloric acid, and 10 per cent. of nitric acid, thus forming a kind of weak aqua-regia by which the gold is dissolved. The hydrochloric acid causes the formation of—when silver exists in the solution—insoluble silver chloride, which separates out from the liquid when it comes to be diluted with the water. If, however, the silver should not all leave the oil of vitriol (a supposition not likely to happen), it may be made to do so by immersing in the liquid a sheet of metallic copper, which will cause the precipitation of the remaining silver. A copper sheet may be substituted for the copperas, in the first instance, and thereby reduce the process to a single operation, as it has the power to precipitate both gold and silver from this mixture.

The Sheet Copper Process

Gold is by some operatives stripped from articles by placing them in strong nitric acid, to which 10 per cent. of dried common salt is added. The gold may be recovered from such mixtures by diluting them with three times their volume of water and immersing therein a sheet of copper, when the gold will be thrown down into the metallic state. Soluble gold and silver in these solutions are both precipitated by metallic copper as metallic gold and silver. To precipitate gold from such mixtures by a solution of iron sulphate (copperas) an excess would have to be added, as the gold will not be precipitated until the nitric acid is destroyed, and this causes a quantity of iron to fall down and mix with the gold in the form of ferric oxide. The gold precipitate would then require to be treated with hot hydrochloric acid to dissolve out the iron (hydrochloric acid dissolves all oxides of metals precipitated by adding a solution of sulphate of iron to a solution of gold), or the gold powder, when dried and burnt, would contain too much iron oxide to enable it to be satisfactorily melted. The sheet copper method as a single operation is therefore preferable with *nitric acid* solutions.

By other operatives aqua-regia stripping solutions of special composition are much employed, and one of these may be mentioned as consisting of a 4 and 1 mixture of hydrochloric acid and nitric acid. Such mixtures like the preceding one require to be diluted with about four times their volume of water, and then have the liquid gold reduced to a brown metallic powder by means of a sheet of copper. The action depends on a simple exchange of metal, the copper dissolves and takes the place of the gold in the solution as nitromuriate or chloride of copper, while the gold at the same time as it is leaving the solution is being reduced to its natural state as a metallic product, and nitrosyl gas ($NOCl$) is liberated from the liquid. Residues reduced by copper are easily melted.

The acid stripping solutions for gold are generally of such a composition, that if any silver is removed from the articles submitted to the mixture for stripping off the gold, it is always thrown down in the form of insoluble chloride of silver to the bottom of the vessel in which the stripping of the gold has been effected, by the hydrochloric acid or common salt of which they are partly composed. Some small portion of silver may, however, be held in solution in a concentrated mixture of these substances, but when the stripping liquid comes to be diluted with water, all of the silver will be precipitated of its own accord, and silver chloride is formed as an insoluble substance, which, however, may easily be reduced to the metallic condition by means of a few pieces of iron or of zinc placed in contact with the silver chloride to extract the whole of the chlorine and release the silver from its chemical combinations.

It is not necessary to remove the silver from the stripping vessel until the gold residue is removed. Stripping oxides, or *removing the green*, as it is called, from rolled gold plated articles, and from those of gold, from 6-carat to 12-carat in quality, have special dipping mixtures, and these, along with the dilute oil of vitriol pickling solutions, and their wash waters, should be saved and used in diluting the strong acid stripping mixtures to neutralise the nitric acid because this oxidises some of the ferrous sulphate without any gold being precipitated, and by mixing all together the gold can then be precipitated by a solution of ferrous sulphate (copperas), if added slightly in excess, without any detrimental effects. The precipitate formed being allowed to subside for a few hours, when the clear liquid may be syphoned off and thrown away, after which the precipitate may be dried and burnt to powder, mixed with two-fifths of its weight of soda-ash, and melted in a refractory crucible until the whole of the gold has collected together into a button, when it may be remelted in the manner before described, and sold as a rough bar by assay to the refiners, or the burnt powder may be sent to them direct, who would undertake to reduce it and allow cash for same. If in any of these mixtures the gold exists in the form of a double salt, the oil of vitriol used in some of the mixtures, being the strongest mineral acid, restores the gold to a single salt, easily precipitated by a strong solution of copperas to the metallic state. Also, if such mixtures are sufficiently diluted with water, a sheet of zinc will bring about the same results, but as a rule they are too strongly acid for this method, and it is not always convenient to find room enough to store tanks sufficiently large for the purpose, to enable the zinc method to be commercially adopted, as they would require a large quantity of water; and, even if the room was available, no doubt the copperas process will be found the most simple and economical if treated in the manner stated, namely, by putting the stripping, dipping, and pickling mixtures with their rinsing waters together, and dealing with the whole as one liquid product.

CHAPTER XIII

RECOVERING GOLD FROM PHOTOGRAPHER'S OLD TONING SOLUTIONS

THE toning of photographer's pictures consists in giving them a rich colour by replacing a part of the silver used in previous processes with gold, from a neutral solution of chloride of gold ($AuCl_3$), and when the gold solution ceases toning, and a quantity of this liquid, along with the washing, has accumulated in the collecting vessel so as to nearly fill it, pour into a hot solution of protosulphate of iron (copperas), gently stirring the solution with a clean rod of cane, or other substitute, not metallic, during the time you are pouring in the hot copperas solution. This reagent will rapidly precipitate the whole of the gold to the metallic state, which, in a few hours, all will have parted from the liquid and fallen to the bottom of the precipitating vessel, with other extraneous matter in the form of a muddy sediment, and after the liquid standing above it has been carefully withdrawn, the residue is removed and thoroughly dried in the drying-furnace until it is reduced to a black powder, when it may be melted, or mixed with other burned wastes and sold to the refiners in that condition, after two or three assay trials have been made. The gold precipitate (other metals, if in the toning solution, are not precipitated by the hot copperas solution) may be washed several times with water, allowing it each time to settle well before pouring the water off; the last washing should be done with dilute oil of vitriol or spirits of salts to dissolve and remove any oxide of iron that may have gone down with the gold, in case the toning solution was not sufficiently acid to hold the whole of the iron salt in solution. Fig. 12 is a very suitable vessel in which precipitates of this kind can with safety be washed, if the operation is carefully performed. The pure gold left in the bottom of the vessel may then be dissolved in aqua-regia, forming chloride of gold again, which is evaporated to an oily consistency to drive off the nitric acid, when it can be dissolved in water and used in the preparation of a fresh toning solution.

If any of the solution of hyposulphite of soda (thiosulphate or hyposulphite), $Na_2S_2O_3$, used in the process called fixing to remove the chloride of silver from sensitised paper used by photographers to enable the plate to stand exposure to the light without change, gets into the chloride of gold toning solution to form a double salt, neither $FeSO_4$ (protosulphate of iron), $SnCl_2$ (stannous chloride), nor $C_2H_2O_4$ (oxalic acid) will precipitate the gold therefrom in a commercially satisfactory manner, but on adding a little hydrochloric acid and passing nascent sulphuretted hydrogen gas (H_2S) through the solution it throws down sulphide of gold (Au_2S). Ferrous sulphate is convenient and effective in extracting the gold from most acid solutions, as also is it suitable for the gold baths used for gilding by dipping with single acid salts, such as the bicarbonates, and to most of the stripping solutions, when diluted with water, the copperas method is perfectly adaptable, but ferrous sulphate is not applicable to the recovery of gold from solutions holding it in the form of a double alkaline salt. For example, copperas does not cause the gold to leave the liquid from the following mixtures of double salts: sulphocyanides of gold, hyposulphite of soda and gold, sulphites of the alkalies and gold, cyanide of potassium and gold, baths used for gilding by dipping when holding a cyanide, and all other solutions which hold gold in the form of a double salt, unless they are first reduced to single acid salts by adding a quantity of oil of vitriol. This method is dangerous, on account of the unhealthy vapours which are evolved, and to the frothing which arises through the addition of the acid to solutions composed of double alkaline salts, and this treatment may cause an overflow of the sides of the vessels if not provided sufficiently capacious to prevent it.

The "zinc-acid method" devised by the author will most successfully throw down the gold and silver from all these solutions of double alkaline salts into a metallic sediment, and will require very little or no attention in its application from the time it is started. For, by slightly acidifying the liquids with oil of vitriol or spirits of salts, and suspending therein a zinc plate in a vertical position, all the gold will be precipitated in the course of from 12 to 24 hours. Put the liquid or liquids into a large vessel (see fig. 13), pour into it a small quantity of acid of either kind, sufficient to turn blue litmus paper red, and immerse therein a zinc plate, about the thickness of the strongest house spouting, or old battery plates may be utilized for the purpose. The zinc will decompose the cyanide and other double salts, causing the gold and other metals that may be in solution to leave the liquid with which they are in chemical union, and to fall down to the bottom of the vessel in which the operation is being performed, in the metallic condition. After the zinc plate has been in the solution a while, it is advisable to add a further small quantity of oil of vitriol, which hastens the process considerably, this addition being repeated at intervals until the solution becomes quite clear, at which stage every trace of metal in the liquid will have become precipitated. The zinc should only be very gradually dissolved by the weak acidity of the liquid, and very little oil of vitriol is required to be added to cause chemical reactions to be set up within the solution to evolve nascent hydrogen gas, which liberates the metal from its solvents, and reduces it to a metallic powder. When the liquid has cleared itself, after remaining unmolested for a time, the supernatant water is syphoned away from the sediment at the bottom of the vessel. The sediment is then dried, slowly at first, the heat being continued until all organic matter has been destroyed, and the product burnt to powder, when the latter is mixed with two-fifths of its weight of soda-ash and melted in a clay crucible, in the manner before stated in dealing with the waste products resulting from gilding by dipping, or of those from electro-gilding, when a button of metallic gold, not pure, but mixed with other metals, is obtained.

CHAPTER XIV

USEFUL HINTS AND TABLES

HAVING now given with some considerable explanatory detail the best and most reliable methods for the recovery of gold from the more prominent of the waste liquids resulting from the manufacture of jewellery and other articles, whereby every particle of gold may be extracted with accuracy and completeness, the necessary formulas for reducing the metal to the solid condition are also given in the details, all of which are thoroughly practical, and with ordinary care and judgment may be successfully accomplished after a little practice, by any thoughtful and studious minded operator, and who may, by perusing these articles, come into the possession of some secret knowledge of such a character with which he has hitherto been unacquainted, and which will no doubt prove of service, for I may state that it has taken me years and years of close personal attention, and deep experimental research into all the recesses of the subject, to acquire. Methods have been suggested in accordance with the requirements of the work that are applicable to all the different kinds of waste liquids, whether large or small in quantities, that may have to be operated upon.

Infallible Tests for Gold in Solution

Protochloride of tin is the pre-eminent test in distinguishing the presence of minute quantities of gold in dilute acid solutions, for it will indicate a smaller portion than any other known substance. The addition of a few drops of this liquid to a dilute solution containing gold will present the following characteristics, varying in the following manner, according to the amount of gold existing therein :—

1. $\frac{1}{500}$th of a gr. of gold in $\frac{1}{2}$ oz. of gold solution, equal to 1 gr. of gold in 120,000 grs. of solution—$1\frac{9}{16}$ gallons, imparts to the dilute solution a deep black colour, its approximate equivalent being two-thirds of a gr. of gold in 1 gallon of liquid, or 48 grs. in 72 gallons.

2. $\frac{1}{1000}$th of a gr. of gold in $\frac{1}{2}$ oz. of gold solution, equal to 1 gr. of gold in 240,000 grs. of solution—$3\frac{1}{8}$ gallons, imparts to the dilute solution a brown-black colour, its approximate equivalent being one-third of a gr. of gold in 1 gallon of liquid, or 24 grs. in 72 gallons.

3. $\frac{1}{2000}$th of a gr. of gold in $\frac{1}{2}$ oz. of gold solution, equal to 1 gr. of gold in 480,000 grs. of solution—$6\frac{1}{4}$ gallons, imparts to the dilute solution a red-brown colour, its approximate equivalent being one-sixth of a gr. of gold in 1 gallon of liquid, or 12 grs. in 72 gallons.

4. $\frac{1}{4000}$th of a gr. of gold in $\frac{1}{2}$ oz. of gold solution, equal to 1 gr. of gold in 960,000 grs. of solution—$12\frac{1}{2}$ gallons, imparts to the dilute solution a faint brown colour, its approximate equivalent being one-twelfth of a gr. of gold in 1 gallon of liquid, or 6 grs. in 72 gallons.

Beyond this diluteness the change in the colour of the liquid is only very faintly distinguishable.

Protosulphate of iron is also a conclusive test in distinguishing the presence of gold in dilute acid solutions, but it does not indicate the presence of *quite so small a portion* as the protochloride of tin. The addition of a few drops of a saturated solution of this salt to a dilute solution containing gold will present the following characteristics, varying in the following manner, in conformity with the amount of gold present:—

1. $\frac{1}{500}$th of a gr. of gold in $\frac{1}{2}$ oz. of gold solution, equal to 1 gr. of gold in 120,000 grs. of solution—$1\frac{9}{16}$ gallons, imparts to the dilute solution a dark brown colour, its approximate equivalent being two-thirds of a gr. of gold in 1 gallon of liquid, or 48 grs. in 72 gallons.

2. $\frac{1}{1000}$th of a gr. of gold in $\frac{1}{2}$ oz. of gold solution, equal to 1 gr. of gold in 240,000 grs. of solution—$3\frac{1}{8}$ gallons, imparts to the dilute solution a dark blue colour, its approximate equivalent being one-third of a gr. of gold in 1 gallon of liquid, or 24 grs. in 72 gallons.

3. $\frac{1}{2000}$th of a gr. of gold in $\frac{1}{2}$ oz. of gold solution, equal to 1 gr. of gold in 480,000 grs. of solution—$6\frac{1}{4}$ gallons, imparts to the dilute solution a pale blue colour, its approximate equivalent being one-sixth of a gr. of gold in 1 gallon of liquid, or 12 grs. in 72 gallons.

4. $\frac{1}{3000}$th of a gr. of gold in $\frac{1}{2}$ oz. of gold solution, equal to 1 gr. of gold in 720,000 grs. of solution—$9\frac{3}{8}$ gallons, imparts to the dilute solution a faint blue colour, its approximate equivalent being one-ninth of a gr. of gold in 1 gallon of liquid, or 8 grs. in 72 gallons.

Beyond this diluteness the colour is very faint, and perceived with difficulty. It will thus be observed that the protochloride of tin test is a more delicate one than the protosulphate of iron. Both tests have to be very carefully prepared; they must be freshly made and in a quite clear state, to bring about the above changes in the colour of the gold solution, when testing for the presence of gold in infinitesimal quantities. The gold solution also has to be in a very clear state to bring about results that are satisfactory and complete. The colouration that takes place in the gold liquid resulting from the reactions of both these chemical tests, when employed in seeking for the presence of gold in very dilute solutions, always appears *deeper tinted* if the liquid is held with your back to the light, so that the light is reflected from it, instead of held between the eye and the light and passing the light through it.

Table showing the weight of crystallised copperas salt and the quantity of water required to dissolve it, to precipitate from 1 gallon of jeweller's waste liquids the following proportional parts of gold, when prepared in accordance with the stated directions.

Waste Liquid.	Copperas.	Water.	Gold Precipitated.
1 gallon	¼ oz.	1¼ ozs.	1 dwt.
1 "	½ "	2½ "	2 dwts.
1 "	1 "	5 "	4 "
1 "	2 ozs.	½ pint	8 "
1 "	4 "	1 "	16 "
1 "	6 "	1½ "	24 "
1 "	8 "	1 quart	32 "

The above table of proportions of the various substances will be as near, approximately, as my experience can dictate, after very careful observation and experimental research into the subject of every kind of liquid wastes resulting from the manufacture of jewellery, and also from other artistic work in which the precious metals form a part.

Table showing the different percentages of loss of gold in stated volumes of solution, occurring daily and weekly in jewellery factories, including the relative money value of the respective portions of gold lost.

1 Percentage of Gold per gallon of water.	2 Water used daily.	3 Percentage of Gold lost daily in grs. and dwts.	4 Water used weekly.	5 Percentage of Gold lost weekly in grs. and dwts.	6 Value of Loss weekly.
					£ s. d.
1 gr.	72 galls.	72 grs. = 3 dwts.	432 galls.	432 grs. = 18 dwts.	3 16 6[1]
⅔ ,,	72 ,,	48 ,, = 2 ,,	432 ,,	288 ,, = 12 ,,	2 11 0
⅓ ,,	72 ,,	24 ,, = 1 dwt.	432 ,,	144 ,, = 6 ,,	1 5 6
⅙ ,,	72 ,,	12 ,, = ½ ,,	432 ,,	72 ,, = 3 ,,	0 12 9
1/9 ,,	72 ,,	8 ,, = ⅓ ,,	432 ,,	48 ,, = 2 ,,	0 8 6
1/12 ,,	72 ,,	6 ,, = ¼ ,,	432 ,,	36 ,, = 1½ ,,	0 6 4½

To ascertain the loss of gold in any intermediate quantity of solution is comparatively an easy matter of calculation. For example, column 2 represents 72 gallons of liquid passing away daily, and column 3 represents the average daily loss of gold. But if only that quantity of liquid is being used weekly, column 3 will then show the average weekly loss of

[1] Pre-war value at 85s. per oz.

gold. The money value of the lost gold is then arrived at by dividing the amounts in column 6 by the number of working days in the week; and as many times that volume of liquid is used so will the loss of gold and its value be increased.

When the loss of gold is greater than any of the portions given in the table, it will be noted how serious a matter, financially this part of the working loss of a large manufactory employing from 50 to 100 workpeople becomes in the course of a year, unless very great attention is constantly given to the waste liquids, and the best available means are employed to extract the gold therefrom, before allowing the waste waters to leave the premises and run into the drains. Innumerable waste-saving devices have been invented and brought into the factories with the view of saving the gold in solution, but most of them have emanated from persons who have had no experience inside a jeweller's workshop, and know nothing of chemistry; or from some foreign adventurer, trusting to make a little capital out of unsophisticated employers, before leaving this country for their foreign homes, and before time proves their contrivances to be utter failures and worthless. Pounds and pounds of money have been spent in purchasing useless kinds of apparatus, described as waste-saving, by trustful, though inexperienced goldsmiths, on the strength of the praise bestowed upon them, but alas! their hopes have been shattered, and their money lost, for the want of a little due consideration and investigation into the merits of the contrivances, and their applicability to the work required to be performed in each individual case—that is to say, to the number of hands employed, and to the quantity and kind of liquid that is allowed to run to waste daily, for unless large quantities of liquids can be treated in factories as they ought to be, only little hope of extracting the whole of the gold is possible.

Commercially, it is not advisable to treat the whole of the solutions separately in the manner I have described, unless the accumulations are very large, the results of a great quantity of work having passed through the different liquids. The several solutions may be collected together and treated as one substance, by putting them into the general waste-water tubs, a contrivance varying in construction and employed in every jeweller's workshop, into which is allowed to run the different liquids (used in the various processes of manufacturing the wares special to the firm) when done with, instead of direct into the drains.

It is my intention to supply my readers with full particulars of two waste metal-saving appliances, which, by an arrangement of tanks being adopted that I have designed, are suitable for treating all kinds of liquid waste in one operation. They are self-acting, absolutely reliable in working, inexpensive, and are practically "unbeatable" in effecting the recovery of the gold from the wash-hand waters and all other waste solutions in one process, by throwing down the gold, and duly filtering away the waste water at the same time, for the whole of the gold (unless it is floating gold) will have left the solution before it reaches the filtering vessel in which every portion of floating gold will be retained, thus nothing of any value passes away with the filtered liquid as it leaves the last of the series of receptacles, and it is then in a perfectly fit state for running off the premises, even if they consisted of the largest kind of manufacturing establishments.

CHAPTER XV

GOLD PRECIPITATING REAGENTS

THE methods for the precipitation of gold from the liquid by-products resulting from the manufacture of goldsmith's work and jewellery, to which I have hitherto alluded specially, are practical commercial plans for the precipitation of, and for the detection of the presence of gold in the liquids used in manufacturing establishments, and as much minute account has been brought into the narration as I conceive will be particularly useful to those operators in factories who may wish to educate themselves in these matters. Practice and close observation will alone produce skill in the department of gold waste saving, as erroneous results must necessarily tend to heavy pecuniary losses to employers.

Before proceeding to describe the methods for recovering silver from the various solutions holding that metal in the liquid condition, I will devote short space to the consideration of some other gold precipitants, taken from different authorities of high standing in the scientific world. These, however, are more suitable to the laboratory in experimental research work. They are all, or very nearly all, unsuitable in manufacturing establishments, where the liquids, when discarded, are in a murky condition, and require special substances to clear them, as well as to precipitate the gold. In laboratory research work the liquids are mostly free from murky materials, and are therefore more easily dealt with. For solutions of chloride of gold, and for weak aqua-regia liquids of different strengths holding gold in suspension, the following reagents are applicable to determine its presence and for its precipitation therefrom :—

Sulphurous Acid (H_2SO_3) in solution, or a current of the gas (SO_2) will precipitate the gold completely, as a black precipitate of metallic gold from chlorine and weak aqua-regia solutions to the insoluble state in nitric acid. When, however, the solutions are strongly impregnated with nitric acid, this acid must be destroyed before the gas is passed through the gold solutions. Mr G. H. Makins advises a small quantity of potash to be added, and then an excess of sulphurous acid, when precipitation will immediately commence, and ultimately the whole of the gold is thrown down in a scaly metallic powder !—*Manual of Metallurgy.*

Cyanide of Potassium (KCy) precipitates yellow cyanide of gold in neutral solutions of chloride of gold, soluble in excess of the precipitant, but when hydrochloric acid is present in excess, no precipitate occurs until it is neutralised.—Dr G. Gore, *The Art of Electro-Metallurgy.*

Liquid Ammonia (NH_4OH) precipitates fulminating gold as an oxide from weak aqua-regia solutions, soluble in cyanide of potassium and the alkaline cyanides, carbonic acid being evolved. *Carbonate of Ammonia* (NH_4CO_3) if added to a solution of gold chloride precipitates a similar compound. It should never be allowed to become dry, for in this condition it is liable to explode with great violence, but so long as it is kept wet, there is no danger attending its use.—Mr W. G. M'Millan, *A Treatise on Electro-Metallurgy.*

Magnesia, or Magnesium Oxide (MgO) precipitates gold in the form of an oxide, insoluble in dilute nitric acid, from solutions of gold chloride, and without evaporating the acid to render the liquid neutral. This oxide of gold may be reduced to the metallic state by the action of light, or by heating to a temperature of 245° F.—Mr W. Jago, *Inorganic Chemistry.*

Tartaric Acid ($C_4H_6O_6$), one of the strongest of the organic acids, precipitates gold from its solution in nitro-muriatic or nitro-hydrochloric acid, or more simply aqua-regia, when diluted with water, as a black precipitate when the mixture is boiled.—Mr J. Scoffern, M.B., *The Chemistry of Gold.*

Antimony Chloride ($SbCl_3$) precipitates gold from its chloride solutions, and throws it down in the metallic state.—Mr J. Scoffern, M.B., *The Chemistry of Gold.*

Caustic Potash or Potassium Hydrate (KOH) precipitates oxide of gold, soluble in excess of the alkaline salt, but when hydrochloric acid is strongly mixed with the solution no precipitate is formed.—Mr T. K. Rose, *The Precious Metals.*

Alkaline Oxalates, such as those of potash, soda, ammonia, and calcium, precipitate metallic gold in dilute aqua-regia solutions, especially on the application of heat.—Mr T. K. Rose, *The Precious Metals.*

Sulphide of Ammonium (NH_4HS) and *Sulphu-*

retted Hydrogen (H_2S) precipitates gold as a dark brown and black powder insoluble in excess of the precipitants in acid solutions, soluble in alkaline solutions.—Dr Lyon Playfair, *Lectures on Gold.*

Carbonate of Soda (Na_2CO_3) and *Carbonate of Potash* (K_2CO_3) no precipitate in cold solutions, but when heated voluminous precipitate like oxide of iron.—Dr Lyon Playfair, *Lectures on Gold.*

Bicarbonate of Soda ($NaHCO_3$) and *Bicarbonate of Potash* ($KHCO_3$) no precipitate.—Dr Lyon Playfair, *Lectures on Gold.*

Tannic Acid ($C_{14}H_{10}O_9$), an organic acid, and best obtained from nut galls, precipitates gold from weak chloride solutions.—Mr G. Frownes, *A Manual of Chemistry.* (Tea leaves, after the tea has been brewed, have been recommended as a precipitant for gold in the waste waters of jewellers, but they are of no earthly use, for the tannin they originally contained is exhausted in the brewing of the tea.)—G. E. Gee.

Sulphide of Potassium (K_2S) precipitates gold from its solutions, soluble in excess of potassium sulphide. This alkali throws down finely divided gold in a cold solution of $AuCl_3$.—Mr G. H. Makins, *Manual of Metallurgy.*

Protonitrate of Mercury ($Hg_2N_2O_6$) throws down gold in a very finely divided state in the form of a dark blue or black powder from the chloride of gold solution.—Mr G. H. Makins, *Manual of Metallurgy.*

Sodium Sulphite (Na_2SO_3), *Hyposulphite of Soda* ($Na_2S_2O_3$), and *Potassium Sulpho-cyanide* (KCyS) will precipitate gold from a clear neutral solution of its chloride when acidulated with a few drops of hydrochloric acid. Soluble in excess of the precipitants.—Messrs C. and J. J. Beringer, *Text-Book of Assaying.*

Oxide of Zinc (ZnO) and *Oxide of Magnesium* (MgO), if put into a neutral solution of gold chloride and heated, will throw down nearly all the gold as oxide, if the precipitants are used in excess.—Mr G. H. Makins, *Manual of Metallurgy.*

Sulphuretted Hydrogen (H_2S) precipitates metallic gold at 212° F., but at lower temperatures mixtures of gold, sulphur, and sulphides of gold are thrown down. Pure hydrogen, *unless nascent*, does not appear to act on chloride of gold.—Mr T. K. Rose, *The Precious Metals.*

According to Mr G. H. Makins and other scientists, many organic compound substances precipitate gold from a solution of the terchloride when assisted by light and heat. Included in these are gallic acid, which, when added to a dilute acid solution, throws it down as a yellow precipitate.

Citrate and acetate of potash will each precipitate it; charcoal, which acts best in boiling it with the mixture; tartrate of potash precipitates gold from the terchloride, as will sugar boiled in it, and also tartrate of soda when the mixture is heated, but the action of all these salts will be prevented by the presence of hydrochloric acid and nitro-muriatic acid in excess in the mixtures. There are other precipitants given in various "text-books," but sufficient has been said to show that they are not eligible for the uses of the workshop, and I will thus leave the matter where it now stands, and proceed to describe the methods to be adopted for saving the waste silver existing in the liquids resulting from manufacturing processes in which that metal forms the principal material.

CHAPTER XVI

RECOVERING SILVER FROM SILVER-PLATING SOLUTIONS

THE methods employed for the recovery of the silver from old silver-plating solutions and the residuary liquids in electro-plating works, and in manufacturing workshops engaged in working this metal, have never been of a satisfactory kind. Many silver-plating solutions are often spoiled by unsuccessful endeavours to improve them when the working is unsatisfactory. The chief causes being attributed to impurities, such as dust and carbonate of potash, getting into the solutions; to adding excess of cyanide of potassium or other substances; to the articles not being freed from grease; to too much "brightening liquid" being added, and to numerous other causes which I need not enumerate further, and when it is found impossible to correct the faulty working of a solution it is necessary to recover the silver therefrom and make a new solution.

There are three general methods of recovering the silver from worn-out or spoiled solutions and the waste residuary liquids resulting therefrom, but none of them are quite satisfactory. They are as follows:—

1. By the "dry process" of evaporating the solution, from which it is desired to recover the

silver, to dryness in a boiler furnace if large in quantity, and further heating the residue to redness, in order to destroy all organic matter, and then melting the residue in a clay crucible to recover the silver in a metallic button. No flux is required with this method to cause the metal to fuse and collect together in the bottom of the crucible, as the salts of potash, combined with the carbonaceous matter of the burnt residue, are sufficient for fluxing purposes. This operation is slow, even if you have the necessary apparatus, and will depend largely upon the volume of the liquid, and then it requires considerable attention to prevent boiling over the sides of the vessel unless of sufficient capacity to avoid the difficulty. The operation is also somewhat dangerous, as cyanogen fumes are given off during the evaporation. Small quantities of solution may be evaporated over a water bath with less danger of boiling over, and there is no waste of silver in performing the operation, but it takes a long time to evaporate the whole of the liquid away. The method usually practised by gold and silver refiners for recovering the silver from electro-plating solutions is to evaporate the liquid to dryness in specially constructed drying pans, and then melt the residue remaining, but is expensive on account of the time it takes.

2. By the "wet process" of precipitating the silver from electro-plating solutions, which have to be abandoned, by means of muriatic acid, is another method frequently followed by workers in the electro-plating trades. The solutions require to be diluted with an equal bulk of water, and if they contain an excess of cyanide of potassium, twice their

bulk of water should be added to weaken its effects, as sufficient muriatic acid is required to be poured in until the liquid exhibits a distinct acid reaction, when a white precipitate of chloride of silver is obtained, if the solution is not too strongly contaminated with foreign impurities; but if it is, then it may appear of a colour between a reddish white and dark brown. Dangerous hydrocyanic acid fumes are given off by this operation, as sufficient acid must be employed in small portions at a time with stirring to throw down all the silver present in the solution in the form of silver chloride, and the greater is the excess of cyanide of potassium the more acid will be required to decompose it, thus causing more and stronger fumes to be evolved. This method is perfectly practical, and is very simple in its application and easily performed, and it has this advantage over some of the processes adopted, that every trace of silver contained in the solution of cyanide of potassium is extracted by the aid of muriatic acid. The operation should, however, only be performed in the open air, or in a place where there is perfect ventilation to allow for the escape of the poisonous gases, which are disagreeable and dangerous to health. The solutions should be treated in very large utensils, because the addition of the acid causes the liquid to rise, and produces a large amount of froth. The sediment produced is dried and then melted with one-fifth to three-fifths of its weight of an alkaline flux, such as potash or soda. Take flux in the following proportions:—

$\frac{1}{5}$ for chloride of silver reduced to metal and unwashed
$\frac{1}{10}$ " " " " " washed
$\frac{3}{5}$ " " not reduced to metal.

Muriatic acid added in excess to a silver cyanide solution causes all the silver that is present in the solution to be precipitated, for the chlorine of the HCl unites with the silver, which then falls down to the bottom of the vessel in the form of chloride of silver. The acid in excess of this requirement combines with the potassium of the cyanide of potassium, liberating the cyanogen in the form of gas, which then escapes, and chloride of potassium is formed in the solution. To avoid too rapid an evolution of cyanogen gas the muriatic acid should only be added a little at a time with stirring, and care must be taken not to stand directly over the solution, nor to breathe the poisonous vapours which arise. Sufficient acid must be poured into the solution to decompose the cyanide and extract the whole of the silver. When blue litmus paper is turned red by dipping it into the liquid enough muriatic acid has invariably been added to destroy the alkalinity of the cyanide solution and render it sufficiently acid to cause every trace of silver to leave the liquid and become precipitated as silver chloride. When these results have been achieved, the solution is allowed to rest for a sufficient time to enable the precipitate of silver chloride to settle completely free of the liquid, after which the clear liquid is poured off, or otherwise withdrawn by means of the syphon. The chloride of silver is then dried and reduced to the metallic state by melting with soda-ash in the manner already described. It may also be reduced to the metallic state by treating with a little dilute of oil of vitriol and adding a few scraps of wrought iron, or of zinc, to extract the chlorine and enable it to combine with the hydrogen gas evolved, when the silver will separate out in the metallic condition, in the form of very small metallic grains having a dark grey appearance.

Silver chloride takes more fluxing salts to melt it, in proportion to its weight, than most of the other residuary substances of precious metal workers, about three-fifths of the gross weight of the product being the minimum quantity required, and the crucible must not be filled too full on account of the frothing that takes place, and this may cause the contents of the crucible to boil over from the escape of chlorine and carbonic acid gases, if you are not very careful.

Silver chloride when reduced to metallic grains or finely divided powder by means of hydrogen will not require more than one-fifth of its weight of flux to melt it down into the solid condition, if freed from most of the organic matter associated with it. This is done by washing the sediment produced in reducing the silver chloride to the metallic condition.

Iron or zinc in the presence of dilute muriatic acid or of dilute oil of vitriol rapidly reduces silver chloride to the metallic state at the ordinary temperature. The chloride of silver should not be dried, when it is desired to perform this operation, as it does not reduce the silver so readily when in a dried condition. The acid mixtures consist of 10 parts of water to 1 part of acid, whichever kind is used. The objection to this method is the length of time that it takes to reduce all the silver and the liability of losing some of the chloride of silver (if it is not all reduced to finely divided silver), or of metallic silver, during the washing operations to free it from impurities.

Chloride of silver is insoluble in water and in acids. It is, however, readily dissolved by strong solutions of ammonia, cyanide of potassium, hyposulphite of soda, and partly so by a concentrated solution of common salt.

3. By the "battery process" of electro-depositing the silver on to a metallic cathode by means of an electric current from a powerful battery, or from a dynamo or other sources of electricity, is another method for treating spoilt cyanide solutions to recover the silver. If the battery is made use of for recovering the silver, a piece of sheet iron is attached to both wires proceeding from the battery —negative and positive—to act for both the anode and the cathode. The silver solution should be diluted with an equal bulk of water, when, after the battery has been in action for some time under a powerful current, the silver, or the greater portion of it, will be deposited on the cathode as a loose powder during the time the electric current is passing through the solution, and this continues to drop down into the bottom of the vessel in which the operation is being performed, and may be recovered in the same manner as any other precipitate. The cyanide solution, when diluted, does not attack the iron anode which remains passive in the solution. A strong current is used so as to cause the silver to deposit upon the sheet iron cathode in a non-adhering form, so that it is constantly falling off as a metallic powder to the bottom of the receptacle, and the portion that does slightly adhere is easily removed by brushing with a stiff bristle brush. It is difficult to remove the last traces of silver by this method from the solution

without continuing the operation for a long time, and as time is of value, it is not the best method to adopt in electro-plating establishments for the recovery of the silver from worn-out and spoilt solutions.

The precipitation also of large amounts of silver solution by the employment of muriatic acid is a difficult operation except in experienced hands, and even then it is not an operation that can be safely recommended for general adoption. For very small solutions it can be more satisfactorily carried out.

The method about to be described will greatly simplify the operation of recovering the silver from spoilt cyanide solutions and their attendant rinsing waters. The process is almost free from danger, and no special apparatus is required in carrying it out to a successful issue. I shall introduce it as the zinc-acid process.

CHAPTER XVII

THE ZINC-ACID PROCESS

THE principle of this process is based upon the fact that if a large piece of sheet zinc is hung in an ordinary cyanide silver solution, or one diluted with an equal bulk of water, the silver is precipitated, or completely thrown down into a metallic powder if a very small quantity of mineral acid is added at short intervals; the cause being, a kind of galvanic action is set up between the metals in the mixture, which evolves nascent hydrogen, and this, while it is assuming the gaseous form, decomposes the solution and effects a reduction of the silver to the metallic state. The zinc is an electro-positive metal to silver, and soon displaces that metal when in combination with acid, and also in alkaline solutions, if they are first made *slightly acid*, leaving the supernatant waters standing above the precipitates perfectly clear and quite free from the precious metals. The zinc decomposes the cyanide of potassium, and the weakness of the acidulated liquid gradually attacks the zinc and prevents the silver adhering to its surfaces. Besides the zinc plate, the only chemical needed to be added to the silver cyanide solutions is a little oil of vitriol or spirits of salts, to cause the necessary reactions to be set up to convert the liquid silver to the solid metallic condition, but the solutions must not be made too strongly acid to corrode the zinc too rapidly. The zinc plate has the property of gradually decomposing all the cyanide or other salts of silver into their metallic state from solutions slightly acid. Sulphuric acid (oil of vitriol) is the most convenient acid to use along with the zinc plate, as it causes a quantity of hydrogen gas to escape from its surface, and this, in its endeavours to come to the surface of the liquid, searches through its interior, displacing some of the oxygen of the water in its course, which fastens on to the zinc sheet, forming oxide of zinc, and the effect of this oxidation is to prevent any of the silver depositing on the zinc, which otherwise it may do, and stop the continuousness of its precipitating action on the silver in the solution. The oxide of zinc, as fast as it is formed, is gradually dissolved by the little free acid present in the liquid into the sulphate of the oxide of zinc, a very soluble salt, which remains in the liquid form taking the place of any silver which the solution may hold in a dissolved condition. These reactions are constantly occurring until the end of the operation is reached. Sulphate of zinc taken in the form of a crystallized salt and dissolved in water, will not precipitate the silver from cyanide solutions as there is then no hydrogen generated. The precipitation of the silver is entirely due to the electrical action of the zinc sheet being immersed in the alkaline solution in conjunction with a very little oil of vitriol, which causes the development of *nascent* hydrogen gas, and also to its being an electro-positive metal to the metal in solution—

silver, which so perfectly throws down the silver, leaving no trace of it in the clear liquid standing above the precipitated silver. When this process is well performed, the supernatant liquid to be drawn off will be as clear as spring water. Any number of gallons of solution may be operated upon by this method, and no special apparatus, extra skill, or costly chemicals are required to recover therefrom the whole of the silver.

The silver-plating solution from which the silver is to be recovered may be placed in a stoneware vessel similar to fig. 18, or any other receptacle that may be convenient for the purpose. Large solutions may be dealt with in a disused wooden paraffin barrel (see fig. 13). A good surface of zinc should be exposed to the action of the liquid, and sheet zinc is the best form in which to employ it, for the reason that it can be readily withdrawn after the silver has been precipitated from the solution, and that the necessary surface can be provided without requiring any after-treatment to be adopted to remove the fragments of zinc that would exist in the precipitate if it is used in the form of powder or fine turnings, or in any other form of finely divided zinc. In using sheet zinc, when the solution is slightly acidulated with oil of vitriol, it is not necessary to trouble about the cleanliness of the surface, for if it is coated with oxide, the zinc will work quite satisfactorily, and any deposit of silver will be prevented from adhering to it, the oxide being dissolved almost as fast as it is formed by the acidity of the solution, which sets up a kind of galvanic action between the silver in solution and the immersed sheet of metallic zinc, whereby the silver becomes precipitated, not on to the zinc, but to the bottom of the vessel in which the operation is being conducted.

The larger is the surface of zinc exposed to the action of the liquid the more quickly will the precipitation of the silver be brought about. The proportions I am about to give are only approximate to the largeness of the respective solutions to be treated, and I repeat, that it is not wholly due to the amount of zinc that becomes dissolved that the silver is so completely precipitated, but to the volume of "nascent hydrogen gas" which is forming and being evolved from over all of the surface of the sheet of zinc, and this gas in its endeavours to reach the surface of the solution passes entirely through it, and by an occasional stirring, every part of the liquid comes in close union with the gas, thus completely decomposing the solution, and causing every particle of silver to leave its former associates, and to fall down to the bottom of the vessel in which the gas is being generated, as a metallic sediment.

From ½ an oz. to 1 oz. of metallic zinc to 1 gallon of silver-plating solution, and the same proportions of oil of vitriol, will be ample to effect the reduction of the silver from most of the cyanide of potassium solutions in general use. The latter proportions will precipitate silver up to 4 ozs. per gallon of solution, with excellent effect. To recover the silver, therefore, from 1 gallon of silvering liquid of the strongest kind, take the following :—

Silver-plating solution .	1 gallon	or 10 gallons
Zinc sheet . . .	1 oz.	,, 10 ozs.
Oil of vitriol . . .	1 ,,	,, 10 ,,

For large quantities of silver solution not so much oil of vitriol will be required, and it is advisable to dilute the solution with an equal volume of water, where this can conveniently be done, as it tends to lessen the soluble action of the free cyanide, and then the precipitation of the silver is effected at a smaller cost of the precipitating substances by employing reduced proportions.

When the silver is all reduced to the metallic state, the remaining part of the zinc sheet is withdrawn, and the liquid allowed to rest for a time to clear itself; a sample is then taken and tested with a few drops of pure hydrochloric acid, when, if all the silver has been removed from the liquid, no chemical action of any kind will take place, nor will any change of colour appear visible to the eye through either transmitted or reflected light. The silver solution, after being submitted to the treatment described above, should have a perfectly clear and colourless appearance, like water. It should not show any grey murkiness in its reflections. This colouring may be the result of too much sulphate of zinc getting into the solution, and if that is so, the hydrochloric acid will soon clarify the liquid sample without any precipitate being thrown down, but if silver still existed in the solution, the hydrochloric acid would immediately detect it by showing a white cloud, and later, by the appearance of a white curdy precipitate of silver chloride—hydrochloric acid is the "test," pure and simple, for silver in solution, with which its chlorine has great affinity to combine.

The supernatant water standing above the sediment should not, when the operation is finished, show a milky colouration, but instead, a clear, transparent, water-white liquid should result, or the operation has been imperfectly performed. To prove the accuracy of this statement I desire to make the following declaration, before proceeding to give further particulars of the "zinc-acid process" of recovering the silver from electroplating solutions.

After experimenting for the extraction of the silver from an old silver solution, which, when first made, contained 1 oz. of fine silver to the gallon of water, and the necessary proportion of cyanide of potassium, I found that less than ½ oz. of zinc sufficed to throw down all the silver from an old solution of the above make up. The small zinc sheet which I employed became coated with silver at first, but on the addition to the liquid of a small quantity of oil of vitriol (H_2SO_4) it at once removed the silver from the surface of the zinc. The zinc then did its work, and cleared the solution of its silver, and reduced it to the metallic state, the liquid being rendered colourless at the same time. I next proceeded in the way of experiment to make a new solution in the following manner, after having dissolved 1 oz. of fine silver in dilute nitric acid, then added to it 2 pints of water, and precipitated the silver with hydrochloric acid (HCl) into the form of silver chloride. The silver chloride was next redissolved in a strong solution of cyanide of potassium, and made up with water to 1 gallon of silver-plating solution. It then consisted of the following ingredients :—

Fine silver	1 oz.
Cyanide of potassium . . .	4 ozs.
Water	1 gallon.

A sheet of zinc was immersed in this solution, to which, after a time, a little H_2SO_4 was added to attack the zinc. The zinc threw down all the silver perfectly into the metallic state, at first, without the acid, the silver deposited on the zinc plate, but readily brushed off. On, however, adding the acid, it did not attach itself to the zinc afterwards, but precipitated to the bottom of the vessel. Froth was formed all over the surface of the liquid through the hydrogen bubbles arising from the decomposition of the ingredients within the liquid, but these soon passed away and left the liquid perfectly clear and colourless like water. After a few hours of these reactions, a sample of the solution was tested with a few drops of hydrochloric acid, which turned the liquid slightly cloudy white, but no precipitate of silver chloride appeared visible. The zinc sheet was at once returned to the solution, and a little more acid added, and as this then appeared to act too energetically on the zinc, after 15 minutes of such action it was again withdrawn, weighed to ascertain the loss, and the solution allowed to rest unmolested for two hours, when the solution became quite clear and colourless, and on adding a few drops of hydrochloric acid to some of the liquid withdrawn into a test tube, no change of colour whatever was observable. The zinc must remain in the silver cyanide solution until HCl has no action on the supernatant liquid above the precipitated silver when being finally tested for ascertaining if any soluble silver is left in solution. When the zinc has removed the whole of the silver, the hydrochloric acid testing liquor will not cause frothing or discoloration of the clear liquid formed by the action of the zinc.

If the silver cyanide is not thoroughly decomposed, on testing for silver with hydrochloric acid, a white colouring will take place, probably with foaming. The result of this experiment, which was verified by others, tends to show that 1 oz. of silver per gallon of solution will be completely precipitated to the metallic form by 1 oz. of zinc and ½ oz. of oil of vitriol in 12 hours, in which time all of the silver will have become separated from the liquid. The silver solution may be diluted with an equal volume of water, and the same proportions of the precipitating substances above named employed in extracting the silver from the double volume, with good effect. A little judgment and experience is required to accomplish the best results, and each operator must be guided by the strength or weakness of the different solutions which come under his charge. It ought to be particularly noted that the liquids should, in the end, be quite clear, and without any colour tinge. The solution of silver cyanide only requires to be made weakly acid, just sufficient to barely turn blue litmus paper red, when the hydrogen liberated in the neighbourhood of the zinc will be in a very active form (while assuming the gaseous state), bringing about rapid chemical changes in the component parts of the liquid, and by these reactions the silver is completely precipitated from the solution. It does not deposit on the zinc at all. Another feature of the zinc-acid method is the removal of all cloudiness from the liquid, thus presenting it in a suitable form for testing purposes.

Silver-plating solutions, unlike gold, usually

consist of large bulks of liquid, probably some hundreds of gallons are employed continuously in one receptacle in very many cases in the silver electro-plating industries, and perhaps it may be found a difficult matter to transfer large volumes of solution from one vessel to another, even if the other vessel formed a part of the plant, and was empty at the time a large spoilt solution required to have its silver separated from the liquid. The dilution, therefore, of such large solutions with

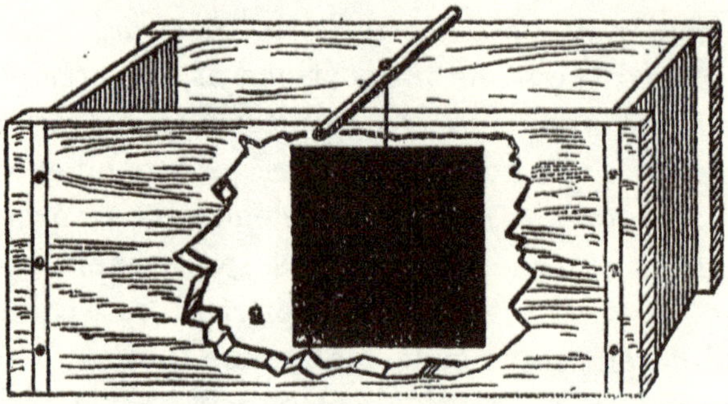

FIG. 19.—Silvering vat as precipitating vessel.

water to reduce the strength seems entirely out of the question on account of the inconvenience and expense. By my process, the only apparatus and chemicals necessary are a few pounds of sheet zinc, a pint or so of oil of vitriol, and a tank for holding the solution (fig. 19), which may consist of the original vat, and its contents may be placed under treatment without removal. The solutions may all be treated cold and need never be warmed. Such solutions if treated over-night by the immersion of a piece of sheet zinc and gentle stirring, will generally have all the silver "thrown down" in the morning. To the vat holding the silver solution when not filled full with liquid, it is always advisable to add as much water as the vessel will conveniently take without causing it to eventually overflow from the slightly arising froth subsequently set up by the hydrogen gas which is being generated within the solution.

The zinc when placed in the solution first turns black owing to a layer of hydrogen over its surface, but this is soon dissolved off by the action of the oil of vitriol, and afterwards the hydrogen escapes from its surface almost as fast as it is being generated, and by its means there is no danger of the silver adhering to the surface of the zinc sheet. It all falls off to the bottom of the vat as a finely divided metallic powder.

When the precipitation of the silver is completed, the clear solution on the top of the sediment (this will contain much organic matter) is syphoned off by means of a rubber tube, which is first filled with water and then immersed in the solution.

MELTING THE PRECIPITATE.

The precipitate which is produced consists of all the metals which may happen to exist in the silvering solution at the time of precipitation. The sediment thus formed will be voluminous, as organic substances are removed from the liquid as well as the metals, and to recover the latter, the sediment will require to be dried and melted in a crucible, assisted by employing a suitable flux, or otherwise sent to a smelting works in the dried condition, and in as fine a state of division as possible. Another method is to mix it with the polishings or other saved waste, and send the resulting product to be smelted, at regular periods of the year, by those who make this part of the business a speciality.

When it is desired to melt a metallic substance of the above kind in a wind furnace provided in a department of your own premises, clay crucibles of various construction are employed in conjunction with proper fluxes. On the small scale this is the most profitable course to adopt. Any material of a metallic nature which contains the precious metals, if dried and pulverised to a fine powder, and then thoroughly mixed with a good reducing flux before melting, may have the metals extracted therefrom.

The flux will unite with the organic matter and form a kind of liquid slag, while the metallic substances will melt and separate from the slag, aggregating in a mass at the bottom of the crucible in the form of a metallic button of mixed metals. On breaking open the crucible and turning out its contents, the slag will be found overlying, and the metallic button underneath. By giving a smart blow with a hammer the slag may readily be separated from the solid metal, and the latter is then ready to be remelted for assaying purposes with the view of turning it into money.

The crucible should not be filled quite full with the mixture requiring to be melted, as the mass has a tendency to rise up during fusion, particularly if the powder has not been thoroughly dried to destroy its contents of combustible organic matter. As an assistant to this end, the sediment, after scooping it out of the vat, may easily have the remaining liquid drained away which you are unable to remove with the rubber tube, by tying or nailing the edges of a piece of finely woven flannel or of a similar piece of unglazed calico to those of a square frame of wood, leaving a concavity to hold the precipitate, and placing this in a manner so that the liquid will drain from it. It is then in a condition to be quickly reduced in the boiler furnace, and the process expedited for melting.

The amount of reducing flux which is needed will vary with the material to be melted. For residues containing much organic impurity, a larger proportion of flux is necessary than for those mostly metallic. For the former products the following proportions should be used :—

Silver plating residue . .	10 ozs. or	100 ozs.
Bisulphate of soda ($NaHSO_4$) .	3 ,, ,,	30 ,,
Fluor-spar (CaF_2) . . .	1 ,, ,,	10 ,,

This is an excellent flux for zinc reduced residues, and for the matter of that, for almost any kind of metallic residues containing gold and silver it will give good results. The first named salt is called *salt cake*, its cost being only 1d. per lb. It possesses excellent dissolving properties for metallic oxides, although not quite equal to the fluor-spar, which quickly dissolves oxides, and also all kinds of siliceous matter when that is present.

Graphite crucibles have been recommended in which the general class of dried sedimentary substances of jewellers may be melted, but it is not found good practice commercially, for besides their being acted upon by the fluxes employed, a longer time is required to effect fusion, and not so clean a melt can be effected, as some of the particles of metal refuse to entirely separate from the slag which is being formed, for detachments of these crucibles become united with the fluxes employed, thus causing the fluxes to thicken and render some portion of the mass within the crucible infusible, so that the smaller particles of metal have a difficulty in draining away from the slag.

There is no fixed time for allowing the pot to remain in the fire. Sometimes the product will melt much more quickly than at others, owing, no doubt, to the temperature of the furnace and to the atmospheric influences surrounding it, and the workman will have to be guided entirely by circumstances and his own experiences as to the completion of the process. But the heat must be continued until the component parts in the crucible cease emitting gaseous vapours through the slag, which is being formed on the top of the fusing mass in the crucible, and until a tranquil condition is arrived at, when the metal will have collected together and settled to the bottom of the crucible and remain separated from the slag after cooling; the crucible may then be broken and the button of metal obtained, or the whole mass, while in the melted state, may be poured out into an open casting mould; but it is advisable to allow it to remain in the crucible until cool, and then break the crucible so as to obtain the silver in the form of a button, for if the melted mass is poured out, a portion of the metal is apt to become entangled with the flux, and a difficulty will be found in effecting a clear separation, unless the slag consisted of a very thin fluid after being poured from the crucible into the open casting mould, so as to enable the fused metal to drain out of the slag again, when the silver will be found underneath in an aggregated solid form.

The slag is required to be rendered so fluid that all metal must settle from it after the whole mass has melted. In a thick slag, the metal does not separate out sufficiently, and numerous shots remain in the slag. A good melt is therefore not obtained. A small portion of nitrate of soda (Chili saltpetre), if added to a thick slag, will quickly give increased fluidity to the mass, and thus enable the particles of metal to freely unite and settle rapidly out of the slag.

Chloride of ammonia (salammoniac) has been recommended for the same purpose, and to assist the fusion in melting the waste products of goldsmiths. This advice can only be acceptable to those without any practical experience of the work I am dealing with. To those experienced at the work it comes as a bolt from the blue, for salammoniac is a volatile salt, and burns away in the dry form in the nature of a gas, giving no liquidness whatever to the slags formed as the result of melting waste residues, but rather, on the contrary, it tends to absorb moisture from the fusing substances in the crucible instead of giving the required fluidity.

A summary of the steps in the zinc-acid process may be enumerated as follows :—

(1) Precipitating the silver with a sheet of zinc in union with a small quantity of oil of vitriol.

(2) Removing the clear liquid standing above the sediment thrown down.

(3) Scooping the sediment out of the vessel and the drying and burning of it to powder.

(4) Mixing the powder thoroughly, with a good reducing flux of nearly half its weight.

(5) Melting the mixture in a clay crucible to

collect together all the fine particles of silver into the solid condition.

(6) Remelting the button of silver obtained into a rough bar ready for refining.

(7) Obtaining two or three assay trials to ascertain its fineness, and then disposing of the bar to the refiner who makes the best offer.

The results of this process will serve to show how nearly accurate it is (when properly carried out) in reclaiming the silver from all kinds of silvering solutions, for there is very little loss of silver. It is also very simple, and is easily put into action.

CHAPTER XVIII

THE SULPHIDE OF SODIUM PROCESS

THIS is another simple method of recovering silver from silver-plating solutions. Silver readily unites with sulphur, and is precipitated by such reagents as sulphuretted hydrogen, and the following soluble sulphides, namely, potassium, ammonium, calcium and sodium. Silver sulphide, unlike gold sulphide, is barely soluble in either cyanides of sodium or of potassium, or of their sulpho-cyanides, and an excess of sulphur, if allowed to get in the solution, does no particular harm. Sulphide of soda (Na_2S) dissolved in water is probably the best of these chemicals to use, as it is cheaper, and has been found to answer admirably. The sulphide of soda should not be added to the silver solution in crystallised form, but dissolved into a liquid in the following proportions, *pro rata* to the quantity of silver solution requiring to have the silver extracted therefrom.

Sodium sulphide . . .	1 oz. or 8 ozs.
Water	5 ozs. or 1 quart.

The cyanide of potassium silver-plating solution is more speedily acted upon by this reagent if it is first diluted with water, for the silver sulphide which is formed is rendered insoluble in weak cyanide solutions, and less of the precipitant is required to effect the reduction of the silver, as the water reduces the strength of the free cyanide. The more water that can conveniently be added the less danger will there be of any particle of silver remaining in the soluble state in the solution after treatment. About its own, or, at the most, twice its volume of water will prevent any loss of silver. Add the sodium sulphide solution little by little to the diluted silver-plating solution and stir up the contents with a smooth stick of wood. On first adding the precipitating reagent there is no action, but very soon, as you continue adding the sodium sulphide, a brown colouration begins to show itself, which afterwards rapidly proceeds to a much darker colour if the solution is strong. The black sulphide of silver (Ag_2S) which is thus forming will soon settle to the bottom of the vessel, leaving a clear liquid on the top. The sulphide of sodium solution should be added with great care, so that precisely enough may be added to throw down the whole of the silver to the insoluble state, and no more, although there is a wide margin between the insoluble and soluble condition of the silver sulphide when precipitated with this reagent. I intend to give further on the proportions which I have found to answer perfectly for the precipitation of the silver to the insoluble state. The liquid in the precipitating vessel is known to be ready for removal when an addition to the sulphide of soda does not produce a dark colour, after having allowed time to clear itself, for this substance will produce a dark colour when only a trace of silver is left in the solution.

If a dark colour is produced, it is a proof that

silver in its soluble state is still in the solution, and more of the precipitant must be added to remove it, and until no reaction of sulphide of sodium is produced. When this point is reached, all doubt that the whole of the silver has been precipitated is removed by the application of a still more delicate test—hydrochloric acid—to prove the presence or absence of silver, for 1 part of this acid in 200,000 parts of solution will show opalescence—a grey colouring. It is necessary that the solution should be stirred while the precipitating liquid is being poured in, in order that every part of the solution may be brought within its reach. After allowing the solution to stand unmolested for a time, the operator withdraws a little of the clear liquor into a clean glass test-tube, and adds a few drops of hydrochloric acid; if no precipitate or change of colour takes place, it is sufficient proof that all the silver has been removed, but the solution should be allowed twelve hours' rest to enable all the precipitate to settle to the bottom of the vessel, when the liquor standing above it is drawn off by means of a syphon or rubber tube, and the black sulphide of silver, which exists with other matter as a slime, is scooped out of the vessel and deposited in a concavity formed in some good filtering material, and hung in a wooden frame for the remaining liquid to drain away. It is then further dried and burnt to powder in the presence of air in the boiler furnace, when most of the sulphur will be got rid of by escaping in sulphurous acid vapours, and metallic silver will result, if wholly freed from sulphur.

Both sulphurous acid and the soluble sulphides are reducing agents of great value in gold and silver liquids when the metals are required to be reduced, because under proper treatment they liberate *nascent sulphuretted hydrogen*, and the action of this gas causes the metals to leave the liquid in which they are in close union, and to be precipitated into the metallic state, or into sulphides of the respective metals, readily reduced to metal by exposing them in an air furnace to a burning heat to get rid of the sulphur by converting it into sulphurous acid gas.

For a silver-plating solution containing 1 oz. of fine silver to the gallon of water holding the usual proportion of free cyanide, it will take 3 ozs. of sodium sulphide dissolved in water to precipitate it, even if treated when only very slightly diluted with water. If it contained 2 ozs. of silver per gallon, 6 ozs. of sodium sulphide will be required, and so on in proportion to the amount of fine silver there has been dissolved and used for making up the solution. If you prefer to treat the silver solution after diluting it with an equal bulk of water, then the 3 ozs. of sodium sulphide will precipitate the silver from the resulting 2 gallons of solution; and if the solution is to be treated after adding two equal volumes of water, the 3 ozs. of sodium sulphide will still suffice for throwing down the silver from the 3 gallons of solution which will result from the dilution, though probably a little less will be sufficient when the solution has been so much weakened. Large quantities of silver in the solution will, of course, require proportionate increases in the sodium sulphide compatible with the weight of silver and free cyanide supposed to be in the silvering liquid. But the above is a perfectly accurate basis to work from.

When the sulphide of sodium has thrown down the whole of the silver to the insoluble state in the liquid, the addition of hydrochloric acid will not cause frothing or discoloration of the clear liquid formed through the action of the sodium sulphide.

To commence the operation of separating the silver from a cyanide silver-plating solution containing the following proportions of substances :—

Fine silver	2 ozs.
Cyanide of potassium	8 ,,
Water	1 gallon

a solution of sodium sulphide is prepared consisting of these proportions :—

Sodium sulphide	6 ozs.
Water	1½ pints

Add this mixture by degrees to a silver-plating solution of the above consistency which you have previously diluted with 1 gallon of cold water, stirring during the time the precipitating liquid is being poured in. The solution will soon begin to turn a blackish colour; allow to rest for a few hours, when the black silver sulphide will have settled to the bottom of the vessel and a clear liquid remain on top of it. When this has been accomplished, take some of the clear liquid into a test tube, and add a few drops of pure hydrochloric acid. If the liquid still remains quite clear, it indicates that all the silver has been removed and no more sodium sulphide need be added, but sufficient of the precipitating solution must be used until this result is brought about, and no more silver sulphide is thrown down. Both solutions are treated cold, but if the sodium sulphide solution is used hot, the silver will be more quickly separated from its combination as a soluble substance and converted into an insoluble sulphide of silver, which readily separates out from the liquid. If the silver-plating solution is treated over-night with sodium sulphide, in sufficient quantity to the amount of silver and free cyanide existing in the solution, it will be found in the morning that all the silver has left the liquid and taken up the form of a black sediment at the bottom of the receptacle in which the operation has been performed.

The various operations in the sodium sulphide process may be briefly recapitulated as follows :—

(1) Precipitation of the silver as sulphide of silver by means of 3 ozs. of sodium sulphide dissolved in three-fourths of a pint of water (15 ozs.) and adding this proportion for every ounce of silver there is supposed to be in the solution.

(2) Testing the clear liquid standing above the precipitate thrown down with a few drops of pure hydrochloric acid.

(3) Decanting the water from the precipitate.

(4) Drying and heating the precipitate in the presence of atmospheric air to convert the silver to the metallic state and liberate the sulphur in the gaseous form.

(5) Melting the silver in a clay crucible by means of a suitable reducing flux, or the burned residue may be sold direct to silver smelters, after grinding to fine powder, so that a satisfactory sample can be obtained for assay.

CHAPTER XIX

REDUCING SILVER SULPHIDE TO METAL

OWING to the great affinity silver has for sulphur, the two substances, when combined, cannot be completely separated and resolved into metallic silver and sulphur by simply melting alone. Silver sulphide when melted alone without any previous preparation yields a button of sulphide of silver, for the compound as then existing would become mixed, and an *alloy* of the two substances formed; while *sulphide of gold* has the *gold* completely separated from the sulphur by melting, the result being metallic gold in a button and sulphur as a slag. Gold, unlike silver, has a feeble affinity for sulphur, and it is readily separated therefrom by heat alone, without having to first oxidise the sulphur, which is necessary with silver sulphide, for sulphide of silver is unchanged by the highest temperature, when heated without access of atmospheric air.

The ordinary common reducing fluxes, such as those principally used by precious metal workers to reduce their waste residues, do not appear to cause the silver to reject the whole of the combined sulphur, for while a solid button of metallic silver may be made to result, the slag will contain some unseparated sulphide of silver, even after long fusing, and all the silver will, therefore, not be obtained from the precipitate by the ordinary reduction process. The workman without any scientific or chemical knowledge may not be aware of this fact, and often be surprised at the smallness of the solid metallic button of silver he has recovered from combined products of silver and sulphur, giving no thought whatever to the slag retaining some portion of the silver.

The processes for the recovery of the silver from sulphide precipitates into its real solid form, by decomposing its compounds and eliminating the sulphur, capable of adoption by the manufacturer, are the following :—

(1) Roasting with free access of air.

(2) Melting with oxide of lead.

(3) Melting in union with an oxidising flux.

The first method is probably the most economical one to adopt, for as a general rule, sulphur may be expelled from silver when it is combined with it, by exposing the compound to prolonged action of heat and atmospheric air; which treatment first causes oxidation of the sulphur by its uniting with the oxygen of the air, and most of this oxide is afterwards destroyed and evolved in the condition of *sulphurous acid gas* by prolonging the heating operation, the whole of the silver being completely reduced to the metallic state, if the heating is done gradually and extended up to a red-heat without melting the product. The sulphur is generally got rid of, either wholly or in part, by the roasting operation, as it is commonly called; that is to say, the silver sulphide on being exposed to a high temperature in contact with atmospheric air, most of the sulphur burns away and escapes in the gaseous form, but the heating must be prolonged to eliminate the whole of the sulphur, and care must be taken not to heat it too hot at first, for if heated above

FIG. 20.—Silver sulphide roasting furnace.

a certain temperature alone, the substance simply melts, without the separation of the silver being effected.

The most useful roasting furnace for workshop practice is that designed by the writer and shown in the illustrations. Fig. 20 is the plan, and that of fig. 21 represents a sectional view of a very excellent apparatus for reducing silver sulphide precipitates. This furnace is capable of giving a good heat, and of admitting plenty of air into the chamber. It consists of a square bricked body in which the heat is produced, over which a rather large shallow iron pan is fixed, in which the product is roasted and burnt, four iron columns or pillars are secured, one to each corner of the brickwork of the furnace, supporting at some height above a hood or cover which contains the chimney by which the heated air and gaseous vapours are carried off. The open spaces, as seen between the iron pillars,

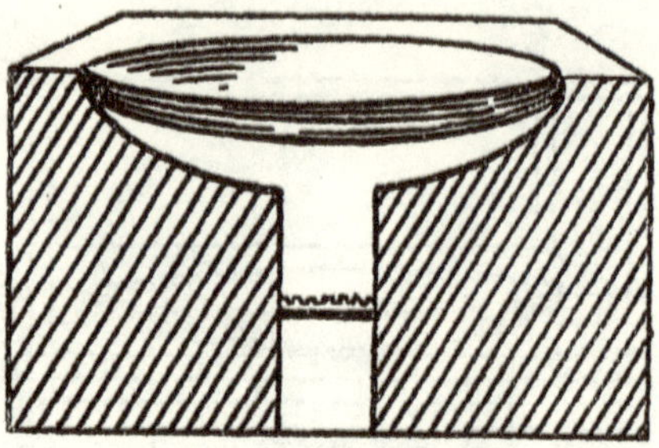

FIG. 21.—Section of furnace.

are for admitting atmospheric air to the burning mass in the shallow iron pan, which can be easily stirred without inconvenience, when this operation is being performed.

In *roasting*, the precipitate must not be fused, but kept in a powdery form, and repeatedly stirred with a round iron rod, about $\frac{1}{4}$ inch in diameter, having a flat bent end for the distance of about 1 inch, until no fumes are given off. The heat, in this operation, requires to be nicely regulated for some time, and should not, when finished, exceed a dull red-heat. The stirring is an important point requiring attention in calcining in a furnace, in order to present the largest possible surface to the action of the oxygen of the air, and prevent fusion; for silver sulphide, when heating, will melt very readily at a low temperature, unless the surface be continually exposed over and over again to the air. The product must not be allowed to agglutinate.

When the moisture from the wet precipitate has been eliminated in the form of steam vapour, the resulting product will rise and ferment, and the drying must be completed at a very low heat to prevent the fusing of the sulphureous substance, and to avoid this happening the addition of a few handfuls of coarse sawdust will assist the process. Constant stirring at a certain stage of the heating is essential, that fresh parts of the product may be exposed to the combined action of heat and air in turn, and as the sulphur becomes burnt out, the resulting material or burnt dry powder—as it will then be—will stand a much greater heat being given to it without its melting, and the heating may then be continued, without much danger, until the residue left is burned to a red-heat, when all the sulphur is burned or oxidised away and the operation may be considered at an end. The product now only requires to be removed from the furnace, pulverised to a fine powder, mixed with a suitable carbonaceous reducing flux, and melted into a metallic button of silver, for silver sulphide when well roasted with free access of atmospheric air, for about three or four hours, the silver is gradually separated from the sulphur and reduced to the metallic state, without the formation of even a trace of sulphate of silver; for all of the sulphur becomes oxidised, the greater portion escaping in the form of sulphurous acid gas (SO_2); and even, through imperfect treatment, *should* some of the sulphur remain behind, unoxidised, and convert a portion of the residue into sulphate of silver, when the residue comes to be melted, practically the whole of the silver leaves the slag and is reduced to the solid metallic condition, the remaining sulphur either escaping in fumes or entering into the slag as sodium sulphate.

In melting the powder resulting from the above operation, probably the best desulphurating reagent will be dried carbonate of soda (soda-ash), to which may be added a little common salt, both reduced to fine powder and carefully mixed with the silver residue, in the following proportions :—

Burnt silver sulphide . .	10 ozs.	or 100 ozs.
Soda-ash (Na_2CO_3) . . .	3 ,,	,, 30 ,,
Common salt (NaCl) . .	1 oz.	,, 10 ,,

The mixture is placed in a clay crucible, which should not be quite filled, as there will be a little boiling up when fusion commences to take place, but the common salt is useful in checking the action of products that cause much bubbling up, and in this case it will assist in effecting the reduction of the silver as well, by converting the sulphur into chloride of sulphur, a volatile liquid. The carbonate of soda is an oxidising flux, and tends to separate the last remaining traces of sulphur from a well-roasted metallic compound through the carbon it contains, and to isolate the silver from its state of combination, thus preventing any loss of silver in the slag. Carbonate of potash (pearl ash) may be substituted for carbonate of soda, but it is much dearer.

Scrap iron is sometimes added in melting direct

a metallic compound containing sulphur, for the special purpose of oxidising and combining with any sulphur which may be present in the product, and prevent loss of metal in the slag. Iron is used to assist in the separation of the sulphur as it combines with sulphur better than with any other metal. It is introduced in the form of clean wrought iron nails or of a piece of iron hoop. The sulphur and iron unite and the production of iron sulphide is the result, which mixes with the slag or forms into a matte above the silver; for the sulphur, on account of its greater affinity for the iron, is separated by it from the silver, and the latter is consequently reduced to a nearly pure metallic state; but prolonged fusion, frequent stirring, and sufficient iron is necessary. The roasting method is, however, the most simple.

Litharge (oxide of lead (PbO) may be used as a desulphurating flux in melting sulphide of silver, and if employed in sufficient proportion, the metallic sulphides acted on are completely decomposed, the sulphur being mostly oxidised or set free in the gaseous state, when, after fusion, all the contained silver will be found alloyed with a portion of the lead. The larger portion of the litharge as it becomes reduced to metallic lead by the sulphureous matter in the product (sulphur reduces fused litharge to metal), passes through the fluid mass in the crucible, alloys with all the silver it finds in its passage downwards, and so concentrates it in a mixed button of silver and lead at the bottom of the crucible. That portion of the litharge which unites with the sulphur remains in the slag as lead sulphide. This method is one more to be associated with the smelting industry, as the argentiferous lead button can be passed on for direct cupellation for the separation of the silver into its pure state. The silver sulphide residue is finely powdered and intimately mixed with its own weight (or more according to circumstances) of litharge, and melted in the usual way in a clay crucible. The operation is simple enough, but it has its disadvantages, for the litharge attacks and permeates earthen crucibles, and in the event of their destruction loss of metal may result; it is also a much more expensive flux to employ, and for these reasons it is one I cannot recommend for factory purposes, for separating the silver from its combination with sulphur by melting in a crucible.

In melting silver sulphide with oxide of lead it is necessary to add to the charge enough oxide of lead, not only to oxidise the sulphur present, but to collect the whole of the silver into that portion of the lead which is reduced to the metallic state by the reactions set up on fusing the two substances together, in which case the litharge oxidises the whole of the sulphur and causes it to enter the slag on the surface of the silver-lead button formed at the bottom of the crucible.

Sulphide of silver, even if melted along with an excess of *metallic lead*, is not completely reduced to the metallic state, the sulphide of lead slag formed as a result of the sulphur combining with some of the lead contains some sulphide of silver, while the silver which is liberated from the sulphur alloys with the excess of lead in the metallic condition; all foreign materials of an organic nature are, however, either volatilised or turned into the slag, and an unadulterated alloy of lead and silver obtained.

Saltpetre (nitrate of potassium, KNO_3) is a desulphuriser, and completely reduces silver sulphide (Ag_2S) to the metallic state; it, however, should not be used in large quantity in the introductory stage when the latter is melted with it, as it causes loss of silver by oxidation and volatilisation as well as sulphur, and sulphate of potassium is formed in the slag as a result of some of the sulphur combining with the potash of the saltpetre. Nitrate of soda, $NaNO_3$ (Chili saltpetre), may be used as a substitute for the saltpetre, if preferred, as the action is similar. For oxidising the sulphur, these are the best substances to employ, as both possess great oxidising properties in melting metallic sulphides rich in silver, although a loss of silver always takes place through the cause above stated. The nitrate salts of potassium and sodium, when heated with metallic substances, give off oxygen which oxidises out the most easily oxidisable impurities, and when deprived of their oxygen the decomposed nitrate salts of potash or soda, as the case may be (owing to the facility with which they pass into sulphates), then acts as a reducing agent for the metal. The action of the nitrate salts is entirely one of oxidation, as it is supposed they first attack the baser impurities in a melting charge more strongly than any other flux, and then commence to attack the crucibles and the purer metals. It is advisable to add the saltpetre by degrees to the charge, and not all at once, or an explosion may take place, and a portion of the contents of the crucible be thrown out if an excess of saltpetre is used, but by using a moderate quantity in the manner stated, just sufficient to burn out all the sulphur and leave the silver unoxidated, the latter will be more successfully obtained. The decomposition of the oxidising fluxes by the sulphur withdrawing therefrom the oxygen, leaving an inert fluxing substance which collects together the silver into a metallic button, is the crux of the whole secret; and as to the resulting reactions of sulphate of potash, or of sulphate of soda on sulphide of silver, it is worthy of remark that if employed in that chemical form alone, neither would have any oxidising action upon silver sulphide when melted with it, but yield a mixed button of metallic and sulphide of silver, so that it is the oxygen in the flux that plays so important a part in the complete separation of the silver from the sulphur. (Sulphur and oxygen combine, forming sulphurous acid at a red-heat, which escapes.) Both sulphate of potash and sulphate of soda are powerful reducing agents, and possess dissolving properties for oxides, which they dissolve into the slag under the process of fusion; thus the sulphur is first oxidised by the nitrate salts, and then dissolved by the sulphate salts, owing to the chemical change the former have undergone, while the silver is set free, and completely reduced to the metallic state.

Wrought Iron Crucibles are sometimes employed for melting down silver sulphide residues, as they are very useful by their action in removing the sulphur and separating the silver therefrom. By this plan the iron, instead of being added to the melting charge in separate pieces to cause the reduction of the silver, is supplied by the sides of the crucible. Wrought iron crucibles are formed by bending a thick plate of iron round a mandril, welding the edges together, and then attaching a thick iron bottom, also by welding. The flux consists as follows :—

Silver sulphide residue . .	10 ozs.	or 100 ozs.
Soda-ash (Na_2CO_3) . .	3 „	„ 30 „
Saltpetre (KNO_3) . . .	1 oz.	„ 10 „

A wind furnace is employed in which the melting is performed; a tall chimney is necessary in order to ensure a good draught. The product should be allowed to enter into a complete state of fusion before it is withdrawn from the fire, and when the crucible is withdrawn its contents should be poured into an open casting mould. The silver collects by reason of its high specific gravity at the bottom, and is recovered by turning over the mould, when it falls out. The accompanying slag is broken away with a few smart blows from a hammer. Wrought iron crucibles, when carefully welded together, will last for from 25 to 30 melts before their sides are eaten through by the sulphur. Carbonate of potash (K_2CO_3) may be wholly or partially substituted for the carbonate of soda (soda-ash), but it has the disadvantage of deliquescing, and also of being much dearer to buy.

Sulphide of silver may, by another method, have the silver separated from the sulphur and reduced to the metallic state. This process consists in reducing the silver sulphide to the metallic form by means of iron or zinc in a solution of sulphuric acid, or one of muriatic acid, whereby *hydrogen sulphide gas* is given off, and this gas while in the *nascent state* (that is before it has passed into the gaseous form) on coming in contact with the metallic solution decomposes the component parts, and sets free metallic silver. The hydrogen that is evolved appears to possess the property of reducing the sulphide of silver to the metallic condition. Sulphuric acid (the commercial oil of vitriol) is principally used to effect the reduction of the silver as it is stronger and cheaper. After the supernatant liquid has been poured off the precipitated sulphide of silver, diluted oil of vitriol (1 part acid to 8 parts water) in sufficient quantity to more than cover the precipitate is poured on, and pieces of thin clean wrought iron placed in contact with it.

The acid should always be poured into the water in making the mixture, and not *vice versa*. The iron becomes dissolved, and, uniting with the sulphur, enters into solution as sulphide of iron, while the silver is being gradually reduced to the metallic condition. When zinc is employed to effect the reduction of the silver a weaker acid mixture may be used (1 part of acid to 16 parts of water) to well cover the precipitate, as the zinc is more vigorously acted upon than the iron. The action of the acid on either the iron or the zinc commences at once; this is allowed to go on for some time, and the mass should be frequently stirred, in order to promote uniformity of action and to ensure the complete reduction of the whole of the silver. When this is done, the undissolved iron or zinc is removed and the liquid syphoned off, fresh water poured on to the sediment, well stirred, and allowed to settle. This is repeated three or four times with fresh water to remove the acid and sulphate of iron or of sulphate of zinc (formed as a result of the particular metal used) that may have been produced from the liberation of sulphuretted hydrogen gas. The silver is then dried, and is ready for melting, and about one-tenth the weight of the dried silver sediment of soda-ash as flux will usually be found sufficient to collect the particles of metal into a solid button. If the action of the acid on the iron or zinc subsides before all of the metal appears to be reduced, the liquid should be poured off and a fresh supply of the acid mixture added to the sediment. The operation does not require any watching, and may be allowed to take its own course, as some length of time is necessary to reduce the whole of the silver. The washing of the reduced silver is not imperative (it may straightway be transferred to the boiler furnace and burnt to powder), but if done, the metallic silver must be safeguarded against loss during the washing operation; while on the other hand the sediment is apt to contain a little iron or zinc, but these may be got rid of by adding a little nitrate of soda occasionally to the mass in the crucible during the progress of the melting.

Sulphide of silver melted with cyanide of potassium (KCy) yields a button of metallic silver covered with slag-like matter of sulphide of silver. The whole of the silver is, therefore, not reduced, a portion remaining in the slag. The supposed action of this flux is that it removes oxygen and sulphur from metallic compounds and forms sulphocyanide of potash, which combines with some of the silver. Sulphates of potash and soda have no oxidising action upon sulphide of silver when melted together, but yields principally a button of sulphide of silver.

Carbonates of potash and soda when melted with sulphide of silver always leave a portion of the silver in the slag formed as a result of the operation.

Sulphide of silver is acted upon and dissolved by strong nitric acid (HNO_3) with the formation of nitrate of silver ($AgNO_3$) and the separation of the sulphur which remains at the bottom of the vessel as a sediment. The silver being in the liquid state, can then be precipitated from the solution (after it is decanted from the sedimentary product) with hydrochloric acid or common salt, into chloride of silver, or by means of a copper sheet into metallic silver.

I have pointed out for the reader's guidance a few of the difficulties that are to be met with in dealing with silver sulphides and its precipitates in manufacturing establishments, that are not commonly known to operatives engaged in dealing with waste products of a particularly special kind, and which knowledge may prove of service to them hereafter. Some of these processes are rather complicated for the inexperienced person, and the best and safest method of adoption in the workshops of precious metal workers that I can suggest for metallic sulphides when treated separately, will be the roasting and the subsequent melting of the residue by means of the special fluxing mixture I have indicated at an earlier stage of these observations, which method has the merit of being simple, and therefore more easily performed. Another safe method is to put the product with the polishings and other rich residues, and dispose of the collection to the smelters.

CHAPTER XX

RECOVERING SILVER FROM ELECTRO-SILVERING RINSING WATERS

IN electro-silvering it is only natural that a considerable amount of solution of silver should be carried with the work from the solutions into the rinsing or swilling waters, and these should be preserved and treated at regular intervals for the recovery of the silver. It is not sufficient to transfer these waste liquids to the general waste water tanks, such as are commonly employed, without some preparatory kind of chemical treatment, for some loss of silver may thereby be incurred, and, for the reason that the cyanide of potassium would still hold the silver in the liquid form, and thus enable it to pass away with the flowing waters from these receptacles, or, at all events, it is found more difficult of recovery by such a method, unless extra precautionary measures are adopted in the treatment so as to decompose the cyanide and prevent some of the silver passing away. The rinsing waters will pay for a separate treatment as the removal of the silver therefrom is very simple, and it is very desirable that the working electro-plater should know how to recover the silver from the rinsing waters made use of for swilling the work after it leaves the silvering vats, as well as from the silvering solutions which have to be discarded. In all electroplating establishments it should be made a rule to save all the rinsings by placing them in a separate tank or reserve tub sufficiently large to take all the rinsings resulting from a day's work at the least, and after the day's work is over proceed to remove the silver from the liquid by adding thereto a reagent which will "throw down" the whole of the silver into the insoluble form. Now, one of the simplest plans of doing this will be by means of a solution of common salt, which is not only inexpensive, but has the virtue of being quick and certain in the reduction of the whole of the silver from these liquids, while very little extra labour is attached to the day's work, so that there is no reasonable cause why it should be neglected.

The Common Salt Process

Common salt will precipitate silver from cyanide solutions as silver chloride (AgCl), if diluted with plenty of water, otherwise some of the silver chloride will redissolve (strong brine dissolves chloride of silver), but the rinsing waters are already sufficiently diluted to destroy the effects that any free cyanide may have in checking the action of the common salt on its addition to the liquid that is carried over with the work from the solution to the swilling waters. If the precipitation of the silver is not absolutely complete by this method in the morning, and on traces of silver being found in the liquid above the precipitate upon testing it with hydrochloric acid, it should not be thrown away, but removed from the tank and emptied into the general waste water tanks, so as to again make the former tank available for the incoming day's rinsings. When, however, this is found to be the case, a stronger solution of common salt must be poured into the day's rinsings in the evening. It will take from 2 ozs. to 2½ ozs. of common salt, dissolved in water, to precipitate about 1 oz. of silver (this weight indicates metallic silver) from these rinsings into 1⅓ ozs. of silver chloride, and this can be taken as a safe guide to work to. Pour the solution of common salt into the rinsings by degrees, stirring during the time with a smooth wooden stick. The common salt is a compound of *chlorine* and *sodium* (NaCl), the chlorine unites with the silver, forming chloride of silver, which falls to the bottom of the tank as an insoluble whitish powder, sufficient of the salt solution is required to be added to the rinsings until all the silver has been "thrown down"; allow to settle thoroughly, which, during the night, it will have ample time, when in the morning the water above the precipitated silver will have become clear and ready to be syphoned off, which must be done without disturbing the powder. The latter need only be scooped out of the tank at stated intervals, and not every time the liquid is removed from it. An excellent plan for ascertaining whether all the silver has been thrown down or not, is to take up a small portion of the clear liquid in a glass test tube and add a few drops of pure hydrochloric acid, when, if it produces cloudiness in the liquid, all the silver is not thrown down; if, on the other hand, no cloudiness is perceived, you may take it that all the silver has been precipitated. A strong common salt solution should be made in accordance to the following scale:

1	Common Salt	2 ozs.	Water	10 ozs. or	½ pint.
2	,,	4 ,,	,,	20 ,, ,,	1 ,,
3	,,	6 ,,	,,	30 ,, ,,	1½ pints.
4	,,	8 ,,	,,	40 ,, ,,	1 quart.
5	,,	12 ,,	,,	60 ,, ,,	1½ quarts.
6	,,	16 ,,	,,	80 ,, ,,	2 ,,
7	,,	24 ,,	,,	120 ,, ,,	3 ,,
8	,,	32 ,,	,,	160 ,, ,,	1 gallon.

Any of these mixtures, after solution has been effected, may be employed in proportion to the quantity of rinsings requiring treatment and to the supposed amount of silver contained in the liquid. The sediment at the bottom of the tank will contain the silver in the form of chloride, and if it is desired to reduce it to the metallic state, all you will have to do is to put a few pieces of scrap zinc at the bottom of the tank and stir up the contents occasionally, so as to bring different portions of the sediment into contact with the metallic zinc to take out all the chlorine and form chloride of zinc, a very soluble liquid. As an aid to the more rapid conversion of the silver, after you have drawn off the liquid standing above the precipitate in the morning, add a small portion of oil of vitriol, or a little of the liquid from an old oil of vitriol pickling solution will have a like effect and be more economical. Both operations, namely, the precipitation of the silver chloride and the latter's conversion into a dark grey powder of metallic silver go on at the same time, with eminently satisfactory results being achieved. The contents of the vessel will then consist of metallic silver as a powder, and chloride of zinc and sulphate of soda as liquids. The sediment is collected at stated periods and well dried in the boiler furnace, when it is pulverised to a fine powder, and is then ready for the melting pot, or for sale to the refiners, after trial by assay to ascertain its value.

Silver chloride is of little value in that form to the manufacturer of silver wares, and in order to know what value it is he needs to have it in the metallic condition, but to the electro-plater it serves a useful purpose, as he can make use of it in the preparation of a new silvering solution and thus effect a saving, but in this case it will require to be well washed. The chloride of silver precipitate obtained from silver cyanide rinsings by means of common salt is usually of high grade silver, as it becomes freed from other metals which do not form insoluble chlorides.

There are two general methods at present employed commercially for reducing the silver from its chloride to the metallic state, which I will describe. The first method is by melting it with either carbonate of potash or carbonate of soda as the reducing flux in a crucible, and the second method is by reducing the chloride of silver to the metallic condition by means of iron or zinc in a weak oil of vitriol solution, and then melt down the metallic silver powder, either after washing with one-tenth part of its weight, or without washing with one-fifth part of its weight of soda-ash as the fluxing substance.

Melting Chloride of Silver

Dry the chloride of silver, so as to evaporate the moisture and liberate the chlorine, in the boiler furnace, or, if only in a small quantity, in an iron ladle or in an iron pan with the sides and ends turned up to prevent loss during the drying process, some portion of which is liable to roll off when it arrives at the loose dry state, unless provided against. The iron instrument of whichever kind is used should be coated with a moist chalk mixture and then dried, to prevent the powdery silver from adhering to the sides of the iron pan. When the silver chloride has been reduced to a fine dry powder (to effect which stir frequently), mix it intimately with the reducing flux in the following proportions:—

Chloride of silver . . .	10 ozs.	or	100 ozs.
Soda-ash	6 ,,	,,	60 ,,

Put the whole into a good-sized *clay* crucible, not a *plumbago* one, and place this in the furnace and heat until complete fusion takes place. The fusion is continued until vapours cease to be emitted, the mass being occasionally stirred with a thin iron rod until a perfectly fluid condition is reached, care being exercised to avoid the boiling over of the contents of the crucible from the escaping carbonic acid gas (CO_2). When the whole is in quiet fusion the crucible is withdrawn from the furnace, allowed to cool until the silver has settled into solid form with the flux remaining above it; it is then broken, when a solid button will fall out. The supposed action of the soda-ash on the chloride of silver is that the chlorine existing in the chloride of silver unites with the sodium and forms chloride of sodium, while the carbon unites with the oxygen upon the action of the heat forming carbonic acid and oxygen gases, both of which escape in vapours, leaving metallic silver behind, which passes through the limpid flux to the bottom of the crucible, where the fine particles collect together and are recovered in solid form. The silver can then be used for any desired purpose of utility. The flux forming slag should and must be sufficiently liquid to allow the globules of silver, as they form under fusion, to pass readily through it and collect together into a united lump of metal. Chloride of silver requires plenty of flux to recover the whole of the silver, but under the common salt process of throwing down the silver some of that substance will be combined with the silver chloride (when unwashed), and will assist in the melting, so that there will be more available flux than that given in the formula, and this proportion has been found sufficient to reduce all the silver if sufficient heat is given, the more so when the substance has been well burnt in the boiler furnace.

CHAPTER XXI

REDUCING CHLORIDE OF SILVER TO METALLIC SILVER WITH METALLIC IRON

INSTEAD of reducing the chloride of silver to the metallic form by melting it with soda-ash, the metallic silver may be obtained by means of iron or zinc in a weak oil of vitriol solution, which converts the substance into finely divided metallic silver. This method is perhaps the more economical, when carefully carried out, as it saves a large quantity of soda-ash in the subsequent melting operation. The choride of silver is placed in any suitable vessel and dilute oil of vitriol (1 part acid to 8 parts water) is poured on (sufficient to well cover it) and pieces of clean wrought iron placed in contact with it. The acid acts on the iron and dissolves it, and in doing this nascent hydrogen gas is given off which eliminates the chlorine, and by uniting, hydrochloric acid is formed, which action possesses the property of reducing the silver to the metallic condition; this is known by the grey powdery form which the chloride of silver assumes when all the silver is reduced. The iron employed should be free from grease and rust; galvanised iron scrap cuttings is a very suitable material to use for the purpose, as they are usually clean and the zinc-coated surface aids the process. The action of the iron through being attacked by the acid commences at once to generate hydrogen, and this is allowed to go on for some time, the whole contents of the vessel being frequently stirred, in order to ensure the complete reduction of the whole of the silver. When this has taken place the iron remaining is removed and the silver settles readily in the bottom of the vessel and collects in a mass. The supernatant liquid is then poured off and water is added, the whole stirred and allowed to settle; this is repeated two or three times to remove the acid and sulphate of iron, or other soluble ferric salts which may be present in the sediment. The silver is then dried in the manner before directed, and is ready to be melted down with a reducing flux in the following proportions:—

Dried metallic silver . .	10 ozs.	or	100 ozs.
Soda-ash	1 oz.	"	10 "

The melting is performed in a clay crucible, known as a "London round," and as you are now dealing with metallic silver the operation of melting is rendered more simple than the previous one of melting the chloride with soda-ash to obtain metallic silver. The washing operation may be dispensed with when it is intended to melt the product into a rough bar and sell to the refiners after trial by an assay, but in this case a little more flux will be necessary, as the silver is liable to be contaminated with a little iron and other organic matter; about one-fifth the weight of the silver powder will be found quite ample to effect a good melt. The iron does not melt along with the silver unless the heat is very high, but remains in the slag, and is easily separated from the silver when cold, and after the crucible has been broken to recover the metallic button of silver. There is always present the risk of losing some of the precious metals during washing operations if entrusted to careless or inexperienced workmen in manufacturing establishments. If there is found any difficulty in melting a product of this kind when it contains a little oxide of iron it is easily overcome by using the following special fluxing mixture devised by the writer :—

Burnt-silver residue . .	10 ozs.	or	100 ozs.
Soda-ash	$1\frac{1}{2}$,,	,,	15 ,,
Fluor-spar . . .	$\frac{1}{2}$,,	,,	5 ,,

The fluor-spar removes the iron from the silver by dissolving it into the slag, and if there are any stray particles of chloride of silver that may have escaped reduction in the conversion process they are quickly reduced to metal by means of this flux, while there is no likelihood of any silver remaining in the slag, as the latter takes up all the foreign substances and melts into quite a thin fluid. Nitrate of soda may be used in place of the fluor-spar when iron has first to be oxidised and afterwards dissolved into the slag, but in products like I am now treating of, the iron will be in the oxidised condition, from its having been reduced from the sulphate of iron or other form of ferric salt, such as the chloride, in which state it would exist when put into the boiler furnace, and, as a result of the burning, oxide of iron is produced, and this is completely dissolved into the slag by the fluor-spar in melting the burnt silver product, in which operation the silver then becomes freed of its presence.

REDUCING CHLORIDE OF SILVER TO METALLIC SILVER WITH METALLIC ZINC

This method is commonly supposed to be the simplest way of obtaining metallic silver from its chloride, as it is more rapid than with iron, and equally as effectual. Crinkled pieces of sheet zinc are placed in the vessel in contact with the silver chloride after its precipitation, and dilute oil of vitriol (1 part acid to 16 parts water) is then poured on to it in sufficient volume to a little more than cover the chloride. The acid immediately begins to attack the zinc, which gradually dissolves and sulphate of zinc is formed; this remains in the solution, while the hydrogen gas that is liberated from the pieces of zinc withdraws the chlorine from its combination with the silver, and metallic silver is produced in a finely divided powder, and by these reactions some hydrochloric acid and zinc chloride is formed in the liquid. The mixture requires to be stirred occasionally to expose fresh surface to each other of the ingredients within the vessel in order to promote chemical action, and thus bring about a more rapid reduction of the silver. When chemical action ceases, either by the dissolving of all of the zinc, or by the acid having become exhausted, add more zinc, or more acid mixture in the above proportions if it is suspected that all of the silver has not become reduced. When the silver is all reduced it will be observed that the white flocculent precipitate has changed into a dark and finely-divided metallic powder, without showing any clots of unreduced chloride of silver intermingled with that which has become reduced. When the chloride of silver is completely decomposed the whole mass will appear of a dark grey or black colour, and is found to settle readily out of the liquid standing above it, and it is advisable to continue the operation until this result is brought about in order to make sure that all the chloride of silver has been reduced to the metallic state. After this has been accomplished, the silver is dried and melted in a clay crucible with one-fifth its weight of soda-ash, if melted without washing; or with the *special fluxing* mixture, as recommended for melting the silver, after it has been reduced from its chloride by means of metallic iron. Well-washed sediments of this kind will only require one-tenth their weight of flux to melt them in a perfectly satisfactory manner. It is, however, not customary to wash these precipitates in the best regulated workshops, neither is it good practice to do so, for some of the very fine metal is often washed away and lost. The right method is to put the different sediments, as they present themselves for treatment, directly into the boiler furnace (see figs. 9 and 16) in which they are first dried, and then burnt to powder; in this manner all the metal will be saved.

CHAPTER XXII

RECOVERING SILVER FROM DIP-SILVERING SOLUTIONS AND THEIR RINSING WATERS

SILVERING by dipping is a process by which small brass and copper or brass-plated iron wares may quickly be silver-plated without the use of a battery or a dynamo electric current, and is somewhat extensively employed for small metal articles of the following kind :—Buttons, hat pins, eyelets, buckles, purse and bag frames, knobs, hooks, brackets, comb ornaments, and a large variety of other small metal goods of upholstery hardware, for the reason that it is quickly and cheaply done, and that the deposit of silver thus effected is in the form of a white and firmly adhering nature.

Dip-silvering solutions which are replenished with silver last a long time, and do not come up for treatment every time they present the appearance of having become exhausted of their silver, but after being replenished with silver are made use of for further service. There is, however, a considerable amount of rinsing or wash water made use of, and this should be saved and the silver precipitated from it; but as this involves a little labour and expense in providing tanks for its reception, and also standing room, it is frequently not considered worth while undertaking the cost of a separate treatment, as the amount of silver may be small. In some establishments the amount of silver which goes into the rinse waters is considerable, and a suitable appliance for receiving the waste rinsing waters and thus save the silver, which is wasted, would make an appreciable difference to the credit side of the trading account in twelve months, and thereby prove a valuable asset to the firm.

The Metallic Zinc Process

The method by which the silver may be saved from dip-silvering solutions and their rinsing waters

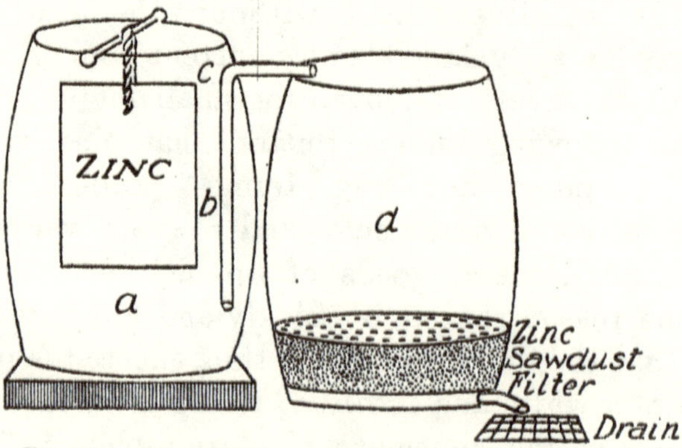

Fig. 22.—Apparatus for recovering silver from rinse water.

consists of a very simple contrivance, and into the first receptacle of this apparatus the dip-silvering solution is poured when it has to be discarded, as well as all the rinsings employed throughout the series of operations. The apparatus is divided into two compartments, as shown in the illustration (fig. 22); they are easily divided when it is found necessary to clean out the silver, which will be in the metallic condition.

The receptacles or tanks may consist of wood, which should be quite water-tight; old beer barrels, or those which have contained paraffin, make excellent utensils, after being well coated with pitch to prevent the destruction of the iron hoops, and to render them impervious to the action of the liquid within. These are both cheap and take up little room, even if they are allowed to remain in the plating room in cases where there is no outside accommodation.

When the tank (*a*) becomes filled with liquid it rises upwards and flows through the tube (*b*), and this lowers the tank again quickly to the level of (*c*), its normal height, so that it *cannot* overflow, unless too large a quantity of rinsings is put into the tank all at once, and at a faster rate than it can be carried into the tank (*d*) by the leaden tube (*b*). There will, therefore, always be an empty space in tank (*a*) down to the outlet where the top of the tube is shown at (*c*), conveying the liquid into the tank (*d*) when no water is being poured in. The water in tank (*a*) is that which contains most of the silver, as it will be understood that all the rinsings are put into this tank, in which is also placed a piece of sheet zinc. This metal has the property of precipitating silver from alkaline solutions into the metallic state. A quicker precipitation is brought about by making the liquid very feebly acid. The silver falls down to the bottom of the tank, and the end of the leaden tube is extended low down in the tank (*a*) so that the water has an upward flow and is conveyed from a tranquil part of the bulk; this is a valuable feature of the contrivance, preventing the direct passing of the argentiferous liquid into the second tank without the opportunity of decomposition being accorded to it in the tank (*a*). When the waste liquids are allowed to flow from vessel to vessel by means of outlets fixed close to the tops of a series of tanks, the liquid is mostly in a state of agitation in those places, and no time is given for the metal to become converted and sink downwards, thus causing the precipitation of the silver to be incomplete, for the slightest agitation causes the light precipitate as it is forming to float and pass through the outlet passages along with the liquid. Without the addition of a small quantity of oil of vitriol a precipitate of hydrate of zinc is liable to form on the surface of the zinc sheet, and this would stop its precipitating action on the silver.

In the tank (*d*) is fixed a perforated false bottom 1 inch from the actual bottom of the tank, over which is placed some good filtering material made to fit well, and above this a layer of coarse deal sawdust, about 8 inches or more in thickness, and over all a circular disc of zinc, pierced with small holes, to enable the fluid to pass through (see illustration, fig. 22). The disc of zinc prevents the sawdust from rising, and also converts any soluble silver, that may be carried that far, into the metallic condition, and by these means only clear water-white liquid, free from silver, is allowed to pass away into the drain. In this tank it is unnecessary to have an extended leaden tube like the one in tank (*a*), as the fluid will be more tranquil, and flow from the bottom through the filter-bed into the drain. When sufficient sediment has accumulated it is burnt in the boiler furnace and melted with reducing fluxes in a crucible in a wind furnace, or otherwise disposed of.

The wooden vessels (*a*) and (*d*) are so arranged that the liquid flows by means of the piping from nearly the bottom of one to the top of the next, wherein the liquid is allowed to rest unmolested until it passes slowly away through the specially constructed filter-bed at the bottom, which allows of only clear solutions to flow through. A stop-cock may be fixed in the outlet pipe in tank (*d*) to enable the liquid to have a longer rest. This contrivance is simply used as an extra security against any loss of silver that might occur by all not being precipitated in the first vessel. It will now be seen why the liquid is brought from nearly the bottom of one vessel through a leaden tube which terminates a few inches below the top of the next vessel.

It is absolutely essential that some of the liquid (which should be water colour) flowing from the discharge-pipe is caught in a clean glass vessel and tested occasionally with a few drops of pure hydrochloric acid, when, if no grey cloudiness is produced, the apparatus is acting with perfect safety, and shows that all the silver has been removed from the liquid, and that the precipitation is being practically effected without any wastage.

This method is applicable for the recovery of the silver from nearly all solutions and their rinsings employed in such processes as dip silvering, contact silvering, galvanic silvering, hyposulphite of silver solutions, electric silvering by means of the dynamo, and cyanide of potassium or prussiate of potash salts—fire dips for the removal of oxides, pickling solutions, etc., and will in each individual case, if properly carried out, give satisfactory results. The solutions should occasionally be stirred in the operations.

Zinc has the property of reducing most metals from acid solutions, and also alkaline solutions if these are rendered slightly acid to blue litmus paper by barely changing it to red. It therefore precipitates silver from alkaline cyanide solutions completely, by employing it with the addition of a little oil of vitriol, so that not a trace of silver is left in them in the liquid form. The best way to use the zinc is in the form of sheets, as they present a large surface to the action of the liquid, and are easily removed from the solution and brushed, if required, at the end of the operation, or at any other time. The amount of zinc needed is usually small, about ½ oz. per gallon of solution is generally found sufficient, and the same quantity of oil of vitriol, or even a less portion may do in weak solutions, its object being to prevent a layer of hydrogen or zinc hydrate attaching itself to the surface of the zinc and obstructing the action. The silver is thrown down in the metallic form in a fine state of division along with any other metal that may be in the solution.

The theory of the process is supposed to be as follows :—Hydrogen is freely generated by sulphuric acid acting on zinc, and the hydrogen on coming in contact with the metallic solution while in its *nascent state* (that is before it has assumed the gaseous form), decomposes the solution ; the oxygen which is liberated has greater affinity for the zinc than for hydrogen, and combining with the zinc is then dissolved by the free acid along with the zinc into the solution as $ZnSO_4$, and metallic silver is precipitated, while the uncombining hydrogen escapes as a gas, along with other gases formed as a result of the decomposition of the different ingredients.

The sediment produced is usually in the form of a voluminous black slime on account of the dirtiness of the liquids which are presented for treatment. The whole mass at cleaning-up time—sediment, saw-dust, and filtering material—is collected together, dried and burnt to powder in the boiler-furnace, then melted down in a clay crucible, the fine dry powder being previously well mixed with the following fluxing mixture:—

Dip-silvering residue . .	10 ozs.	or 100 ozs.
Salt-cake	4½ „	„ 45 „
Fluor-spar	1½ „	„ 15 „

The use of this flux will soon render the mass quite liquid in a good furnace, so that all the metal the residue contains passes readily through the slag to the bottom of the crucible, where it settles in a button of adulterated metal, and after the pot has cooled sufficiently to enable the metal to become solid it is broken and the button recovered. It is then remelted and cast into a rough bar, and finally disposed of to the refiner whose trial report is the most satisfactory.

CHAPTER XXIII

RECOVERING SILVER FROM OLD STRIPPING SOLUTIONS

Acid stripping solutions in which oil of vitriol and aquafortis have been used in stripping silver from old spoons and other wares of silver-plated manufacture, in which the base metal is either copper, brass, or German silver, it is found necessary to remove therefrom all the silver before commencing the process for renewing the goods, otherwise, when they are replated, the second coating may strip or peel off such parts as may have portions of the old coating adhering to them, and it is the common practice to use the so-called acid stripping solution to effect the removal of the silver. The stripping solution, after it has been employed for some time, becomes impregnated with silver, and this should be recovered at stated intervals. There are several methods of treating old acid stripping solutions in vogue, but the following is one of the simplest. It is by no means a difficult operation to recover the whole of the silver by those experienced at the work.

The Common Salt Process

The stripping solution should be diluted with water and the silver precipitated with a strong solution of common salt—one part of salt to five parts of water—in the form of silver chloride (AgCl). For every gallon of the stripping solution to be treated, put into a large stoneware vessel (see fig. 18) about 3 gallons of water, and then pour the stripping solution into it, little by little, stirring during the time with a strip of wood. When all the acid has been added to the water, allow to cool; it is then ready to receive the solution of common salt, which will completely throw down the silver, not in the metallic condition, but in the form of chloride of silver, leaving all other metals in solution that are likely to be present, and this, after being well washed, may be used for making up a new silver solution, or the chloride may have the silver reduced to the metallic state by immersing in the vessel several pieces of zinc (after decanting most of the liquid), in direct communication with the chloride, and adding a small quantity of oil of vitriol, or of spirits of salts, as described on pages 203–7, which will convert the chloride into metallic silver in the form of a dark grey or blackish powder. The common salt should not be added to the diluted stripping solution in the dry form, but in a state of solution, as the precipitation is then more complete. Sufficient salt solution must be added to clear the liquid (when copper is present in the liquid it will give it a bluish tinge) on the top of the precipitate—common salt does not precipitate copper—and when no further precipitate is produced all the silver will have parted company with the liquid and have been "thrown down." A good stirring of the precipitate with the wooden stick will serve to further the union of the chloride of silver by enabling it to cohere and settle well to the bottom of the vessel.

When the chloride of silver has thoroughly settled, the clear blue or greenish blue solution on the top is withdrawn by means of the syphon or rubber tube, the sediment removed from the vessel, and next dried and burned in the boiler-furnace, with repeated stirrings to prevent its melting; when, if desired in the solid metallic condition, it will have to be afterwards melted. Use a clay crucible and mix the powder with about three-fifths of its weight of soda-ash. Do not fill the crucible more than two-thirds full, as the mixture is apt to rise in the earlier stage of melting. The soda-ash causes the particles of silver to enter into the metallic state and to unite together when they become melted, and it is a good plan, when the crucible has become fully heated, and the mixture of soda-ash and chloride of silver begins to melt, to gently stir the mass with an iron rod so as to cause the crust of unfused matter at the top to be brought in contact with the lower portion, which usually enters into a more liquid condition at an earlier stage of the heating. Should the contents of the crucible rise up and appear likely to overflow, a very little *dried common salt* thrown into the pot will speedily cause the mixture to subside. When the collecting together of the globules of silver is complete, two or three small crystals of nitrate of soda, or of saltpetre, may be added at intervals, to the fused mass, which will destroy any traces of iron or other base metals that may be present, although, under the common salt method of precipitating the silver no other metal is supposed to be thrown down with the silver chloride. The bar of silver obtained from silver chloride is pure, or nearly so, and may be alloyed and used over again instead of selling to refiners.

When chloride of silver is melted down with soda-ash, and the operation is completed, the pot is removed from the fire and placed aside in a safe place to cool, when it is broken at its lowest part by a few blows from a hammer, and the solid metal falls out; it is next remelted, poured into a bar, and afterwards assayed as to fineness. If found practically pure silver, it may be employed for any of the chemical purposes requiring pure silver, and if intended for other operations of a mechanical nature it may be alloyed down to the desired standard and manipulated in the regular manner.

The Metallic Copper Process

Another method for treating acid stripping solutions for the recovery of the silver is by first putting the stripping liquid into an acid-proof stoneware vessel (see fig. 18) and diluting it with an equal volume of water (the latter must be put into the vessel first and the stripping liquid added to the water in a gentle stream with stirring), and then precipitating the silver by means of metallic copper in the form of a stout sheet, or in strips being immersed therein; the silver becomes reduced by electro-chemical or voltaic exchange as a finely divided metallic powder which falls to the bottom of the vessel, and may be recovered in the usual way, dried in the boiler-furnace and melted with one-fifth to two-fifths of its weight of flux, in a fire-clay crucible; the fused metal is then remelted and poured into an ingot mould as a rough bar, with the view of sale to silver refiners. Heating the mixture assists the decomposition of the silver and shortens the time it takes the copper to effect the entire reduction of the silver, but where time is of little consequence it is unnecessary for this to be done, as it does not take long to perform the work in a cold solution, and in order to ascertain if there is any silver remaining in the clear bluish green liquid, test a portion withdrawn from the bulk into a test-tube with a few drops of hydrochloric acid at stated intervals, and as soon as no cloudiness or curdy precipitate is formed in the liquid the operation has proceeded far enough and the precipitation of the silver may be regarded as completed, for all of the silver will then have been "thrown down" in a finely divided metallic powder. The copper sheet, or strips, should occasionally be shaken in the solution to remove any sediment that may attach itself to the copper, and if any deposit of crystallised sulphate of silver should adhere to it, it is advisable to brush it off, as this, by exposing a clean surface, helps to facilitate the process. This is a very simple method for the recovery of the silver from stripping solutions composed of oil of vitriol and a small portion of aquafortis, and may be carried out by anybody with very little experience of the operative treatment. The copper exerts a powerful chemical action on the liquid, a portion of the sulphuric acid becoming decomposed, by part of its oxygen being transferred to the copper, forming copper oxide, while this, along with the metal is gradually dissolved by the free H_2SO_4 into sulphate of copper, and sulphurous acid gas is given off in its nascent state, which precipitates the silver into the metallic condition—the sulphate of copper remaining behind in the liquid, and imparting to it a bluish tint. The reduced silver, if required for use in the factory, should be well washed, to free it from organic matter and any sulphate of copper or other soluble cupric salts that may have gone down with the silver, and afterwards dried and melted as before with one-tenth of its weight of soda-ash; or the metallic silver powder may be dissolved in dilute nitric acid, after well washing with hot water, to form nitrate of silver, which can then be used for making up a silver solution, if the operative treatment is being conducted in connection with a silver-plating establishment; or the complete precipitate, without any washing, may be dried and burnt to dry powder and mixed with two-fifths its weight of soda-ash and melted, as before directed, and afterwards sold to the refiners, on the results of two or more assay trials. When washing is adopted, the wash waters should always be preserved and passed through the general waste-water tanks, as they usually contain a little silver, which may there be precipitated, when in the soluble state, with muriatic acid or a solution of common salt as a final treatment. The solid particles of silver settle down by the act of gravity alone when the liquid is at rest.

The Metallic Zinc Process

There is still another method for treating acid stripping solutions for the recovery of the silver in the metallic condition, and this consists (after diluting the strip with about five times its volume of water, in the manner before directed) in placing in the diluted liquid a stout plate of zinc, and this will precipitate the silver as a finely divided metallic powder. More water is required to be added to the stripping solution in the zinc process in order to destroy some of the corrosive action of the acids and render them less attackable on the zinc plate—zinc being much more attackable than copper. The chief point for consideration by the zinc process is to see that the stripping liquid is not too strongly acid to dissolve the zinc too rapidly, but it is open for silver workers to consider which method is best suited to their individual interests to adopt, as I am giving in detail all the practically commercial plans most likely to lead to economical and successful results. The zinc is an electro-positive metal to silver, and soon displaces that metal when in solution in all kinds of liquids in the most perfect manner, leaving the liquid standing above the precipitate thrown down a clear water-white colour and quite free from the precious metal. Zinc precipitates all the metals likely to be present in stripping solutions. If the acidity of the solution is of just sufficient strength to dissolve the oxide of zinc as it is formed, and also prevent a layer of hydrogen adhering to the surface, the silver will not be deposited thereon, but fall to the bottom of the tank in the metallic state, and this electro-chemical action continues until the whole of the silver is thrown down. The theory of the process is, that owing to the immersion of the zinc plate the oil of vitriol is decomposed into its component parts, the free acid attacking the zinc, the oxygen of the water uniting with the zinc, forming ZnO, and then dissolves into a liquid, whilst the nascent hydrogen gas which is liberated becomes dispersed, and in its endeavours to escape from the solution reacts on the soluble silver, separates it from the liquid, and reduces it to the metallic condition; thus an exchange of metals takes place in the solution, the sulphate of silver being replaced by sulphate of zinc in the liquid form. Iron may be employed in obtaining metallic silver from acid stripping solution instead of copper or zinc, but it is a slower method and not so clear and certain in its action. The sediment resulting from the metallic zinc method may be finally dried, burnt and melted in a fireclay crucible in the ordinary manner, with two-fifths its weight of soda-ash as flux, into a button of mixed metals, or with one of the special fluxing mixtures with which I have already provided the reader for this kind of waste. An alternative method is to put the precipitated product into other saved wastes and sell the lot to gold and silver smelters at stated periods of the year. When it is desired to obtain pure silver, to be made use of again in the workshop, the same kind of treatment must be adopted for this purpose as that outlined in narrating the metallic copper process for the reduction of silver from acid stripping solutions.

In the case of acid silver stripping solutions, when they become saturated with silver, it begins to crystallise on the sides and bottom of the vessel as sulphate of silver (Ag_2SO_4), and when the water is mixed with the liquid in preparing it for the precipitating operation these crystals do not entirely redissolve, sulphate of silver being almost insoluble in water, unless very much diluted, but, notwithstanding this, the liquid will still be impregnated with dissolved silver, which is thrown down by either of the methods I have described above, in the stated form applicative to each particular process. No notice need be taken in the factory of the undissolved sulphate of silver in recovering the silver from stripping solutions as regards changing its form purposely, that may be safely left to the ordinary reactions taking place in the mass of mingled ingredients within the vessel; it will be sufficient to collect the whole of the intermixed sediment after the precipitation of the silver has taken place and to dry and burn it to powder; the residue may then be transferred to a crucible and melted in the usual way into a solid lump of metal, or mixed with other wastes rich in metal and preserved till clearing-up time, when it can be sampled by refiners as a part of the ordinary waste, and treated by them to a process of smelting for the recovery of the precious metals.

Copper, zinc, and iron will precipitate metallic silver from solutions of the sulphate of silver.

Sulphate of silver (Ag_2SO_4) may be formed by boiling precipitated silver, or silver filings, in sulphuric acid; it consists, as a chemical salt, of white crystals, only slightly soluble in cold water, and to a solution of nitrate of silver it may be precipitated by adding a solution of sulphate of soda (Na_2SO_4), commonly known by the name of Glauber's salts. Precipitated silver sulphate is converted into metallic silver and sulphurous acid gas (SO_2) which escapes by gradually heating the product up to a red-heat.

CHAPTER XXIV

RECOVERING SILVER FROM NITRIC ACID SOLUTIONS

The Metallic Copper Process

Solutions of silver in which nitric acid has been used should be diluted with about twice their bulk of water (the water can be added *to* the nitric acid solution, as there is no danger of injurious results following, such as those of adding water to sulphuric acid), and then precipitate the silver with strips of copper or sheet copper. The silver is precipitated in a finely divided metallic condition, and after resting to enable the liquid to become clear, the latter can be drawn or decanted off, the sediment well washed with three or four separate lots of water, if you desire fine silver, and melted down with one-tenth its weight of bicarbonate of potash into soft fine metal. The clear blue solution, after the precipitating operation has been performed, contains nitrate of copper, and this may be precipitated, if desired, by the immersion of a few scraps of clean wrought iron and recovered again in the metallic form. The silver solution should be, to the light, quite blue-clear when the copper has done its work, and in order to ascertain if there is any silver remaining in the liquid test a portion withdrawn from it into a test-tube with hydrochloric acid, and if no grey cloudiness is observed it will be a complete indication that all the silver has been extracted from the solution.

The Common Salt Process

When a mixed alloy of silver has been dissolved in nitric acid, and it is desired to recover the silver therefrom in the form of pure chloride of silver, the liquid should be diluted with about three or four times its volume of water, and a solution of common salt—one of salt to five of water—poured in by degrees with gentle stirring; this will precipitate the silver in the form of *silver chloride*, leaving all the other metals in the solution in their liquid state, except lead or mercury, and these are not likely to be present. Hydrochloric acid will also precipitate the silver in a like condition to the common salt. The common salt solution causes the following reactions to take place in the nitrate of silver solution, which, when mixed together, will consist mainly of the following mass of mingled ingredients: silver, nitric acid, chlorine and sodium; the nitric acid leaves the silver and combines with the sodium, for which it has greater affinity, forming nitrate of sodium and liberating the chlorine, and the latter in turn combines with the silver, forming chloride of silver, which gradually falls to the bottom of the precipitating vessel as a white curdy flocculent powder, which does not adhere to the sides of the vessel in which it is thrown down, but falls directly to the bottom.

Sometimes it may be desirable to use only just as much common salt solution as will completely precipitate the silver. In such cases great caution has to be exercised, and it is advisable to add the salt solution by degrees, especially towards the end of the operation, until all the silver has been thrown down. Frequent stirring and allowing the liquid to clear before another addition of the salt solution is made will enable you to judge whether any more of the precipitant is required; also by withdrawing a little of the liquid into a clean glass test-tube and pouring into it a few drops of pure hydrochloric acid, when, if it produces a grey cloudiness in the liquid, all the silver is not thrown down; if, on the other hand, no cloudiness is perceived, all the silver has been precipitated, and the bluish green liquid may safely be poured or syphoned off the precipitated silver. This is a very delicate mode for ascertaining the existence of silver in nitrate solutions. The liquid standing above the chloride of silver, when the process is completed, is not clear water-white, it will show a greenish tinge when other metals are present besides copper, but it will be sufficiently transparent to perceive if any turbidness is set up when the testing liquid is applied. Hydrochloric acid will precipitate chloride of silver from its nitrate solution in a condition consisting of finer flakes than does common salt. The same reactions as with the latter are produced in the solution, with the exception, that instead of nitrate of sodium being formed and remaining in the liquid, nitric acid takes its place.

CHAPTER XXV

WASHING PRECIPITATES

WHEN pure chloride of silver is the object to be obtained, after drawing off or decanting the supernatant liquid, you will next proceed to wash the sediment remaining from all trace of acid, or of nitrate of sodium that may be contained therein, with plenty of water (hot preferred); it will require three or four separate washings, with sufficient time between each to allow the water (if hot) to cool and the sediment to settle, before pouring the water off, which at the finish should be clear, and you will then have pure silver chloride theoretically containing 15 dwts. of fine silver in each ounce of the dried compound mixture. The vessels in which precipitates are to be washed by decanting or pouring off, should have a smaller top than bottom, and be rather tall, when the clear supernatant liquid may be poured off by gently inclining the vessel (see fig. 12), which will enable the water to run away without carrying any of the precipitate along with it. When the first lot of water has been poured off, the vessel is nearly filled with water again, the precipitate stirred with a glass rod or other clean substitute, allowed to settle, and the water poured off as before; this is repeated until all foreign matter is removed and the washings come off clear, when the operation is finished. Any copper or other metal that may have been in the silver alloy remains in the nitric acid solution in the liquid form after the silver has been "thrown down" by either hydrochloric acid or the solution of common salt, and is poured off in the first liquid. It is not worth while trying to recover the copper. The pure chloride of silver may, according to circumstances, be used to make up into a silver-plating solution, or dried, and melted into a bar of pure silver and used for alloying purposes. An alternative method, which finds the most favour in manufacturing establishments in dealing with precipitated products, is to add them, without washing, to other preserved wastes and sell the lot to refiners of precious metals at regular periods, and this, no doubt, when not being scientifically treated, is by far the most economical and best course to adopt.

Silver chloride is perfectly insoluble in water and diluted acids, and only sparingly soluble in concentrated acids; it is, however, freely soluble in liquid ammonia, cyanide of potassium, cyanide of sodium, and hyposulphite of soda.

CHAPTER XXVI

RECOVERING SILVER FROM GLASS MIRROR MAKERS' SOLUTIONS

GLASS mirrors or looking-glasses are silvered, according to the modern method, with a solution composed of nitrate of silver, Rochelle salts, water, and a small portion of liquid ammonia, and after the ammonia has escaped a brilliant mirror of silver is obtained; and as a large amount of silver is used in these trades, it is only natural that a considerable quantity of silver should find its way into the waste products; and many small firms make no attempt to save their waste liquids, but regularly throw them away as not being worth saving for the recovery of the silver. This is a mistaken notion of economy, and the method herein given for recovering the silver from glass mirror makers' solutions is so simple and inexpensive, when carried out according to the directions briefly outlined below, that no difficulty is placed in the way of its recovery. There are several ways by which the silver may be recovered, but the following is perhaps the simplest of all:—

THE MURIATIC ACID PROCESS

The waste silver liquid is put into a prepared wooden tank or barrel of sufficient size to take all that is made during a full day's working; and after the day's work is over, pour some muriatic acid, or spirits of salts, into the receptacle in sufficient quantity to "throw down" the whole of the silver that has gone to waste during the day. Stir the solution with a wooden stick while you are steadily pouring in the acid. It will take about 2 ozs. of muriatic acid to "throw down" 1 oz. of silver in the form of silver chloride; this will consist theoretically of 75 per cent. of silver. In the morning the whole of the silver will have left the liquid and fallen to the bottom of the tank as a grey flaky substance, when the major portion of the water may be drawn off with a rubber tube, but before doing this it will be advisable to test some portion of the supernatant liquid with a few drops of *pure hydrochloric acid*, to make sure that no silver is retained in the liquid. The liquid at this stage should be quite clear, and when free from silver no chemical action of any kind should be observable to the craftsman on adding the hydrochloric acid testing fluid. When this is the case, the clear solution at the top may be syphoned off and thrown away. The chloride of silver on the bottom may be left there, and the ensuing day's waste put on the top of it, and in the evening again treated as before, and repeated day by day until it is necessary to remove the chloride of silver from the tank and start afresh. If preferred, a solution of common salt may be used as a precipitant in place of the muriatic acid.

The chloride of silver, after being removed from the tank, is dried and melted in order to obtain metallic silver. There are two methods employed in effecting this result, namely—(1) by melting the dry chloride with soda-ash, and (2) by reducing the chloride of silver to the metallic state by means of metallic iron or zinc in a dilute oil of vitriol solution, and then melting down the metallic powder and casting into an ingot mould. The latter method only requires one melting operation, whereas the former will require two, for the chloride of silver requires a considerable amount of flux to collect together all the fine particles of silver into a button or solid lump; thus a large quantity of slag is formed which necessitates the silver being allowed to cool in the crucible, and then breaking the crucible to recover the metallic button of silver; and this has to be remelted and cast into a bar before it can be assayed, for the reason that no refiner of metals will take an assay from the lump form as it leaves the crucible after being broken, and presenting a corresponding shape to the inside bottom of the crucible, therefore it is necessary for his test to again melt the button and cast the metal into ingot form. Unless plenty of flux is used in melting chloride of silver, the slag does not become sufficiently liquid to allow the tiny globules of silver as they form under fusion to collect together and fall to the bottom of the crucible. A small quantity of flux will not reduce all of the silver. From one-half to three-fifths the weight of the silver chloride precipitated from hydrochloric acid will be needed of reducing flux to separate the whole of the silver from acid-formed precipitates of this kind.

The chloride of silver must have all moisture eliminated before the melting is undertaken; this is best done in a shallow iron pan (see fig. 20), coated with a chalk or lime paste and well dried, before putting in the silver chloride. Do not heat the iron pan too hot, or the silver chloride may melt into what is known as *horn silver* without changing the nature of the fused body. If the heating is done carefully, and at a low heat, there should be no risks of melting the chloride into horn silver or fused chloride of silver, and it is quite possible to make the iron pan almost red-hot without doing this by constant stirring of the contents. After the drying and burning, so to speak, has been effected, and time allowed for the mass to become cold, it is removed from the iron pan, and pulverised into a fine state, mixed with soda-ash in one or other of the aforesaid proportions, and melted in a clay crucible. The action of the soda-ash is to bring the silver together into one solid compact metallic lump at the bottom of the crucible by searching through every part of the mixed mass where the silver is, and assisting in its complete collection and fusion. If the boiling mass, when that point is reached, shows any signs of boiling over, add some *dried* common salt in powder, and this will check it at once. When the contents of the crucible assumes the form of a thin liquid, and quietness exists, without any kind of upheaval, all the silver has left the slag, and the pot can be taken out of the fire. When cool, it can be broken, and the whole of the silver will be found at the bottom, which only requires remelting and casting into bar form, when it will be ready to be sent to a refiner for an offer, after trial by an assay. The reason why refiners will not make an offer from the lump, or button, as it is often called, is because it is not uniform in quality—a feature caused through the slow cooling and the high specific gravity of all precious metals, which force their way through lighter metals to the bottom, thus making it richer in that part than at the top.

In manufacturing establishments, where a large accumulation of precipitated substances have to be dealt with, the best and cheapest plan is to sell them to smelting firms, who, having large appliances, execute their work on a large scale at the lowest possible cost, and will either pay the value of the precious metals extracted in cash or exchange the amount in new gold or silver, whichever may be required.

CHAPTER XXVII

RECOVERING SILVER FROM PHOTOGRAPHERS' OLD HYPO SOLUTIONS

THE method herewith described for recovering the silver from photographers' solutions applies principally to the alkaline hyposulphite of soda solutions, as these are the only ones used by photographers which contain sufficient silver to make it worth while adopting a separate process of treatment for the recovery of the silver. The hyposulphite of soda ($Na_2S_2O_3$) has the property of dissolving bromide of silver ($AgBr$), iodide of silver (AgI), chloride of silver ($AgCl$), and nitrate of silver ($AgNO_3$), and this led to the "hypo" solution being employed for the removal from the photographic picture either of the above-named salts of silver unacted on by the light. Nitrate of silver, as a salt, consists of colourless crystals, easily soluble in water, and is the basis of the other silver salts. As very few photographers make their own plates, these being prepared in large manufactories, the principal solutions, therefore, of the photographer, to which importance is attached so far as this discourse is concerned, are the "developer" and the "fixing" solution. The "developer" is the solution employed to bring up the likeness of the picture, and before using contains no silver; after being used traces of silver are found to exist, but not in sufficient amount to warrant the separately saving of the solution. The developing solution reduces the silver salt to metallic silver on that portion of the plate which has been exposed to the light through the lens in the camera, and which then constitutes the image or likeness of the object taken. The portions of the plate unacted upon by the light are not reduced by the "developing solution," and are, therefore, left in the form of bromide of silver, or in the form of any other salt of silver that may have been employed in making the plates. After the "developer" has been used, the plate is sometimes dipped into a solution of alum to harden the prepared gelatine coating which holds the silver and prevents it from peeling off; when it is finally soaked in a solution of hyposulphite of soda, or "hypo," as it is usually called by photographers. This solution removes the silver salt that the light has failed to act upon, but does not dissolve the black metallic silver reduced into the form of the picture by the developer. This process is called "fixing," and the plate may thereafter be exposed to the light without injury. All the silver unacted upon by the light is dissolved by the hyposulphite of soda, leaving the black, reduced silver comprising the likeness untouched. The hyposulphite of soda, or "fixing solution" is the only one containing sufficient silver to call for separate treatment in the photographer's studio. All the other solutions may be put into the general waste-water tanks and there dealt with in the ordinary manner for the recovery of the silver.

The silver in hypo solutions, if large, cannot be "thrown down" in the same manner as is applicable to cyanide and other solutions of silver—that is to say, with hydrochloric acid to form silver chloride; but if the solution of hypo is a small one and could conveniently be boiled for some time, and then adding to it an *excess* of hydrochloric acid, or by using slightly diluted sulphuric acid, the whole of the silver is quickly precipitated as silver sulphide in admixture with free sulphur, while sulphurous acid gas is at the same time abundantly discharged; and this process, although not one of the best, may be adopted for the recovery of silver from some old photographic hyposulphite solutions, if preferred.

Waste liquids holding silver in the form of a *single salt* by nitric acid are easily treated, for, by adding to them an excess of common salt, or hydrochloric acid, the silver will be precipitated into a curdy, white, spongy substance of chloride of silver, which after washing, to remove the slight amount of soda or acid in admixture with it, may be employed for the preparation of a new solution—though some may not be in favour of this method—or reduced to the metallic state in the manner already described. Common salt, however, is without action upon strong liquids holding silver in the state of a *double salt*, and will rather aid its solution than effect its precipitation. The hyposulphite of soda and silver is one of these compounds. By first adding a moderate quantity of oil of vitriol to an "hypo" solution the reactions set up are such as to neutralise or displace other acids and salts, and cause the silver to be restored to a single salt like sulphate of silver, which is easily precipitated by common salt.

The three mineral acids—nitric, hydrochloric, and sulphuric—do not precipitate silver directly from the double hyposulphite of silver solutions in a commercially satisfactory manner. Hydrochloric acid, if used to great excess, will precipitate nearly all the silver from these solutions, but it is not completely successful; it is also expensive, and sulphurous acid gas is evolved in large quantity, the silver being precipitated as sulphide of silver (Ag_2S) in admixture with free sulphur, and not as one would have contemplated, as silver chloride ($AgCl$). Similar reactions occur when excess of sulphuric acid is employed in place of hydrochloric acid to precipitate the silver from hyposulphite of silver and soda solutions, namely, the silver is precipitated in the form of sulphide of silver together with some sulphur, sulphurous acid gas (SO_2) being at the same time liberated in large volume. This method is, therefore, not one that can conveniently, or with safety, be adopted commercially for the recovery of the silver from old hyposulphite solutions of photographers. The whole of the silver may, very simply, be precipitated from its solution in hyposulphite of soda by sulphide of sodium (Na_2S), and to this method I shall now direct the reader's attention.

The Sulphide of Sodium Process

The hyposulphite of soda solutions are used until nearly exhausted. They should then be poured into a capacious vessel or tank and preserved until it is nearly full, when the precipitation of the silver may be brought about by adding a solution of sulphide of soda. This is poured into the hyposulphite of soda and silver solution, with gentle stirring, and sufficient added until all the silver is supposed to be thrown down, afterwards allowing it to stand for a few hours to enable the sulphide of silver which is forming, to fall down and settle to the bottom of the precipitating vessel. The sulphide of sodium will precipitate the silver in the hypo solution as silver sulphide (Ag_2S), and the precipitating liquid must be added until no further reactions are set up, which is easily ascertained by the absence of black flakes in the body of the solution after the formation of the sulphide of silver has taken place, as these almost immediately settle down to the bottom of the vessel, leaving the top of the solution clear. The flocky precipitate having settled, add some more sulphide of sodium liquor to make sure that all the silver has been removed from solution, and if it is found that a sufficient quantity has not been added to effect this, add some more of the sodium sulphide solution to precipitate all the silver, but do not add more than is necessary, although a moderate excess is not detrimental to the success of the operation. When enough sulphide of soda has been added to "throw down" or precipitate all of the silver to the insoluble condition in the hypo solution (hyposulphite of soda does not redissolve sulphide of silver), the contents of the vessel must be allowed to stand unmolested for a reasonable time to give the sulphide of silver an opportunity to completely settle free of the hyposulphite of soda liquid, when the black sulphide of silver will all be found on the bottom of the vessel. A little of the clear solution standing on the top of the sediment is now taken in a clean glass test-tube and tested with a few drops of sodium sulphide. If a dark colour is produced, it is proof that some silver is still in solution, and more precipitant is required to be added to remove it; if, on the other hand, no dark colour appears in the test-tube, all the silver has been extracted, and the liquid may be syphoned off the sediment and thrown away.

Silver in hyposulphite of soda solutions is wholly precipitated from its solution by sulphide of soda. The precipitating solution should consist of one part of sulphide of soda dissolved in five parts of water, and a sufficient quantity of solution prepared upon this basis to satisfy the different requirements, such as those dealing with small or large volumes of discarded hypo solutions.

The sulphide of silver, when the precipitation is completed, will all be on the bottom of the vessel contaminated with other matter as a slime, and if you want to obtain fairly pure sulphide the precipitated mass should be washed with water, by well stirring it up and then allowing it to settle. The water on the top is decanted or syphoned off and the washing operation repeated. This is renewed several times until the water comes off clear, when nearly pure sulphide of silver will remain in the moist state, and when dried contains about 17 dwts. per oz. theoretically of metallic silver. If you wish to reduce this to metallic silver, the simplest way is to dissolve it in dilute nitric acid —1 of acid to 4 of water—when the silver will go into solution, leaving the sulphur unacted upon; filter this off the insoluble sulphur residue, dilute the liquid with water, and precipitate the silver (which is contained in the liquid that passes through the filter) with a few strips of sheet copper into a pulverulent powder of metallic silver. The filtering is adopted for the purpose of extracting particles of floating sulphur that may remain in the silver nitrate solution in a flocculent or flaky condition; in this manner all the sulphur is separated from the liquid and remains on the filter. You may precipitate the silver from the nitric acid solution in the form of silver chloride by using a strong solution of common salt dissolved in water in the proportions previously stated, or change into metallic silver by means of a thin sheet of copper, which is better.

The sodium sulphide process of precipitating the silver from hyposulphite of soda solutions is so simple that photographers can do it at frequent intervals without much trouble or difficulty of any kind, and may gradually allow the sulphide of silver to accumulate, so that when it has reached a sufficient amount to pay for sale, it can be sold to silver smelters and refiners, who readily purchase this kind of waste as it is usually rich in silver. This procedure will be found the best to adopt financially, for the reason that it is not advisable for those inexperienced in chemical operations to enter into operations of an intricate nature without taking into consideration the ultimate results that are likely to follow defective operations.

The recovery of silver from hyposulphite of soda solutions used for photographic purposes, by means of hydrochloric or sulphuric acid, is extremely faulty, as traces of the metal is invariably to be found in the liquid standing above the precipitated silver sulphide, after it has been given time to settle down. For precipitating the silver from these solutions the sodium sulphide process is much to be preferred, excepting, perhaps, the one to be described hereafter. When the sodium sulphide fails to precipitate any more silver sulphide, stop making any further addition, although a moderate excess is not a detriment, the idea being to avoid adding too much and cause the operation to be more costly. When a quantity of sediment has accumulated in the precipitating vessel — which should have a very smooth surface, else it is difficult to remove therefrom all the finely divided wet powder which appears in the form of a slimy substance — this slime, after the water standing above it has been drawn off, may be at once removed and placed in a filter bag of closely fabricated material, prepared for such operations, and hung in a wooden frame to drain away the remaining water. The accompanying drawing shows a suitable kind of arrangement for this purpose.

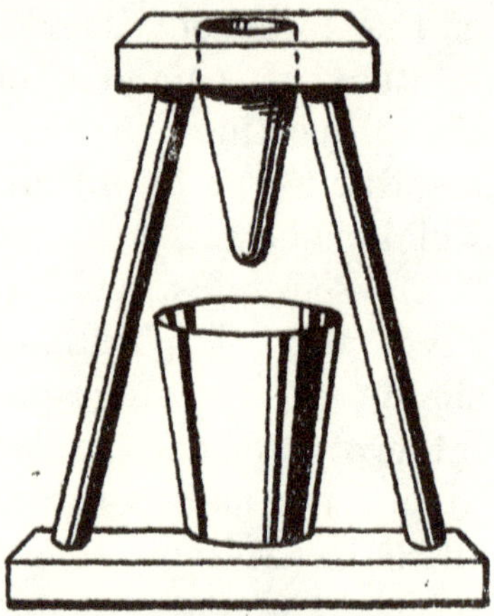

FIG. 23.—Filtering bag in wooden frame.

The mass in the bag when drained, which, beside sulphide of silver, will be composed of sulphate of soda; but this may be washed out along with other soluble salts by pouring water over the mass and letting it drain into the bucket placed beneath the filtering bag. The washing operation being repeated several times—but this is unnecessary—it will be sufficient to well dry the powder at a low heating at first, in the roasting furnace (see fig. 20), and to a further and higher heating afterwards to which atmospheric air has access, to prepare it for the melting pot, or for placing among other preserved wastes. Ordinary hyposulphite of soda solutions do not react on silver sulphide (Ag_2S) like alkaline cyanides, which dissolve it.

If it is desired to melt down the dry burnt powder into a solid lump of metal, it is mixed with a reducing flux, put into a clay crucible, the crucible put into the fire of a wind furnace, and when the whole has fused, it is taken out of the fire and allowed to cool. After a time it becomes congealed, and then with a few blows from a hammer the crucible falls apart and the button of silver rolls out. This is weighed and carefully tested by assay, and notes are usually made of the results of the various operations in the order that they proceed. The following is a good fluxing substance for this kind of residue :—

Burnt silver sulphide . .	10 ozs.	or 100 ozs.
Carbonate of soda (soda-ash) .	3 ,,	,, 30 ,,
Nitrate of potassium (saltpetre)	1 oz.	,, 10 ,,

The saltpetre is best added by degrees as the melting progresses, and not all at once at the commencement, as this causes the mass in the crucible to rise, and such action may lead to the overflowing of its contents. The object in employing saltpetre is to separate any remaining sulphur from the silver by oxidation, when, owing to the changed condition of the fluxing salts, it is easily dissolved by them into the slag, and the whole of the silver set free.

The most important operations in treating the double hyposulphite of soda and silver solutions by the sulphide of sodium process may be briefly enumerated as follows :—

(1) Precipitating the liquid silver as sulphide of silver by means of a strong solution of sodium sulphide.

(2) Drying and roasting the precipitated silver sulphide in the roasting furnace to oxidise and eliminate the sulphur in the form of sulphurous acid gas, which, by prolonged heating, escapes in smoky fumes.

(3) Melting the dried silver powder with soda-ash and saltpetre to obtain metallic silver in the form of a button, then remelting this and casting it in the form of a bar.

(4) Weighing the bar of silver and having it tested by means of assay, to ascertain its commercial value.

(5) Selling the bar of silver to the highest bidder for purposes of refining, as it needs purification before it can be used to good effect for most industrial purposes.

If everything has been done well in carrying out this method to the end, there will be obtained more silver, and less loss will result to the firms adopting it, than by some other of the methods which are employed for the recovery of the silver from the double hyposulphite of soda and silver solutions used for different purposes in the arts and crafts.

There is another method by which the operation of precipitating the silver is so simple that it must not be left out of this discourse, besides, it has the advantage of "throwing down" the silver from hyposulphite of soda solutions to the *metallic* state, and, therefore, the operations for the recovery of the silver are somewhat simplified. It is as follows:—

The Zinc-Acid Process

The recovery of the silver from all the liquid salts and slops of photographers may be effected, by a simple apparatus designed for small workers, by precipitation into metallic silver with zinc scraps or shavings. Collect all the liquids holding silver in a large cask, tank, or tub—this should be large enough to hold all the liquids resulting from a day's work. The zinc scraps may be suspended in a perforated earthenware dipper (fig. 24), or a willow basket will answer for the same purpose.

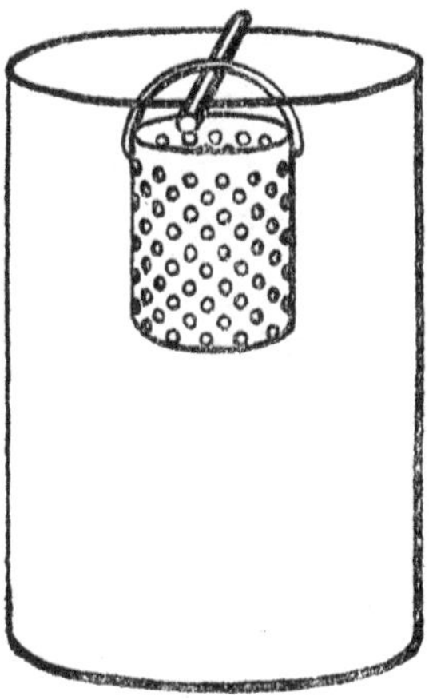

Fig. 24.—Precipitating vessel with contrivance showing how the zinc shavings are suspended.

This is placed in the upper part of the liquid, and by occasionally agitating the solution with the contrivance by shaking it about, the metallic silver will fall to the bottom of the vessel into a finely divided powder. A chemical reaction takes place without stop or hindrance, if a small quantity of either spirits of salts or oil of vitriol is poured into the waste liquid now and then with gentle stirring, the zinc is substituted for the silver and remains in solution, taking the place of the silver as the latter departs from it. Thus the zinc is reduced from a metallic substance to a soluble salt in the solution, while, during the change, the silver leaves the solution and assumes the metallic condition. An entire change of places, and of metallic forms, therefore, of the two metals is being produced as long as the operation lasts. A canvas bag filled with coarse zinc filings may be used for *throwing down* the silver, in place of the earthenware dipper or willow basket; it should be fixed to the top of the precipitating vessel so as to hang in the upper portion of the liquid. Metallic zinc "throws down" *metallic silver* from the double hyposulphite of soda and silver solutions in a commercially satisfactory manner, and may be employed to extract all the metal. The precipitate produced is rather voluminous, but after removal from the vessel and well burned in the furnace provided for the reduction of saline sediments, or for other similar combinations containing the precious metals, its bulk is considerably diminished, and this preparatory treatment is a necessary step to the operation of melting the mass in a crucible. No washing of the precipitated metallic silver need be put into practice—in fact, it is advisable not to attempt doing this commercially; if the best results are to be achieved, it will be quite sufficient to mix the burnt dry powder, in a thorough manner, with a suitable fluxing substance, to recover the whole of the metal in the solid condition by melting in a crucible, if sufficient heat is given, and dispose of the product to a firm of smelters.

The nature of the flux is of some importance, because there is sure to be some oxide of zinc (after the product is burned) combined with the powder when it is ready to be melted, and this is difficult of fusion, unless appropriate fluxes are employed to assist the melting of unwashed precipitates of this kind.

The chemical reactions which occur in hyposulphite of silver solutions in contact with zinc evolves sulphuretted hydrogen; the zinc liberates hydrogen, and the sulphur on being displaced from the silver salt unites with the hydrogen, forming sulphuretted hydrogen or hydrogen sulphide (H_2S), which escapes in the gaseous form, leaving sulphate of zinc in the solution, and metallic silver as a sediment. The zinc sulphate, when the "hypo" solution is rendered acid by H_2SO_4 is not precipitated along with the silver by hydrogen sulphide gas. This compound is also known as hydrosulphuric acid. The quantity of zinc required to precipitate a given quantity of silver may be approximately estimated at from $\frac{1}{2}$ to 1 oz. of zinc for every ounce of silver the solution contains.

After the operation of precipitating the silver is completed, and before the decanting of the clear liquid standing above the sediment takes place, a portion of the solution should be withdrawn by means of the pipette for the purpose of testing if any traces of silver remain in the liquid. The water withdrawn should appear to the sight through the clean glass test-tube into which it is put as transparent as water, when, if there is any silver in it a drop or two of pure hydrochloric acid will immediate indicate its presence, by showing a grey colouration, no matter how infinitesimal the portion may be ; a little clear solution of common salt will present the same effect. This method is most perfect in its results, and I have devoted considerable time in experimental research in order to ascertain whether these waste liquids would become cleared of their murkiness by this method.

CHAPTER XXVIII

RECOVERING SILVER FROM SUNDRY SALTS OF SILVER

A SIMILAR manner of procedure to the foregoing may be adopted with safety for the recovery of the silver from its various compound soluble salts, for by this method there is less risk of the precious metal being lost than by some of the different methods which have been suggested. The metal is obtained in the metallic condition, and the silver residue, after drying and burning, is presented in a suitable form ready for the melting pot. Solutions of any bulk may be treated directly by the zinc-acid method in a simple and convenient manner, as there is little heaving up, swelling, or effervescence arising, if the operation is well conducted. Silver compounds of the following kinds may have the silver extracted and recovered from its various salts in the metallic form, or from small or large volumes of liquids by means of metallic zinc, namely, nitrate of silver, bromide of silver, iodide of silver, sulphide of silver, sulphate of silver, cyanide of silver, hyposulphite of silver, chloride of silver, and sulphite of silver. Zinc decomposes all these salts and solutions of silver by separating the silver from the compounds and reducing it to the metallic state.

When, however, the silver is in the form of a crystallised salt it must have some acidulated water added. Any one of the aforesaid silver salts may be separately treated, or a combination of them (according to the nature of the business being carried on) can be subjected to one operation, as the result will be the same, metallic silver being "thrown down" in consequence of the constituents of the different salts becoming decomposed by the reactions of the zinc, in the presence of water and a very little acid. The solutions, when strong, or nearly saturated with chemical salts, should be diluted with water—an equal volume, or more, according to circumstances—and this will be best determined by the operative in charge of the particular solution requiring to have its silver extracted. There should always be kept a tank into which silver salts and their liquors can be thrown when of no further use, for whenever compounds of any of the precious metals are being employed, it is necessary to save all the materials, liquid or otherwise, and recover therefrom the precious metals, so that nothing is allowed to pass away to waste without an attempt being made to extract the metal.

The most important salt of silver is the nitrate ($AgNO_3$), and from this, or the oxide of silver (Ag_2O or AgO), most of the salts of silver are formed. The nitrate of silver is obtained by dissolving fine silver in dilute nitric acid; and, by diluting this with water, and adding either caustic potash, magnesia, caustic soda, or lime water, the silver is precipitated, and oxide of silver is formed, which is insoluble in water, but readily soluble in mineral acids, in liquid ammonia, and in the alkaline cyanides, forming silver salts again in the liquid condition. Liquid ammonia added to a solution of nitrate of silver will not precipitate the silver therefrom as an insoluble oxide of silver, but rather aids its solution than otherwise. Nearly all the crystallised salts of silver become converted into cyanide of silver by dissolving in a strong solution of cyanide of potassium forming a double salt of cyanide of potassium and silver.

The single cyanide of silver is freely soluble in liquid ammonia, in solutions of carbonate of ammonia, nitrate of ammonia, chloride of ammonia, hyposulphite of soda, the cyanides of the alkalies, ammonium, sodium, and potassium, also by ferrocyanide of potassium. Boiling sulphuric acid added to a *small* quantity of water dissolves silver cyanide with escape of hydrocyanic acid gas, and produces sulphate of silver.

In the manufacture of silver wares for commercial purposes, many of the discarded solutions will contain organic matter, and when mixed together decomposition invariably sets in, and the silver is in several cases separated from its solvents by the adoption of this method and readily precipitated as an insoluble powder, which falls to the bottom of the vessels provided to receive all the kinds of waste liquids resulting from the different operations conducted in the establishment. Thus, in factories devoted to silver working only, the following construction of an apparatus may be employed for collecting the whole of the watery fluids preliminary to their passing into the drain, or a modification of the apparatus to suit particular operations will probably be found to be most useful, and especially so in the service of the smaller firms. It consists of three receptacles, as shown in the drawing :—

FIG. 25.—Apparatus for precipitating silver from mixed liquids.

This device represents the mode of collecting the disused liquids, and how the displacement of the silver is effected. The collecting tubs should stand out in the yard of the factory, as the action of air and light tends somewhat to decompose some of the salts of silver. The apparatus is self-acting and needs very little attention, as the watery fluid, after being received into the first tub, passes into the second tub, and from that to the third tub by the act of gravity alone, down to the outlet pipes, so that there is no overflowing from the occasional running into it of the disused liquids from the different shops constituting the factory. A filtering substance (which will be described) is placed in the last tub so as to safeguard and prevent passing into the drain any floating silver that may be cast up and remain on the surface of the liquid. A lead or earthenware long-legged funnel is fixed in the top of each tub, with *only* its leg immersed in the water, as the funnel cup has to hold a gradually dissolving chemical substance to act constantly on the liquid silver waste and separate the silver from the fluid. If the funnel cups were to be wholly immersed in the water, the precipitating substance would become reduced too quickly to powder and fall through the leg to the bottom of the tub, and the operation would then be faulty and prove more or less unsuccessful in its working.

The precipitant for the silver is burned lime, or quicklime as it is frequently called after being manufactured from limestone. Chemically, it is known under the name of calcium oxide (CaO), and is a hard white solid substance only slightly soluble in water (1 part in about 700 parts of water at ordinary temperatures), but probably sufficiently so for the purpose of this operation and method of recovering the silver from already partly decomposed spent liquids. A solid lump of lime is placed in the cup of each of the funnels, and as the water flows on to the solid lime and spreads over it, a little becomes dissolved away, and this, reacting on the silver, precipitates it in the form of oxide of silver (Ag_2O), which is insoluble in residuary liquids so largely diluted with water. The lime converts the alkaline salts into hydrates, which conversion greatly assists in the precipitation of the silver.

The three tubs employed for receiving the waste water for treatment are arranged in the manner shown in fig. 25. They consist of wine casks or old beer barrels, well coated with melted pitch or asphaltum, to prevent absorption of some of the silver liquid. The funnels are made of preservable materials, and the liquid-conveying pipes should consist of lead, so as not to be acted upon by the liquid passing through them.

No further construction of the first and second tubs is required than that shown, but with regard to the third tub a perforated disc of wood is placed to act as a false bottom, on the top of which, and covering the rim, is stretched some good filtering cloth. This is tightly wedged down to within 1 inch of the bottom of the tub in order to avoid the liability of some of the water flowing out of the tub without its first passing through the filtering cloth. On the top of the filter is put a thick layer of coarse deal sawdust, well pressed together, and through this the liquid to be filtered passes to the filtering cloth, then through that, and, lastly, through the outlet pipe fixed in the bottom of the tub into the drain.

The wash-waters and spent solutions from all parts of the factory are conveyed through lead-piping and enter the funnel of the first tub, being dispersed over the lump of lime placed therein, then flowing through the leg of the funnel are released at nearly the bottom of the tub, where the silver is deposited in the form of oxide along with other matter of a muddy nature. By the funnelling arrangement the water in the uppermost parts of the tubs is prevented from being in a constant state of agitation, and time is given for the light particles of the precipitating silver to fall down to the bottom of the tubs. The water rises very steadily and gradually upwards to the outlet pipes without any disturbance of the precipitated product, and in consequence of a large volume of water being used in factories devoted to manufacturing purposes, and wisely being allowed to mix with chemical solutions, experience has proved that the deposited silver remains insoluble as a residue, although it may not all have been reduced to the metallic state.

The water flowing from beneath the filter placed in the third tub should be tested at intervals with a few drops of pure hydrochloric acid to determine if the whole of the silver has been freed from it. The inspection should take place in a good light, and if no change of colour is observed in the liquid, not even the slightest grey colouration, the whole of the silver will have been extracted from the watery fluid.

In removing the sediment from the respective tubs, which are easily separated for that purpose, it is necessary to stop the flowing in of the water, to allow the water to drain entirely away from the filtering tub; and, after syphoning off the water above the sediment in the two other tubs, put their accumulated sediments into the filtering tub and let the water completely drain away through the sawdust and filtering cloth into the drain, when the silver material, including sawdust and filtering cloth, is burned in a suitable furnace, and with as light a draught as possible, to prevent the passing away of light fine particles as the product becomes dried to dust. Nothing can be done with the product until it is burned. After this treatment, the most satisfactory method is to grind the material to a fine powder and mix well, so that a fair sample can be obtained for an assay test to be taken by two or more smelting firms, and on a comparison of results sell to the highest bidder.

If it is desired to melt down the burnt residue in the works factory, you will have to do this in a clay crucible with the assistance of a special flux, as there will be in combination a large amount of foreign matter, and some lime, and as this is difficult to melt in a crucible (but in smelting it is often found an advantage) a flux of a suitable kind is necessary, and this is found in "Fluor-spar."

The melting of residuary products of the jeweller and silversmith have never been a satisfactory or profitable undertaking in manufacturing establishments, although several have attempted it time after time, and now that the writer has standardised formulas for every description of waste, better results and a good recovery of metal may probably be obtained. At least it is to be hoped so. The flux for silver-lime precipitate is composed of the following materials:—

Silver residue	10 ozs.	or 100 ozs.
Soda-ash	4½ ,,	,, 45 ,,
Fluor-spar	1½ ,,	,, 15 ,,

The fluxing ingredients should be well mixed with the silver residue before they are put into the pot. When the mass is melted, which is assisted by stirring up the contents towards the end of the operation, the heat being continued until a thin fluid is arrived at, when the pot is withdrawn from the fire, allowed to cool, then broken, and the metal recovered in a solid lump. This flux not only reduces the silver to the metallic condition and collects it into a lump, but separates the lime from the silver by causing it to become liquid and act as a flux. The lump of silver, when melted into a bar, will have to be sold to silver refiners, as it cannot be used without refining.

CHAPTER XXIX

SILVER PRECIPITATING REAGENTS

THE metallurgical processes for the recovery of silver from most of its liquid residues having now been described in plain and simple language, and a certain amount of hitherto unpublished knowledge has, I believe, been imparted, which, when once learned, will tend to prevent the many disappointments usually following the mistakes made (in the manufacturing establishments associated with the working of precious metals) in dealing commercially with discarded liquids. But before leaving this subject to describe the tests most applicable for the recovery of platinum from its various solutions, it is desirable to direct the reader's attention to some other precipitants for silver, and point out their distinguishing characteristics in relation to that metal :—

Carbonate of Potash (K_2CO_3) precipitates silver as a brown powder of silver carbonate (Ag_2CO_3), insoluble in water; soluble in strong solutions of ammonia, hyposulphite of soda, cyanide of potassium, cyanide of sodium, and cyanide of ammonium.

Carbonate of Soda (Na_2CO_3) precipitates silver as a light white powder of silver carbonate (Ag_2CO_3), insoluble in water; soluble in ammonia, hyposulphite of soda, cyanide of potassium, cyanide of sodium, and cyanide of ammonium.

Chloride of Sodium (NaCl) precipitates silver as a white flocculent powder of silver chloride (AgCl), insoluble in water and dilute acids, but readily dissolved in liquid ammonia, carbonate of ammonia, cyanides of potassium, sodium, and ammonium, sulphite of soda, and hyposulphite of soda.

Liquid Ammonia (NH_4OH) precipitates silver as a brown oxide (Ag_2O), readily soluble in excess of the liquid—in fact it is a rather difficult matter to arrive at the precipitating point without causing some of the silver to be redissolved.

Carbonate of Ammonia (NH_4CO_3) precipitates silver as a white powder, readily soluble in excess of the precipitant and in the alkaline cyanides above named.

Caustic Potash or Potassium Hydrate (KOH) precipitates silver as a brown oxide (Ag_2O), insoluble in water and in potash, but readily dissolved in strong solutions of ammonia and the cyanides of the alkaline metals.

Hydrochloric Acid (HCl) precipitates silver as a white curdy precipitate of silver chloride (AgCl), insoluble in water and in nitric acid, but is easily dissolved in liquid ammonia and in a strong solution of its carbonate. It is also dissolved in strong solutions of hyposulphite of soda, sulphite of soda, cyanide of potassium, cyanide of sodium, and cyanide of ammonium. Chloride of silver is dissolved by hot *concentrated* hydrochloric acid, and also by boiling in strong alkaline chlorides, but dilution with water will suffice to reprecipitate the silver again into the form of silver chloride from all these liquids.

Phosphate of Soda (Na_2HPO_4) precipitates silver as a yellow powder, soluble in ammonia and in all of the above-named reagents in which the various salts of silver are dissolved.

Sulphuretted Hydrogen Gas (H_2S) precipitates silver from its solutions as a black powder of silver sulphide (Ag_2S), as also do the soluble sulphides of the alkaline metals. Silver sulphide is insoluble in water, in diluted acids, and in hyposulphite of soda. The alkaline cyanides dissolve it, however; it is also acted on by strong nitric acid with the formation of nitrate of silver ($AgNO_3$) and the separation of the sulphur.

Sulphate of Soda (Na_2SO_4) precipitates silver from its nitrate solutions as sulphate of silver (Ag_2SO_4); this is redissolved in solutions of carbonate of ammonium, liquid ammonium, nitric acid, cyanides of the alkaline metals, and in about 180 parts of water to 1 of the silver salt.

Bromide of Potassium (KBr) precipitates silver from its nitrate solutions as bromide of silver ($AgBr$), insoluble in water, and only slightly soluble in ammonia, but freely soluble in hyposulphite of soda and the alkaline cyanides.

Iodide of Potassium (KI) precipitates silver from its nitrate solutions as iodide of silver (AgI), insoluble in water, practically insoluble in ammonia, and almost so in strong alkaline chlorides. In solutions of hyposulphite of soda and in the cyanides it is readily dissolved.

Cyanide of Potassium (KCy) precipitates silver in neutral solutions as cyanide of silver ($AgCy$), but when acid is present the acid must first be neutralised. The precipitate is easily redissolved by an excess of the precipitant.

Lime Oxide of Calcium (CaO) precipitates silver

as an oxide (Ag_2O), insoluble in water and in dilute solutions, and all the oxides of silver possess the common properties of being reduced to the metallic state by heat.

Metallic Zinc (Zn) precipitates metallic silver from all solutions when slightly acidulated with either oil of vitriol or spirits of salts, owing to the liberation of nascent hydrogen gas, which throws down the silver into the metallic state. Metallic Copper (Cu) possesses similar properties to zinc—that is to say, it throws down silver to the metallic state, although it is a slower process.

Oxalic Acid ($C_2H_2O_4$) precipitates silver as a whitish powder, but only in neutral solutions of a water-diluted nature, and this after long standing for rest.

Sulphate of Iron ($FeSO_4$) precipitates silver in very weak solutions of a neutral kind to the metallic condition, but is not a reliable reagent for this purpose in manufacturing establishments using the precious metal.

Hyposulphite of Soda ($Na_2S_2O_3$) precipitates silver as a powder of silver hyposulphite ($Ag_2S_2O_3$) from neutral solutions of its nitrate, soluble in excess of the reagent.

When the different waste liquids of the silversmith are mixed together in the waste-water tanks it must not always be taken for granted that those compounds in chemical union continue to remain so, or that the metal is precipitated therefrom of the same colour as when a pure chemical salt is being prepared; but on the contrary, the result of a number of different substances being mixed together results in a muddy slime consisting of metallic particles, chemical salts, and organic matter being thrown down. Nor must it be generally supposed that when the metal has been deprived of its chemical union and precipitated, no matter in whatever form, that it is certain to be redissolved by other solvents in large bulks of diluted liquids such as waste waters consist of, for the large volume of water entering into the collecting tanks of manufacturing silversmiths will prevent their acting upon the precipitate when once it is formed, for it can only be redissolved by strong alkalies or acids.

It frequently happens, however, that when several different salts or acids are dissolved and dispersed in the same liquid, they neutralise each other; that chemical decomposition is set up of its own accord, and thus the metal becomes liberated from its solvents and falls down, sometimes in the metallic condition, or more often as a metallic salt, without any special precipitant being added to bring about that result. Silver has strong affinities for chlorine, bromine, iodine and sulphur, and combines with them at the ordinary temperature. Substances, therefore, containing any of these gases precipitate silver to the insoluble state in large bulks of water, and in consequence of this affinity or elective attraction complete changes take place in chemically combined substances, and on this principle the separation or analysis of different compound mixtures is brought about, for it is the power which one body possesses over another, of forcing it from its combination, to allow of that particular body taking its place for which it has a greater elective attraction.

The most certain and easily applied of these reagents, for the precipitation of the silver from the waste waters of the silver worker, is the solution of common salt and hydrochloric acid—commonly called muriatic acid, and spirits of salts—which gives a curdy precipitate of silver chloride, not easily redissolved by any of the substances contained in the general waste waters, after being so freely diluted with water. The only point of consideration in the complete extraction of the silver by the one process operation is, that a sufficient length of time for rest should be allowed for the liquid to remain in a stagnant state without agitation, and for that reason it is advisable to employ a series of tanks for the waste waters to flow through, and in manufactories where a considerable quantity of water is made use of for the dilution of the various solutions, for the rinsings, for the wash-hands, and as requisites for various other operations incidental to the manufacture of silver wares, this is a point of considerable importance.

The precipitate of silver does not wholly fall down at once, no matter what is the reagent employed for that purpose, especially in solutions supercharged with mineral acids, or with the usual alkaline salts employed in cleaning and in plating the work, although the latter are supposed to be sufficient to neutralise the acids used when all come to be mixed together, and also to precipitate the silver. But this process of reduction is not always sufficient, and from the results of a large practical experience has not proved eminently satisfactory, particularly when large quantities of the precious metal finds its way into the waste liquids, and therefore special treatment, including suitable reagents for "throwing down" the whole of the silver, is necessary for the completion of the process.

The waste waters should all be filtered through a good filter bed in every factory in which the precious metals are manipulated, before they are discharged into the drains; and as carbonate of soda is universally employed in one or other of the operations, and as that substance acts as a precipitant for silver, reducing it in the form of carbonate of silver, a light powder having a tendency to float; but this, when a filter is employed, is collected together on the filter and prevented from otherwise passing into the drain and being lost.

CHAPTER XXX

RECOVERING SILVER FROM MIXED SOLUTIONS

WHEN all kinds of silver solutions, all the waste waters, and all the spent acids used in a manufacturing silversmithing establishment are mixed together, decomposition of the silver is generally effected in one form or another—for instance, hydrochloric acid, and also any soluble chloride, precipitates the silver as an insoluble chloride in the watery liquid; caustic soda and caustic potash throws down the silver as an insoluble oxide; carbonate of soda and carbonate of potash throws down the silver into insoluble carbonates of silver; and as this form of silver salt is very light, some of it has a tendency to rise and swim on the surface of the watery liquid, and, as already stated, means should be taken to prevent its passing away into the drain with the wash waters. All the above-mentioned substances, along with others, are necessary requisites in silversmithing operations, and assuming that you are satisfied that a combination of the substances employed is sufficient to decompose the silver, it is imperative for the reason I have stated, before letting the waste liquid run entirely away, to have it filtered first; and where no special reagents are employed for the purpose of precipitating the silver, the following contrivance is probably one of the best for the removal of suspended and floating matter from the watery fluid :—

FIG. 26.—Waste-water filtering apparatus.

The tubs employed for the purpose of receiving the waste liquids are three in number, and are arranged in the manner shown in the illustration, from which the construction of each of the tubs will be easily understood. Midway between the top and bottom of each is fixed a circular disc of wood pierced with small holes to enable the water to run through ; on the top of the wooden disc is laid a layer of coarse deal sawdust 5 or 6 inches thick ; this is compressed to prevent its floating about in the water above. Each tub is arranged in the same way, the sawdust in each tub acting as filters for the liquid to be treated. The waste solution from the different workrooms flows through the conveyance piping into the first tub, drains through the sawdust, falling into the bottom section, then flows out through the outlet-pipe fixed in the bottom into the lower part of the second tub, from which it takes an upward course, passing through the sawdust, and is carried upwards to the outlet-pipe near the top by the force of water standing above it in the receiving tub, from whence it is discharged into the last tub, being there conducted through the sawdust, and, entering the lower compartment, flows from there through the outlet nearly at the bottom and is carried away into the drain. The particles of silver that may remain in mechanical suspension in the liquid being safeguarded by the sawdust in the series of tubs, which allows only the fluid to pass through.

If you find that the whole of the silver is not being separated from the water, after testing a small portion of the liquid passed through the filtering apparatus in a test-tube, with a few drops of pure hydrochloric acid, it can easily be rectified. This reagent will immediately indicate the presence of silver, though the portion may be only infinitesimal. The greater the quantity of acid, or alkaline salts in which the silver is soluble, is the cause, as they are more capable of redissolving some of the precipitated silver. All such liquids will, however, have the silver extracted from them completely as they pass through the filters of sawdust, by mixing the latter with a few granulated zinc flakes, or zinc shavings or turnings, *not filings*, and if sufficient time is allowed for the water to remain in the tubs for the conversion of the silver into the metallic condition (this can be assisted by fixing a stop-tap to the end of the last outlet-pipe), the success of the process largely depending upon the bulk of the liquid made use of in the manufacturing establishment, and of its speed of progress through the sawdust. For small workers it is a very simple and effective apparatus for filtering purposes. At stated periods the silver residue, including the sawdust from each tub, is collected together, dried, and burnt in the boiler-furnace, if you have one on the premises, and then either sold to the smelter, or melted with three-fifths its weight of soda-ash, as the fluxing substance, in a clay crucible, taking care to ascertain that the fused mass above the button of alloyed silver is quite free from metal before taking the crucible out of the fire. The button of mixed metals, when obtained, is remelted and cast into ingot form with the view of sale to the highest bidder, after trials by assay.

After the burning operation, the product must be well pulverised to break up any clotted parts that may not have burned to powder, for it is essential that the product to be melted should be in fine powder, and thoroughly mixed with the reducing flux, also in fine powder, to enable every particle of metal to leave the slag and thus be recovered in solid form. If the product is put into the crucible in lumps, some of these will not yield to the action of the flux, and numerous beads of metal will be found dispersed in different parts of the slag upon breaking it into fragments when cold—the operation, therefore, then shows that it has been imperfectly performed, and loss of metal is the result.

The simplicity of the working of this method for the recovery of the silver from its various solutions will be apparent, for the more water that is allowed to mix with them the more perfect will be the recovery of the silver. The amount of loss being very small indeed—no more, in fact, than would be expected in treating each solution separately for the extraction of the silver. While the cost of its recovery is much less, and this alone will commend the process to the notice of manufacturers generally.

CHAPTER XXXI

RECOVERING PLATINUM FROM PLATINUM SOLUTIONS

As platinum is now used extensively in the manufacture of jewellery, and in some other of the arts and crafts, its separation from its alloys and soluble salts is a matter of no little importance, and is of special interest to manufacturing jewellers, refiners, and other kindred workers employing the precious metal; and taking into consideration its great cost (more than twenty times that at which it could be purchased at thirty years ago) any general information of a commercially practical nature will naturally be sought for with avidity by those concerned in reducing the metal by manufactural operations into different forms suitable for their everyday uses. For platinum is used in the manufacture of jewellery in various ways: it is employed as a separate part of jewellery in conjunction with a separate part of gold, each portion representing the distinct characteristics of the two metals, which, when united, exhibit a most pleasing effect, as both metals can always be seen. It is employed in diamond ring setting as a backing to the stone to enhance and preserve its brilliancy—the pure metal itself being sometimes employed for this purpose, though more often alloyed with silver. It is also used to alloy the gold, and particularly in the manufacture of 18-carat *white gold* it is almost imperative, as nickel makes the gold too brittle, and as silver alone alloyed with the gold imparts to the mixture a greenish tinge, and this is not altogether satisfactory in the construction of some of the designs of high-class art craftsmanship. I have in like manner succeeded in introducing platinum into an alloy to be used as the solder for platinum workers, and which, I believe, some of the more eminent firms are now using for uniting the different parts together. Platinum is used in the manufacture of electrical apparatus, and as pins, sockets, and other requisites of the dentist for artificial teeth.

Platinum is a true metal, insoluble in nitric acid, hydrochloric acid, and sulphuric acid, but soluble in aqua-regia, that is a mixture of hydrochloric and nitric acids; also, it is infusible at the highest temperatures attainable in wind furnaces, but capable of being melted therein when alloyed with some of the other metals, notably lead, tin, and silver. The pure metal is only melted in the oxy-hydrogen flame, and the furnace consists of two well-fitting pieces of quicklime, hollowed out to form a crucible or hearth; an opening at the side serves as a spout for the melted metal, and for carrying off the fumes and products of combustion. The nozzle conveying the oxyhydrogen blowpipe is introduced through an opening made in the centre of the cover. Lime is used as the crucible for the melting, because it is infusible at the heating point of the flame, and because it is porous and absorbs the slags and impurities formed during the operation;

it is also a poor conductor of heat. Platinum can also be melted in an electric furnace.

Chemically pure platinum is quite soft, flexible, and extremely malleable and ductile, like wrought iron, and can, like the latter, be welded at a white heat. Native platinum is usually alloyed with iridium, rhodium, etc., and these make the commercial metal brittle. The use of platinum for jewellery has greatly increased during the last decade, and in view of the great increase in the price of platinum many attempts have been made to find substitutes for it, and in some instances this has been partially accomplished. I have succeeded in preparing alloys equal to all the English standards, at the same cost[1] of each or thereabout, and of the same colour as the platinum metal itself. These are perfectly workable, being only required to be manipulated with the usual safeguards common to gold alloys of the different standards.

The distribution of platinum in the principal industries was estimated in 1902, by a high authority in the industrial world, to be as follows:—For dental manufacturers, 50 per cent.; for chemical and electro-chemical industries, 30 per cent.; and for electrical manufacturers and jewellers, 20 per cent.; but the use of platinum in the electrical and manufacturing jewellery trades has considerably increased since that date, and the percentage is, therefore, now much higher in both those industries.

[1] Taken at pre-war prices.

CHAPTER XXXII

RECOVERING PLATINUM FROM ELECTRO-PLATINATING SOLUTIONS

The principal salt of platinum in the formation of solutions for electro-depositing the metal is the bichloride of platinum ($PtCl_4$), also called platinic chloride; this salt is soluble in water. There is also a protochloride, or platinous chloride ($PtCl_2$), which is produced by overheating the bichloride or prolonging the operation of evaporating away the liquid to form the correct crystallised salt. Platinous chloride is insoluble in water, and further application of heat, when it has been evaporated to that stage, quickly reduces it to metallic platinum. It is, therefore, the platinic or bichloride which is so valuable to the metallurgist, electro-depositor and manufacturer, as it is from this salt that all the platinum compounds are obtained, either directly or indirectly. This salt is prepared by dissolving platinum in aqua-regia, in the proportion of 3 parts of hydrochloric acid to 1 part of nitric acid, the same as is commonly used in the dissolving of gold, but platinum takes a longer time to dissolve than does gold, and usually a larger bulk of liquid is required to effect complete solution. About 6 fluid ozs. of aqua-regia are necessary to dissolve 1 oz.

of platinum. The two acids should not be mixed until they are required to be used, for the reason that they mutually decompose each other, and then have no dissolving power for either gold or platinum. But when mixed together, and used at once, the hydrogen of the hydrochloric acid combines with the oxygen of the nitric acid to form water, and chlorine gas is set free, and this uniting with the water forms liquid chlorine, a true solvent for both gold and platinum, and nitrosyl chloride gas (NOCl), a mixture of gases resulting from the reactions of the two acids, but have no dissolving effects on the platinum, escapes in reddish brown fumes. When all the platinum is dissolved no further action takes place, and heating the liquid greatly hastens the operation. The quantity of aqua-regia given above should be sufficient to dissolve 1 oz. of platinum. The aqua-regia is increased in quantity in proportion to the number of ounces of platinum to be dissolved; if, however, much evaporation takes place while the platinum is being dissolved, more may be required than stated in the formula, as it takes a long time to dissolve large quantities of platinum, and particularly is this so if the acids are not pure and not very strong. After the platinum has completely dissolved, the liquid is evaporated in a porcelain dish until a scum appears on the surface and the liquor becomes thick and almost like a syrup, when nearly all the nitric acid will have been evaporated or driven off, and bichloride of platinum is obtained after cooling, in the crystallised condition.

The crystals of bichloride of platinum may be dissolved in water and used in union with such other chemical salts, in the preparation of the different solutions as have been compounded by various authorities for platinating purposes, as are preferred. Chief among these being phosphate of ammonia, citrate of sodium, cyanide of potassium, phosphate of sodium, cyanide of sodium, pyrophosphate of sodium, platinic ammonium chloride, and oxylates of platinum. I mention all these to show that the method I am about to describe is applicable to one and all of the solutions for the reduction of the platinum, either treated singly, in series, or all mixed together, after diluting with water and adding a little acid.

It is advisable to keep all platinum solutions, and all the waste resulting from its use, separately, as far as is possible, from all similar products of gold and silver, for the reason that when mixed residues containing the three metals are sold to some smelting firms, the vendors may lose the value of the platinum which forms a part of the residue through the unsatisfactory method of testing the samples submitted for trial by assay, which may fail to discover the platinum by its not being sought for, and when inquiry is made, you will be met with the reply that no platinum was found to exist in the samples tested. When, however, this kind of residue is kept entirely separate from all other kinds of waste you will be sure of the fact that platinum does exist, and you will be able to give directions as to what metal the waste residue does actually contain, and there should then be no difficulty whatever about arguing the truth of the matter further. It is always the best plan to submit platinum waste to smelting firms who make platinum smelting a speciality, for an offer, and who are of known reputation for straightforward dealing, for great losses are easily made with platinum at £24 per ounce (its present price), unless extra precautionary measures are taken to recover the value of most of the platinum that goes into the waste as the result of manufacturing operations of various kinds in industrial pursuits. There are several methods practised for the recovery of platinum from its solutions, such, for instance, as evaporating the solutions to dryness and then burning the residue to obtain metallic platinum; the precipitation of platinum by various chemical salts into platinum salts, the insolubility of which, however, is not always certain; and the precipitation by copper into metallic platinum. None of these methods are altogether of the most satisfactory kind when practised in a jeweller's workshop. The former being too expensive if the bulk of liquid to be reduced is large; the second in consequence of its requiring skilled attention; while the third, although it does not require expert knowledge, its action is very slow, unless the solutions are strongly acid to cause the copper to be attacked and enter into solution to take the place of the platinum, which then goes out and is precipitated as metallic platinum. I shall describe two processes for the recovery of platinum from its solutions, which, for simplicity of working, leave nothing to be desired in their application to the work of the factory. The first is by the agency of metallic zinc, and nothing will be found more satisfactory.

THE METALLIC ZINC PROCESS

The principle of this method is the same as those described for gold and silver by the immersion of a sheet of metallic zinc in cyanide solutions of the respective metals, whereby the gold and silver is precipitated (thrown down) in the metallic state, and in this way the platinum may also be completely recovered from all its soluble solutions if sufficient time is given for the separation therefrom of the platinum. The platinum is "thrown down" into the metallic condition, from which any adulterating substances may be removed, by a process I shall describe, leaving pure metallic platinum undissolved.

The solutions from which the platinum is to be recovered should be placed in a glazed earthenware vessel (see fig. 18) or any other receptacle having a smooth inside surface. The solutions should be made slightly acid, if they are not already so, with spirits of salts—the commercial acid will do for the purpose—and when they turn blue litmus paper red they are in a fit state to receive the zinc. The metallic zinc employed may be in the form of a large thin sheet suspended by a cross-bar from the top of the vessel and hanging in the centre of the solution; or it may be used in a series of long strips reaching nearly to the bottom of the vessel, the top portion of each strip being bent into the form of a hook and hung over the rim of the vessel, which will secure them in their places, at certain distances apart, in a circle around the vessel. The object being to expose as much surface of zinc as possible to the action of the liquid, the acidity of which will provide fresh surfaces by removing the oxide as it forms, and thus keeping up a constant supply of hydrogen to precipitate the platinum without further attention, unless it be the addition, occasionally, of a very small quantity of spirits of salts to increase the action of the zinc, and thus hasten the operation by the evolving of more hydrogen; but it is by no means advisable to have too much agitation set up in the solution—it will be found much better to let the operation proceed more slowly, for the reason which will hereafter be amply explained.

The proportion of metallic zinc required to decompose platinating and other solutions is not large, as hydrogen readily precipitates platinum into the metallic state from its soluble compounds. From ¼ oz. to ½ oz. of zinc to 1 gallon of solution holding platinum in solution not exceeding those proportions, if allowed to work slowly, will separate all the platinum from its solvents in a commercially satisfactory manner. The formula is as follows, for solutions holding respectively per gallon of solution 5 dwts. and 10 dwts. of platinum :—

Platinum solution, 1 gallon; Metallic zinc, 5 dwts.
Platinum solution, 1 gallon; Metallic zinc, 10 dwts.

The more surface of zinc exposed, the more quickly is the removal of the platinum brought about, and if the liquid is occasionally stirred during the day, in order to bring all parts into contact with the zinc, it will greatly assist the operation, and by allowing it to remain during the night, it will be found in the morning that all the platinum has been separated from the solution.

The zinc, when first placed in the platinum solution, turns of a blackish colour, and this prevents the platinum adhering to it, and the constant action of the acidulated liquid on the zinc surfaces causes the platinum to fall off and sink to the bottom of the vessel in a metallic powder. The zinc, therefore, gradually becomes dissolved, and zinc chloride ($ZnCl_2$) is produced—a very soluble salt which remains in solution, unless too much zinc has been used, and it has been allowed to remain too long in the solution as to saturate it with zinc chloride—which takes the place of the platinum. When the zinc has remained in the platinum solution a certain length of time, say 12 to 24 hours, it is removed and its surface brushed to free it of adhering matter, and the solution allowed a period of rest to enable suspended matter to well settle. A portion of the clear liquid is then taken up in a clear glass test-tube and a few drops of protochloride of tin (stannous chloride $SnCl_2$) poured into it, when, if platinum still exists in the solution, an intense brown-red colour will be imparted to the liquid, and if no change of colour takes place all the platinum will have been separated from the solution.

After the zinc has remained in the platinum solution for the necessary length of time to precipitate the whole of the platinum, the clear liquid on the top of the precipitated platinum is either poured or syphoned off by means of a syphon or rubber tube; if by the latter, the tube is first filled with water and then one end immersed in the solution, care being taken not to withdraw any of the sediment at the bottom of the vessel, which will contain the platinum. The sediment is then washed thoroughly by pouring water over it—preferably hot—well stirring, allowing to settle again, and then pouring off completely. This operation should be repeated four or five times in the same way, when the platinum will have been washed free from zinc salt, acid, and foreign matter, with which it may be contaminated. The metallic platinum thus obtained may not be absolutely pure, for if the platinum solutions contained any traces of silver or other base metals these would be precipitated along with the platinum, and then the latter needs purification before it can be sold to the best advantage. This is brought about in the following manner:—Put the washed platinum in a porcelain vessel (fig. 11)—large quantities are treated in a cast-iron pot—and pour over it a nearly concentrated solution of sulphuric acid made up as follows:—

Sulphuric acid (H_2SO_4)	5 ozs.
Water (H_2O)	1 oz.

The sulphuric acid is added slowly and by degrees to the water, and, when mixed, it is poured over the impure platinum powder; the mixture is heated over a sandbath, or other convenient heating apparatus, until all effervescence ceases, when all the contaminating base metals will have become dissolved out of the sedimentary powder. The liquid, holding the base metals in the form of sulphates, is then poured off, but if it is suspected that the platinum has not been left quite pure, a fresh portion of acid mixture is added and the whole again heated to a high degree of heat, when the platinum will have become thoroughly purified. After the second boiling has been completed, the liquid is poured off, as before, and clean water added to wash the platinum residue. The washing operation is repeated, with clean water each time, until the water comes away from the platinum quite clear.

The platinum will then be quite pure and in the metallic condition, but in a finely divided form after being gently dried, and may be used for making bichloride of platinum, or any other salts of the metal, or, as an alternative method, sold outright to platinum dealers, or bartered in exchange for new metal, whichever is the most convenient form of dealing with it. Platinum cannot be melted in wind-furnaces, it must be melted in a lime crucible by means of the oxyhydrogen blow-pipe, or by means of an electric furnace. Platinum does not dissolve in sulphuric acid if slightly diluted with water, not even when in a finely divided condition, and platinum residues without any gold in them may all be treated for the recovery of the platinum by means of strong sulphuric acid, in which not a trace of the platinum is dissolved.

The various operations, therefore, by this process for the recovery of the platinum may be recapitulated as follows :—

(1) Precipitating the platinum from its soluble solutions with sheet zinc or zinc strips.

(2) Washing the precipitated platinum with water to free it of all foreign matter.

(3) Boiling the platinum residue in strong sulphuric acid to remove base metals.

(4) Washing the purified platinum powder free from acid to obtain clean metal.

(5) Heating the pure platinum powder to dryness to evaporate away all traces of moisture, when metallic platinum will be obtained in the dry state and in an elementary form.

This mode of treatment for the recovery of the platinum from its chemical solutions is applicable alike to both small and large bulks of liquid, and is less costly than some of the other methods, as it throws down metallic platinum.

The Hydrogen Sulphide Process

Platinum may be precipitated from its solutions by means of sulphuretted hydrogen gas (H_2), and this gives the metal in the form of sulphide of platinum, which has to be heated for some time at a high temperature with or without exposure to atmospheric air to eliminate the sulphur, when metallic platinum will remain behind. From not too large bulks of liquid the precipitation of the platinum with sulphuretted hydrogen gas (a current of the gas must always be used in adopting this method, as a solution of the gas in water is not quite so satisfactory) is preferable to evaporating the solutions to dryness and then separating the platinum from the residue. Sulphuretted hydrogen, or hydrogen sulphide gas, dissolves readily in water, and a solution of this mixture is commonly termed *hydrosulphuric acid;* but a solution of the gas in water is not sufficient, the gas must be *nascent* to be completely satisfactory in effecting the reduction of the whole of the platinum from solutions containing a variety of mixed alkaline constituents. The operation is conducted as follows:—First dilute and acidulate the platinum solution (if it is not already acid) with spirits of salts until blue litmus paper is faintly turned red, after which a current of the gas is allowed to flow into it, and this will precipitate the platinum and any silver or copper that may be present as sulphides of the respective metals. The apparatus for the production of the hydrogen sulphide gas is shown in fig. 27. The gas is produced in the vessel A by dilute oil of vitriol acting on sulphide of iron (FeS), and the vessel B contains the solution holding the dissolved platinum which it is desired to separate from the liquid in the solid form.

For the production of the hydrogen sulphide take a good sized wide-mouthed jar, or double-necked bottle; if the latter, into one neck of which a funnel passing through a cork is fixed, and into the other a bent tube C is passed, also through a cork, and

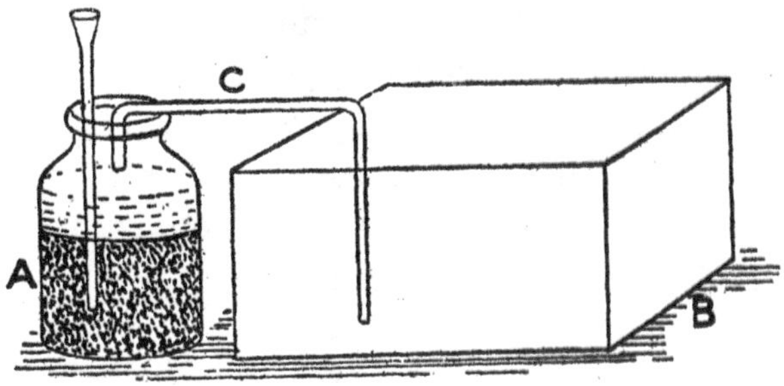

FIG. 27.—Sulphuretted hydrogen apparatus.

terminating in the platinum solution at nearly the bottom of the vessel. When this arrangement is made use of, water and sulphide of iron are introduced into the jar A, the corks containing the funnel and gas-conveying tube are then fixed in the neck, and oil of vitriol very gradually poured in through the funnel until one-eighth the volume of water has been poured in. The substances in the hydrogen generating jar should about half fill it. The gas being evolved almost immediately, and accumulates in the space above the water, then passing through the tube C, is discharged into the liquid holding the platinum in solution, from which, in a short time, the whole of the platinum will be reduced to the condition of an insoluble sulphide, readily converted into metallic platinum by raising the residue to a dull red burning heat with exposure to air, the sulphur and oxygen escaping from the burning mass and disappearing up the chimney. The iron sulphide should be used in lumps about as large as bullets, and the leg of the thistle funnel should reach nearly to the bottom of the generating vessel. The combination of sulphur and hydrogen produces an acid gas which is exceedingly poisonous if inhaled, and it is advisable when employing this method to use every precaution in conducting the operation, and, if possible, perform it in a room where there are no workpeople employed. The conditions which most favour the precipitation of the platinum are, that the solution is only made moderately acid, after being diluted with water, as the acid properties of the gas in its *nascent* form obviates the need of an excess of acid being added to cause the platinum to more readily be thrown down from its alkaline solutions by passing sulphuretted hydrogen gas through them.

The theory of the process of generating sulphuretted hydrogen, just described, is that the sulphide of iron partly decomposes the oil of vitriol and the water by liberating the hydrogen, then the free acid attacks the iron sulphide, liberating the sulphur, and by these reactions the hydrogen and sulphur unite in the gaseous state to form the above named gas, while the oxygen of the water combines with the iron to form oxide of iron, and this oxide unites with the undecomposed acid of the oil of vitriol to form sulphate of iron, which is retained in the generating jar, the gas only passing through the bent tube into the platinum solution. If hydrogen sulphide is passed through a platinum solution of an acid nature till no further precipitation takes place, a precipitate of platinum will be formed which is insoluble in the solution. When the operation is finished the sulphate of iron should be poured out of the gas generating vessel, and if any sulphide of iron remains it should be well washed with water and placed aside for use another time.

When the sulphide of platinum has been obtained, the liquid above it is poured off, and when this is done wash the residue with water several times —the residue left contains the platinum, with probably traces of base metals in commercial products. The sulphide of platinum is next burned gradually to a red-heat to expel the sulphur in the form of vapour, care being taken at first to avoid too rapid heating which would cause the mass to agglomerate; the heating being continued until all the sulphur in union with the platinum is driven off by the heat in sulphurous acid fumes, when finely divided metallic platinum is left as the residue. This is then boiled in slightly diluted pure sulphuric acid, which dissolves away all other impurities, and pure platinum only remains, which is well washed to remove all traces of acid, gently dried to evaporate away every trace of water, when the series of operations are at an end, and the recovery of the platinum is an accomplished reality in the form of a heavy greyish black powder. Absolutely pure concentrated sulphuric acid, at the boiling point, dissolves portions of platinum when its purity is infringed upon, but when it is used slightly diluted (see formula of strength above) the base metals can be removed without loss of platinum.

CHAPTER XXXIII

SEPARATING PLATINUM FROM GOLD UNITED BY SOLDERING IN A MANNER THAT BOTH METALS CAN ALWAYS BE DISTINGUISHED

THIS is a topic in which the manufacturing jewellery trade appears to be possessed of very little knowledge, and now that platinum is used somewhat extensively in the manufacture of jewellery in various ways, this subject presents itself as one of interest when the platinum is employed as a separate portion of the jewellery itself in conjunction with a separate portion of gold, and not used to alloy the gold at all. When the two metals are employed in this manner they are always distinguishable to the eye, and it is well to know how to separate the two metals again, from old articles, scraps, and sundry manufacturing materials, from the solder, without having to send the principal portion to the refiner. The quality of the gold used in conjunction with the platinum is of importance, as the separation will have to be performed by the "wet process," and a low quality of gold may thereby be affected; but when the gold is of good quality, that is to say, 9-carat and upwards, and the solder used in joining the gold and platinum together consisted of either fine silver,

silver solder, low quality gold, dental alloy, or one of the platinum solders devised by the writer, a convenient method of separating the two principal metals again is as follows :—Commence by heating the scraps or other material red-hot, in order to destroy oil, grease, or other organic matter, for unless the heating is done the acid solvent for the solder would be obstructed in its action by the greasy parts. The solvent for the above solders is pure dilute nitric acid. For the proportions of the cold mixture, take—

Pure nitric acid	2 parts.
Water	2 ,,

Place the prepared scraps in a glazed earthenware, porcelain, or glass vessel, and pour over them sufficient of this mixture to a little more than cover the scraps. The nitric acid has no action on the platinum, nor on 9-carat gold, if the copper with which it is alloyed is not too excessive, but the nitric acid must be quite free from hydrochloric acid, or the 9-carat gold will be slightly attacked. Cold nitric acid has no action on gold alloys when they contain more than one-third parts of fine gold in their composition. The process of dissolving the solder may be hastened by using a warm mixture of nitric acid of a more diluted nature. For this mixture take as follows :—

Pure nitric acid	1 part
Water	4 parts.

Do not employ it any stronger than this, and it is as well not to make it too hot if the gold parts to be separated from the platinum are of no higher standard than 9-carat. Nitric acid is decomposed by nearly all the base metals, the weakest being first attacked, by their taking possession of the oxygen from the nitric acid and from the water to form metallic oxides, which ultimately become dissolved into the liquid. In the case with which I am now dealing, the attackable material is the solder mixture, and, if of the composition stated above, it is speedily disintegrated from the integral parts—the gold and platinum—by the liberated free acid converting the substances into liquid nitrates of the respective alloying metals from which the solder has been compounded, while the gold and platinum will remain in their original forms, although separated from each other, and may be used again for any desired purpose.

The operation of parting gold and platinum from each other in ornamental construction work when soldered together in separate parts, and, when the solder employed in their connection is not of high standard, is by this method one of great simplicity, as neither of these metals will be dissolved by the acid mixture used for dissolving the solder, only the solder itself being dissolved. When, however, solder of very high standard, such as 15-carat gold, or 18-carat gold, is used for soldering purposes, this mixture will not suffice to effect the object in view, and another mixture of a different kind will have to be brought into application for their solution.

The liquid containing the dissolved solder should not be thrown away, but preserved and treated for the recovery of the soluble metals contained therein.

Gold and platinum, when soldered together in

separate parts by high-quality gold solder, may be dealt with in the following manner for the parting from each other, and although by this method a little of both the gold and the platinum will become dissolved along with the solder, it is the only process known to be practically successful in its operation if you want to obviate the dissolving of the whole of the material forming this kind of waste, the precipitating of individual metals, washing, filtering, and other operations to separate the respective metals, a kind and form of processes altogether too complicated for persons employed in jewellery establishments, unless possessed of more than average skill and intelligence; and besides, there is the further risk of loss of metal to be added to the extra time and materials required in effecting all the necessary operations to completely separate all the metals from each other and effect their recovery. This information will, however, be given as the discourse proceeds, my object just now being to describe the most suitable, practical, and economical method to pursue in the jeweller's workshop, in dealing with difficulties such as those I am treating of in this chapter.

There is no legal standard for the solder to be used for joining platinum and gold of any standard together, and in instances where the gold is employed of high standard, say as high as 18-carat, the solder may be 9-carat, 12-carat, or 15-carat—seldom higher in the fineness of the gold—as it is an entirely open question for individual manufacturers to adopt which is best for their trade interests and requirements, as the assay offices do not hall-mark this class of work. A good mixture for dissolving the gold solder from scraps of this kind is made up as follows :—

Nitric acid	4 ozs.
Hydrochloric acid	½ oz.
Water	1½ ozs.

The mixture is used hot, and by careful treatment it will dissolve gold solder of inferior quality to the stock metal, without attacking the latter to any great extent. Some portion is, of course, dissolved as the solder is first attacked, and while this is being dissolved the full force of the action of the acid is being withdrawn from the two chief metals, so that by carefully watching and noting when the solder is removed, and when that stage is reached, by pouring off the liquid, the gold and platinum, then in separate pieces, may be removed from the dissolving vessel, rinsed and dried, when it will be found that much less gold and platinum has been dissolved from the waste material by this mixture than would have been the case had the operation been performed according to the general method of dissolving the solder in the usual aqua-regia mixture. The liquid poured off will contain platinum, gold, silver, copper, and probably some other metals, that depending on what the mixture of solder consisted of, and by diluting this liquid with about ten times its volume of water every trace of the silver will leave the liquid and fall to the bottom of the vessel containing the poured off liquid in the form of silver chloride. Some of this (AgCl) will remain in the dissolving flask, and when the metallic pieces of gold and platinum have been removed and washed, the washings, along with the silver chloride found in the dissolving flask is added

19

to the poured off liquid, and after the diluted liquid has remained unmolested for some time, the silver chloride will all have settled, and may at once be recovered by pouring off the liquid standing above it (this contains the portions of gold and platinum which has been dissolved along with the solder), drying, and melting it in a clay crucible, with about half its weight of soda-ash, into metallic silver. The treatment for the recovery of the liquid gold and platinum will be described in another chapter when dealing with the separation of gold from platinum. The watery wastes holding the latter metals in liquid suspension should be strictly preserved and carefully guarded in a separate vessel from the ordinary waste liquids of the establishment for a preliminary treatment prior to being passed on to the general waste-water tubs; for the great cost of platinum makes it imperative that the closest watch should be kept over the ever-occurring waste in manufactories employing so expensive a metal, as such a thing as financial ruin may soon overtake the negligent.

Platinum liquids like those just dealt with, by putting into a large stoneware vessel, may have both the gold and platinum precipitated together in one operation into the metallic condition, by the immersion of sheets or strips of either copper, iron, or zinc; and as the vessel becomes filled with liquid the only attention required will be the testing of the liquid for both gold and platinum (chloride of tin will show the absence of both gold and platinum when no change takes place in the colour of the liquid) each time before it is drawn off. The sediment thrown down may remain until the regulation period arrives for a balancing up of the affairs of trading.

CHAPTER XXXIV

SEPARATING PLATINUM FROM GOLD

When gold and platinum are melted together so as to form an alloy of these two metals only, the gold and platinum may be separated from each other, when it is so desired, by dissolving the compound mixture in aqua-regia with hydrochloric acid in excess. There is no other known method of doing it but this, and heat must be applied until the two metals have gone into solution. Neither pure gold, nor unalloyed platinum, are dissolved by either pure nitric acid, hydrochloric acid, or sulphuric acid when employed separately. The solution of aqua-regia found best to answer the purpose is as follows:—

Hydrochloric acid	4½ ozs.
Nitric acid	1½ „
Platinum-gold	1 oz.

When the gold and platinum has all become dissolved and gone into solution, dilute the mixture with about 8 to 10 times its volume of water (in which has been dissolved a little washing soda to render neutral the nitric acid), so that it will contain about 10 dwts. of gold and platinum per pint of solution. To the liquid, when diluted, add a strong solution of *ferrous sulphate* (copperas) prepared in hot water as follows :—For every 1 dwt. of gold the liquid is supposed to contain add $1\frac{1}{4}$ ozs. of water, in which has been dissolved 5 dwts. of ferrous sulphate; this will precipitate the gold in the form of a fine metallic powder of fine gold. Withdraw a little of the liquid standing above the precipitated gold by means of the pipette into a glass test-tube, add to this a few drops of clear copperas solution (chloride of tin cannot be used to distinguish these two metals when in solution together, as a few atoms of gold gives nearly the same colour as the platinum to the liquid) to ascertain if the gold has all been "thrown down," when, if it has, no further precipitate, cloudiness, or change of colour in the liquid will be visible to the eye. When the latter stage is reached, and the gold has been allowed time to well settle, which does not take long, the liquid is carefully poured away from the gold precipitate. The liquid poured off contains the platinum, still in solution, and is next treated for the recovery of the platinum. The gold powder is washed first with dilute hydrochloric acid to remove iron in case a little had gone down with the gold, and afterwards with water; it is then dried and melted in a clay crucible with one-tenth its weight of bicarbonate of potash, or the same weight of common salt may be employed, when a button of pure gold will be the result. Ferrous sulphate does not precipitate platinum in acid solutions.

The solution from which the gold has been removed contains the platinum, and this is precipitated by passing hydrogen sulphide gas (H_2S) through it for some time, and allowing to stand afterwards for the platinum to settle. Hydrogen sulphide may be employed *for this solution*, either in the gaseous state as *nascent* hydrogen sulphide, or in solution as hydrosulphuric acid, which is a saturated solution of the gas in water. When the latter is employed it is advisable to add a further small quantity of hydrochloric acid to the solution containing the platinum, after the gold has been removed, for when the solutions contain a free mineral acid the platinum is precipitated more readily from its chloride solutions by the application of hydrosulphuric acid. When the gas is employed in its *nascent* state this addition is not imperative. Figs. 15 and 27 represent the kind of apparatus required for making hydrogen sulphide gas, and it is as well to keep one in the factory in case it should be required, the cost being very little.

Hydrogen sulphide gas (sulphuretted hydrogen) does not precipitate iron or zinc in acid solutions, only in alkaline solutions, therefore it is an excellent precipitant for platinum after the gold has been thrown down with ferrous sulphate.

Hydrosulphuric acid added to a solution of platinic chloride will produce a black precipitate of PtS, which is a sulphide of platinum. The theory is that double decomposition takes place—the hydrogen sulphide (H_2S) and the platinum bichloride ($PtCl_4$) mutually decompose each other. The hydrogen takes the chlorine from the platinum, and combining with it forms hydrochloric acid (HCl), while the sulphur takes the platinum with it to form platinum sulphide (PtS), which falls down in an insoluble black powder. When all the platinum has been separated from the liquid, which is ascertained by testing in the usual way with a few drops of pure tin chloride, which produces a red-brown colouring of different degrees of intensity, according to the amount of platinum found in solution, and when no platinum is found therein, no such distinguishing features in the alteration of the liquid will be apparent. Chloride of tin does not precipitate platinum.

The platinum thus precipitated has the water decanted off, washed and roasted in the roasting furnace (fig. 20) to drive off the sulphur, when pure metallic platinum is left in a finely divided condition. This is one method by which alloys of gold and platinum *only* may be separated from each other, and is probably the best and most suitable, although there are other methods which have found much favour with men of science. To some of these I will next direct attention. The undermentioned facts should, however, be known.

Hydrogen-sulphide gas, when dissolved in water, and if not immediately required for use, must be kept in a closely-stopped bottle in the dark, for by exposure to air and light it becomes decomposed, the hydrogen loses its hold on the sulphur, and the latter, as a consequence, becomes precipitated.

Chloride of tin cannot be employed in testing for gold in mixed gold and platinum solutions, on account of a reddish brown colouring being given to the liquid when platinum is present, and this, to the operative in charge of the work, may not be sufficiently distinguishing to determine when all the gold has been thrown down, as this testing liquid in weak aqua-regia solutions, containing if *only a few atoms of gold*, will also give to them a browny red colour, the same as with platinum; it is safer, therefore, to test for gold with a clear white solution of ferrous sulphate. Undecomposed nitric acid existing in the solution will also cause the ferrous sulphate to change the colour to a mahogany tint, and it is then necessary to add carbonate of soda to neutralise the nitric acid.

SEPARATING PLATINUM FROM GOLD BY ANOTHER METHOD

THE following plan may be adopted for the separation of these two metals when alloyed together. Dissolve, after reducing the scraps to small particles, by heating for some time in the mixture of aqua-regia according to the formula before mentioned, and then, after diluting the solution with plenty of water, add a solution of oxalic acid ($C_2H_2O_4$) until all effervescence has ceased to take place, then add a little more to make sure that all the gold has been thrown down. The oxalic acid solution thus prepared, 1 oz. of the salt to 5 ozs. of water, will cause all the gold to leave the liquid and fall down as a dark brown metallic powder, but the precipitation is not brought about so rapidly with oxalic acid as with ferrous sulphate; it acts slowly in cold solutions, but if they could be warmed the precipitation of the gold is brought about more quickly. Oxalic acid precipitates gold to the metallic state without any other metal that may be in solution going down with it. With platinum therefore in the same solution it causes no precipitate of that metal—the gold only is affected; thus all other metals in solution are left undisturbed.

The chemical reactions which take place in the diluted aqua-regia solution containing gold and platinum, when a solution of oxalic acid (the crystallised oxalic acid is composed of $C_2H_2O_4$) is added, may be thus described. The hydrogen of the oxalic acid and of the water taking the chlorine from the gold to form hydrochloric acid, the gold being therefore liberated from its solvent falls down in spangles of metallic gold, while at the same time portions of the oxygen of the oxalic acid and of the water unite with the carbon of the precipitant, converting these elements into carbonic acid gas (CO_2), which rises to the surface as the action progresses and escapes. The following equation will show the change which takes place :—

$2AuCl_3$ (chloride of gold) + $3C_2H_2O_4$ (oxalic acid) = $6CO_2$ (carbonic acid) + $6HCl$ (hydrochloric acid) + $2Au$ (metallic gold).

These reactions do not affect the platinum in the least, which still remains in solution, and is poured off after the gold has all gone down. Oxalic acid is an excellent precipitant for combined platinum and gold solutions, when you want to precipitate the gold independently of the platinum, as in assaying.

The gold residue, when the liquid holding the platinum in solution is decanted away, is washed with water to remove any soluble matter, dried, and melted in a clay crucible, with one-tenth its weight of bicarbonate of potash, or the same weight of common salt, and you will have recovered the gold in a state of purity, and capable of being re-worked when alloyed down to any standard most needed for your requirements. The gold washings should be added to the liquid holding the platinum in order to render the solution more dilute and only slightly acid, as the platinum is thrown down much more completely from such solutions when about to be precipitated with a salt of ammonium.

To the solution containing the platinum, add a strong solution of chloride of ammonia (salammoniac) in water—1 oz. of chloride of ammonia to 5 ozs. of water—and stir the solution while pouring in the ammonium chloride. Enough of this substance must be added to precipitate the whole of the platinum, which will go down in the form of a dark yellowish precipitate of the double chloride of platinum and ammonia (Am_2PtCl_6) not wholly insoluble in water, but sufficiently so if the operation is conducted with care; and if the washing of the residue is performed with methylated spirits diluted with an equal quantity of water the sedimentary substance remains insoluble.

To ascertain whether all the platinum has been precipitated add a few drops of tin chloride, and if this does not change the colour of the sample you are testing, you may conclude that all the platinum has been precipitated, and then may proceed to draw off the liquid above the precipitated platinum. After washing the powder and heating it gently at first, continue the operation gradually till a red heat is reached, and the double chloride of platinum and ammonium will separate, the ammonium chloride being driven off in vapour with the liberation of metallic platinum in a very finely divided condition. This is called spongy platinum, and is in the form of a heavy black powder, which may be sold to the refiner, or employed in any other way, such as making any required salt of the metal, or for alloying purposes, as it cannot be melted in a jeweller's furnace as a single metal.

The above is a better way of separating gold and platinum from dilute aqua-regia solutions than that recommended by some authorities, of first precipitating the platinum by means of salammoniac, and afterwards the gold with ferrous sulphate, as some of the gold is liable to be precipitated along with the platinum by this method, as most ammoniacal salts precipitate gold from dilute acid solutions when the neutral point is reached.

Another method of separating the two metals from their solution in aqua-regia is to firstly precipitate the platinum by means of "chloride of potassium," as a nearly insoluble double salt of platinum and potassium, when the gold is subsequently recovered from the decanted liquid by precipitation with "sulphate of iron" into the metallic state.

In both the latter methods the precipitated platinum (after the liquid containing the gold has been decanted or syphoned off) is dried and gradually heated to a red-heat, whereby metallic platinum is left behind as a black powder; this is sometimes treated with a nearly concentrated solution of sulphuric acid (for which mixture see special formula before mentioned) in order to dissolve any base metals that may incidentally be present, but the operation is not necessary when the latter metals are known to be absent. The platinum being then in both cases absolutely pure, which, after washing and drying, may be sold to a dealer in bullion.

Platinum ammonium chloride (Am_2PtCl_6) and platinum potassium chloride (K_2PtCl_6) are slightly soluble in water, therefore washing operations will have to be conducted with very great care to prevent loss of platinum; but if the washing is performed by means of a weak solution of alcohol the platinum in both cases will remain insoluble, which, on roasting to drive off the ammonium and potassium salts, should leave no residue but platinum in the metallic state.

CHAPTER XXXV

SEPARATING PLATINUM FROM SILVER IN DENTAL ALLOY

A CONVENIENT and perfect method of separating the platinum from the silver when the latter metal is equal to or in excess of the proportion required to form the parting alloy as directed by the "assayers' wet process," and when gold is known to be absent, is the following:—Cut the scraps into small pieces and anneal to a low red-heat so as to destroy all grease or other organic matter, then immerse the scraps in the dissolving vessel, which may consist of a glass flask or glazed earthenware or porcelain vessel for small quantities of metal, and pour over them a nearly concentrated solution of sulphuric acid, and bring to a boiling heat, which will dissolve all the silver without affecting the platinum in the least. Nitric acid will likewise dissolve the silver, but more or less of the platinum along with it, which would then be lost, unless other processes were brought into requisition for its recovery afterwards. According to Dr John Percy, the eminent metallurgist, "an alloy of platinum and silver, containing only 5 per cent. of platinum, dissolves completely in nitric acid, but if the platinum much exceeds that proportion, part of it remains undissolved, but in concentrated sulphuric acid only the silver dissolves." It has, however, since been proved that absolutely *concentrated sulphuric acid* dissolves considerable quantities of platinum from platinum-silver alloys, but by using *slightly diluted sulphuric acid* the platinum can be extracted without loss. The following mixture will effect this purpose :—

Sulphuric acid	3¾ ozs.
Water	¾ oz.
Platinum-silver	1 ,,

Apply heat to the solution until all ebullition of the acid ceases, when as a result with platinum-silver alloys the residue left is pure platinum, after the parting with sulphuric acid. If after the action of the acid ceases to be discernible, and it is feared some of the silver remains undissolved, pour off the liquid and add fresh acid mixture to the residue, expose to heat again until all the silver is dissolved out, then pour off the second lot of acid into the first (this liquid contains the silver), the residue remaining in the dissolving vessel is pure metallic platinum in a fine state of division, which is washed two or three times with hot water to remove every trace of silver sulphate, and the washing waters added to the liquid containing the silver formerly in the alloy, and now in the form of sulphate of silver (Ag_2SO_4).

The separated platinum, after washing, has only to be dried to reduce it to a commercial condition for sale as pure platinum or for its employment in the industrial crafts wherein its requirements are necessary and can be made use of to better advantages without sale to the refiner.

The silver solution is diluted with water to three or four times the quantity, and the silver precipitated therefrom with a strong solution of common salt (1 of salt to 5 of water), which should be added gradually, in small quantities at a time, until all the silver is precipitated, and no more clouding is formed in the water standing above the precipitated silver, after the latter has all gone down. The water above the silver chloride is afterwards carefully removed, the precipitated silver dried, and reduced to the metallic state by melting with about half its weight of soda-ash in a clay crucible.

Platinum and silver is not satisfactorily parted from the wet parting alloy with nitric acid the same as with gold and silver parting, because some of the platinum becomes dissolved along with the silver, and an alloy of platinum containing twelve times its weight of silver has both metals completely dissolved in hot dilute nitric acid. With the sulphuric acid formula I have given, the silver will be entirely dissolved out without affecting the platinum. When, however, the platinum exceeds $\frac{6}{24}$ths the weight of the whole mass no proper solution of the two metals takes place, and the material will have either to be reduced to the parting alloy by melting with some more silver and granulating by pouring the melted constituents into water; or by dissolving in the regular aqua-regia mixture; but the latter is bad practice and not to be recommended, on account of the silver being precipitated into silver chloride by the hydrochloric acid in the mixture, and this gives a strong protecting coating to the metal to be dissolved, on which surface, when deposited, the aqua-regia ceases to act further.

In case the alloy of platinum and silver is such that the mixture of both metals may be dissolved in hot dilute nitric acid without any preliminary preparation, and there may be instances in which this method is applicable, particularly when the platinum does not exceed 7 or 8 per cent. of the whole mass, as both metals are then reduced to complete solution. The platinum is then separated from the silver, after the whole alloy has completely dissolved and the solution has been diluted with twice its volume of water, by first precipitating the silver with weak hydrochloric acid, sufficient of this being added by degrees until every atom of silver has been "thrown down," which is ascertained by the usual means of testing. The silver is thus precipitated as chloride, and the liquid above it on being decanted or syphoned off will contain all the platinum, which can be precipitated out of it by means of salammoniac solution of the strength before stated. The platinum precipitate is dried and heated to eliminate the ammonia, and pure metallic platinum results.

CHAPTER XXXVI

SEPARATING PLATINUM FROM COPPER

THESE two metals may be alloyed together in variable proportions, but the application of platinum-copper alloys is somewhat limited. An alloy of copper and platinum containing 10 per cent. of the latter metal has nearly the colour of 18-carat gold, and can be used for many ornamental purposes. It can be drawn into fine wire, and rolled out into very thin sheets. The separation of platinum and copper may be brought about by dissolving the alloy (if the platinum is in excess of the parting mixture of alloys on which nitric acid and sulphuric acid act) in aqua-regia of the following proportion of constituents :—

Hydrochloric acid	3 ozs.
Nitric acid	1 oz.
Platinum-copper	1 ,,

When the platinum-copper alloy has gone into solution, dilute the liquid with twice its volume of water, and precipitate the platinum with a sheet of copper into metallic platinum. The copper remains in solution and is not precipitated along with the platinum. The platinum cannot, however, be precipitated in this solution by other metals like zinc and iron in seeking its separation from the copper, because the latter metal would be "thrown down" along with the platinum. Any mixture of alloy of platinum and copper may be treated direct with aqua-regia by following these directions.

Platinum and copper may also be separated from each other when only these two metals are alloyed together, by acting on the respective alloys as prepared for parting, or *quartation* as it is called, with dilute nitric acid or sulphuric acid, which will dissolve only the copper. The liquid holding the copper is poured off the platinum powder, and when the latter is washed and dried pure metallic platinum is obtained, which can readily be weighed to ascertain the success of the operation.

Hydrogen sulphide gas cannot be employed as a first application to precipitate the platinum in a mixed copper and platinum chloride solution, as both ingredients would be "thrown down" as mixed sulphides of the respective metals. But as an alternative method, the copper may first be precipitated by a solution of carbonate of soda—1 salt to 5 of water—as copper carbonate. The solution of carbonate of soda has no action on the platinum. Carbonate of potash, however, must not be used, as it would precipitate both the copper and platinum. Thus to the cold aqua-regia solution containing the platinum and copper, after diluting it with water, add the solution of carbonate of soda until no more effervescence takes place. This will precipitate all the copper which may be present, leaving the platinum in solution. Remove the liquid holding the platinum, accidulate it with a little hydrochloric acid, and add a solution of salammoniac, stirring the while—this will precipitate the platinum as the double platinic ammonium chloride; wash this, after very carefully pouring from it the water, with weak methylated spirits to avoid the dissolving of any of the powder, dry, and calcine to a low red-heat to obtain the platinum in the metallic condition.

Solutions of platinum chloride are precipitated by ammonium and potassium salts, but not by sodium salts—for this great discovery the public is indebted to Mr Margraf, the German chemist.

Hydrogen sulphide gas will precipitate the platinum from the solution from which the copper has been removed by means of carbonate of soda, by first acidulating the solution withdrawn from the copper sediment with hydrochloric acid until blue litmus paper is just changed to red, and passing the hydrogen-sulphide gas into it for some time, and then allowing the solution to stand for the platinum to settle. The platinum by this method is precipitated (thrown down) as platinum sulphide, which is washed, and roasted in the drying furnace (fig. 20) to drive off the sulphur, when metallic platinum is left in a finely divided condition.

Aqua-regia will dissolve all platinum and copper alloys, also platinum and other commercial metals when alloyed together in all proportions as mechanical mixtures, nitric acid and sulphuric acid only when the platinum is not in excess of $\frac{7}{24}$ths and $\frac{6}{24}$ths respectively, of the whole mass, in which the base metal alone is dissolved.

CHAPTER XXXVII

SEPARATING PLATINUM FROM LEMEL WITHOUT ANY GOLD IN IT

To recover the platinum from filings, or "lemel," as they are commonly called in the jewellery trades, from all descriptions of work in which gold forms no direct part, either in the mixture of alloy, or in the solder which has been used in the construction of certain miscellaneous articles for utilitarian purposes in other trades as well as those of the jeweller and silversmith, and wherein it is the common practice to dispose of the "lemel" resulting from the filing up of the work to the refiner or to firms who deal in platinum, and from whom the metal was originally purchased. The sacrifice in the cost that has to be made is very great when this method of dealing with waste products containing platinum is considerable, and to know how to dissolve out of the "lemel" any silver, copper, nickel, brass, zinc, or other metal, by a simple and almost costless plan, will be found well worthy of the jeweller's consideration in saving some of the cost.

The method to be adopted for the recovery of platinum from filings of the above description is one of a simple chemical nature. Filings, or lemel, require to be treated separately in a jeweller's workshop. They must first of all be well annealed to destroy grease and other extraneous matter. This is done by placing a portion of the lemel (if large in quantity) in an iron ladle and submitting it to the heat of a clear fire, continuing the heat with a little stirring of the contents of the ladle until the whole is thoroughly heated in all parts. Repeat this operation with another portion until all the lemel has been annealed; and when the whole has cooled, remove with a magnet whatever of iron or steel filings there may be contained therein; then immerse the filings in slightly diluted pure sulphuric acid as follows:—

Filings	1 oz.	or	10 ozs.
Sulphuric acid . . .	5 ozs.	,,	$2\frac{1}{2}$ pints.
Water	1 oz.	,,	$\frac{1}{2}$ pint.

Raise the acid mixture containing the filings to a boiling heat, and continue the boiling as long as effervescence shows itself. After the ebullition of the acid ceases, and the liquid is allowed a period of rest, carefully withdraw the liquid from the residue and pour fresh acid mixture over the lemel residue remaining, heat again until no visible action is perceived on the lemel, then pour off the second lot of acid into the first, and wash the powder remaining, which is pure platinum, two or three times with warm water, and add the rinsings to the sulphuric acid pickle to that before withdrawn, which contains the silver and all other base metals which may have been in alloy with the platinum. It is advisable to have sufficient sulphuric acid mixture, not only to dissolve the base metals, but also hold them in solution when completely dissolved. Boiling *nearly concentrated sulphuric acid* will dissolve the silver and the other base metals from all alloys poor in platinum — such as watch-arbor alloy, Cooper's pen alloy, dentists' mixtures, and the alloy called "Mystery Gold," and numerous other fancy mixtures, without touching the platinum.

The platinum recovered from "lemel" of this kind, after treatment with oil of vitriol and washing, has only to be heated to expel all moisture, and it is then ready for sale as pure metallic platinum.

The acid liquid which has been used in separating the platinum, and to which has been added the platinum rinsings, is diluted with twice its volume of water, and the silver precipitated therefrom by a strong solution of common salt, which is added gradually, until no further precipitation takes place, and the salt solution has no action on the liquid above, after allowing time for the silver to settle, which is easily determined when all clouding or turbidness ceases on a further addition of the salt solution. The precipitation of the silver is then complete. Hydrochloric acid may be employed to precipitate the silver in place of the common salt solution, if preferred, as both substances bring about the same result.

The water above the chloride of silver powder still contains the other base metals which may have been in alloy with the platinum, and after being carefully decanted off the chloride of silver, may be thrown away, as it is never worth while attempting their recovery.

The precipitated silver is well washed with either

warm or cold water, dried, and melted in a clay crucible with, as flux, about half its weight of soda-ash in fine powder, the two being intimately mixed together, and you will have pure metallic silver; or the chloride of silver may be reduced to the metallic state, after it has been washed, but still wet, by mixing with it scraps of zinc and a small quantity of dilute oil of vitriol or spirits of salts, and stirring them about occasionally so as to come in contact with all the silver chloride, and thus convert it into a metallic powder; this is washed, dried, and melted, with one-tenth its weight of soda-ash, prepared as before in a fireclay crucible, and the result will be a button of pure metallic silver. This process of recovering platinum from such of its mixtures of alloy can only be successful, it must be thoroughly understood, when no gold is present to form a part of their composition, for when the latter metal is present it remains undissolved along with the platinum, and a different process will be necessary to recover the gold, silver, and platinum as three separate metals.

The filings being in small particles give plenty of surface for the acid to act upon them, and a richer alloy may be treated by the sulphuric acid process with more success than would be possible if the metal existed in large pieces, as the action of the acid receives more obstruction to its attack when in that condition, and the alloy then ought not to contain much more than 20 per cent. of platinum or $\frac{5}{24}$ths of the whole mass of scraps that are going to be treated by the sulphuric acid process, to satisfactorily remove all the silver, copper, and other metals from the platinum, which latter will remain in the form of a metallic powder at the bottom of the dissolving vessel, all the other metals being in solution when the operation is completed, as liquid sulphates of the contaminating metals, which may be poured off and thrown away.

CHAPTER XXXVIII

SEPARATING PLATINUM FROM LEMEL CONTAINING BOTH GOLD AND SILVER

PLATINUM filings mixed with both gold and silver filings will have to be separated by the usual wet method; there is no other way, as no satisfactory dry method is known. You will have, therefore, to dissolve the whole of the filings in aqua-regia (3 parts muriatic acid and 1 part nitric acid), and dilute the liquid with water until the excess of acid is weakened—about 2 pints of water to each 4 ozs. of acid mixture. The silver will then all settle to the bottom of the vessel in the form of silver chloride, an insoluble flocculent white powder. The solution standing above this contains the gold and platinum, and is drawn off. To this liquid a strong solution of ferrous sulphate (copperas) must be added as long as any precipitate forms. The copperas precipitates the gold in the form of fine scales, which is pure metallic gold, and after the solution above has been removed (this will contain the platinum), is washed with water several times, dried, and melted in a fire-clay crucible with 10 per cent. of either common salt or bicarbonate of potash into a button of fine gold. Oxalic acid may be used to precipitate the gold instead of iron sulphate, but it takes a longer time, in a cold solution, to precipitate the whole of the gold than the iron sulphate—both, however, are easily used, and bring about the necessary result without throwing down the platinum.

The solution from which the gold has been separated still contains the platinum, and this is thrown down by passing hydrogen-sulphide gas into it for some time, and then allowing to rest, when the platinum will separate out of the liquid in the form of sulphide of platinum, a black powder, which is washed to remove extraneous matter, dried, and roasted at a red-heat for some time to volatilise all the sulphur, when metallic platinum remains in a finely divided and pure condition. Chloride of ammonium (salammoniac) may be used to precipitate the platinum from the liquid removed from the gold sediment instead of passing hydrogen sulphide gas into it, the precipitant used being merely a matter of choice or convenience of individuals experienced in the extraction of precious metals from their waste liquid residues. When salammoniac is used to precipitate the platinum a sediment is produced in the form of ammonia chloride of platinum, a combination of the two chlorides—ammonium and platinum. This sediment or precipitate is washed with cold water, or with a weak solution of methylated spirits to prevent the dissolving of any of the platinum precipitate, then dried, and roasted for some time at not too fierce a heat. This will drive off the chlorine and the salammoniac in union with it, leaving behind a dark porous-looking mass. When cold, this is pulverised to fine powder to render it perfectly uniform, washed to remove foreign substances, again dried, and the result is pure platinum in a form commonly known as *spongy platinum.*

CHAPTER XXXIX

SEPARATING PLATINUM FROM GOLD, SILVER, COPPER, NICKEL, AND OTHER METALS

PLATINUM is used in the manufacture of jewellery and other kindred wares in various ways, and will probably be used still more as more uses are found for it, and a better economised method is understood of dealing with its separation from gold, silver, and other base metals with which it is often alloyed. Then there are filings, scrap metal, and other forms of wastes, which result from its many and varied manipulations which have to be employed in the manufacture of commercial wares, and a considerable portion of this waste material accumulates from manufacturing jewellery, and is usually sent in its different forms to the refiner. I have already treated of the separation of platinum from several individual metals in mechanical union, or loosely mixed with it, and pointed out the best means for its recovery in the elementary state, as applicable to manufacturing establishments engaged in working considerable quantities of this metal, so that a saving of some of the loss which is often discovered upon figuring out the financial results of certain transactions, as there are many difficulties to encounter, and the inexperienced and unwary may easily fall into one or other of them unless well acquainted with the various steps that ought to be taken to recover the platinum from every description of waste material, for, unless this is known, only a portion of the platinum in some mixed residues may be recovered, and paid for on selling to the refiner. The method I am now about to describe is not a difficult one, especially to anybody accustomed to preparing chemical solutions for the workshop, such as colouring mixtures, gilding solutions, and other kinds of acid solutions. Such a person ought to perform the necessary operations with sufficient exactitude to accomplish most satisfactory results for those engaged in the manufacture of articles in which platinum now forms so expensive a part.

In separating substances in which gold and platinum form the principal portion of the material, the other metals being present as the base alloy, there is also the solder, which may be gold, to be considered, and the first operation in dealing with such products is melting the gold and platinum with a quantity of silver to form the "parting alloy," so that parting of the metals can take place in an acid mixture. It is necessary that when platinum, gold, and silver are intimately mixed together as in an alloy, that the silver should be present in the proportion of at least three or more parts to one part of the platinum and gold combined, otherwise the particles will be so enveloped and protected by the gold and platinum, that the sulphuric acid mixture used for the parting will be prevented from exercising its solvent action upon them in a sufficient manner as to dissolve out all the other metals, and care should be taken, therefore, to see that the requisite proportion of silver is present before subjecting the material to the action of the parting acid.

The plan to be adopted in the preparation of the "parting alloy" is first to melt the platinum and gold material with a sufficient portion of fine silver so that the parting may readily be effected. Take the following when the amount of the precious metals contained in the scraps to be dealt with are not approximately known.

Platinum and gold material . . .	1 oz.
Fine silver	3 ozs.

Melt the whole together in a plumbago crucible, and when completely melted and mixed together by gentle stirring, pour from some height into a deep vessel nearly filled with cold water, which should be stirred in a circular direction during the operation, in order to granulate or cause the metal to take the form of small irregular particles, in which condition the metal is more effectively acted upon by the parting acid. Another method is to pour the metal into a hot mould, and then roll down into sheet as thin as possible; cut this into small strip pieces, and coil into spirals. This is to allow the parting to be done more easily, by preventing them from lying flat one upon another in the dissolving vessel.

The Parting Mixture

The parting should not be done with *nitric acid* in the same manner as gold and silver parting is done, as some of the platinum is likely to be dissolved along with the silver. This is one of the difficulties that presents itself in dealing with substances containing platinum when alloyed with silver. *Aqua-regia*, likewise, is unsatisfactory for the first operation. It is necessary, then, to use slightly diluted pure sulphuric acid for the parting, which does not act upon either gold or platinum when alloyed with silver in any proportion, but the acid must be chemically pure, as the presence of hydrochloric acid, chlorine, or other impurities would cause both the gold and platinum to be attacked. Put the material prepared for parting in a porcelain vessel if the quantity is small, and if large in quantity a cast-iron pot or kettle can be used. For every ounce of the waste that has to be treated, take about 6 ozs. of the sulphuric acid mixture, as follows:—

Platinum and gold material	1 oz.
Pure sulphuric acid	5 ozs.
Water	1 oz.

Pour this on to the metallic waste substance in the dissolving vessel, and heat hot on any suitable heating appliance until all action of the acid ceases; this takes some time—several hours in fact—to dissolve out all the baser metals from the gold and platinum. This leaves the gold and platinum in fine metallic powders at the bottom of the vessel, the silver and all other metals being solution; and a sufficiency of the acid mixture must be taken to hold these metals in solution in the form of sulphates. Allow the acid to cool, and pour off from the gold and platinum into cold water, and let the water be eight times the volume of the acid which is being poured into it. It is advisable to boil the gold and platinum in a second lot of acid to make sure that all the base metals have been removed therefrom. When all the acid has been poured into the necessary quantity of water, wash the gold and platinum residue remaining in the dissolving vessel with water several times to remove soluble sulphates, and pour the washings into the water containing the previously poured-off acid.

Dissolving the Platinum and Gold

These two metals can only be dissolved with aqua-regia, which is a mixture of nitric acid and hydrochloric acid, and both acids must be quite pure and of the highest strength, or they will not react with one another with sufficient force to dissolve gold and platinum. Aqua-regia is the Latin name for "Royal Water" or "King of Waters," so called because it was the only mixture known to the old alchemists which would dissolve gold, and the name is still retained at the present time. The reason for its solvent action on gold and platinum is, that when the two acids are mixed together chlorine gas is set free from the hydrochloric acid, whilst its hydrogen takes the oxygen from the nitric acid to form water on being heated, and as a result of the great affinity chlorine gas has to combine with water, *diluted chlorine* is formed, and this possesses the power to dissolve both platinum and gold whilst it is forming; thus by the decomposition of the constituents of the two acids a ready solvent is formed for dissolving these metals. The two acids should not be mixed until the aqua-regia is needed, for if they were mixed beforehand the reactions would go on until all the nascent chlorine gas was expelled in vapour, and the liquid would then have no dissolving powers for gold or for platinum.

It usually requires about four fluid ounces of aqua-regia to dissolve one ounce of gold, and a little more to dissolve one ounce of platinum. If, however, much evaporation takes place while the dissolving is going on, more may be required to effect complete solution. The two following mixtures of aqua-regia will be found to give excellent results, in proportion as the gold or platinum predominates :—

FORMULA No. 1

Hydrochloric acid	3 ozs.
Nitric acid	1 oz.
Gold and platinum material	1 ,,

FORMULA No. 2

Hydrochloric acid	4½ ozs.
Nitric acid	1½ ,,
Platinum and gold material	1 oz.

Dissolve the metal in a porcelain dish placed on a sandbath or other suitable heating apparatus, and continue the heat until all the metal is completely dissolved. It will be advisable to use no more acid than is actually necessary, so that, when the gold and platinum are formed into liquid chlorides, there remains very little free nitric acid in the liquid to react on the ferrous sulphate, the precipitant employed at a later stage to "throw down" the gold. In a three to one mixture the hydrochloric acid is much stronger than the nitric acid, and its strong properties are a distinct advantage in driving off the nitric acid during the solution of the platinum and gold material, as this acid is objectionable in the free state, and would cause the solution, when the copperas is added, to change into a mahogany colour, and this tinge remains after the gold has all gone down out of the liquid, making it difficult to test for gold.

Ferrous sulphate or copperas causes no effervescence or frothing to be evolved in cold solutions of aqua-regia containing gold and platinum, but unless the nitric acid is neutralised, a reddish tinge is given to the liquid on the addition of ferrous sulphate, but this is overcome by dissolving a little carbonate of soda in the water into which the aqua-regia is poured after the two metals have become completely dissolved therein.

Recovering the Gold

After the gold and platinum is all dissolved, allow to cool, then add 2 pints of water, in which about 1 oz. of carbonate of soda is dissolved, for each ounce of metal there is in the liquid; after solution has taken place, allow to rest for a time, and if a sediment forms, the clear solution on top must be decanted, or carefully syphoned off, as there may be a slight amount of silver left from the sulphuric acid operation of removing all the baser elements of alloy from the gold and platinum, and this must be separated from the two superior metals. Carbonate of soda does not precipitate gold, nor platinum, but it will silver and all other metals. Its object in the process I am describing is to completely neutralise the nitric acid, so as to prevent the copperas mixture from turning the solution into a lasting ruddy colour on being mixed with it.

Undecomposed nitric acid in a solution of gold, if precipitated by copperas, turns the colour of the liquid a browny red. When the aqua-regia containing the gold and platinum has been diluted with the above quantity of water, add a quantity of ferrous sulphate dissolved in hot water, to precipitate the *gold alone*, and by these means the gold will all be thrown down in the metallic state, and in a very finely divided, but pure condition.

For every ounce of gold there is computed to be in the liquid, add 5 ozs. of copperas dissolved in 1¼ pints of hot water, and this will be found sufficient to throw down the above amount of gold completely, without any action on, or interference with, the platinum, which will remain undisturbed in the liquid above the precipitated gold.

The solution containing the platinum is carefully withdrawn from the gold (after being tested in a test tube with a few drops of pure sulphate of iron), when the gold may be dried, and melted in a fire-clay crucible with 1 oz. of bicarbonate of potassium, or the same weight of common salt, to every 10 ozs. of the metallic gold powder, when a solid button of fine gold will be obtained.

Recovering the Platinum

To the solution containing the platinum add a strong solution of chloride of ammonia (salammoniac) prepared as before directed, to precipitate the platinum. Add sufficient of this liquid to precipitate the whole of the platinum as ammonium platinic chloride. Salammoniac throws down a dark yellow crystalline powder not absolutely insoluble in water, and I should suggest, after pouring off the liquid standing above it, to heat the mass in a good-sized clay crucible (after it has been gently dried) to a red heat to decompose the ammonia and yield metallic platinum, and so avoid the previous washing operation, as some of the platinum may be lost by unscientific washing beforehand. Allow the solution containing the platinum, after the salammoniac has been added, to stand at rest for some time for the platinum to leave the liquid, then add a little more of the salammoniac solution, and test with chloride of tin to ascertain whether the platinum has all been removed from the liquid; allow to rest again unmolested for several hours. The ammonium platinic chloride thus obtained has the water poured or syphoned off, dried, and gradually heated to a red heat. The ammonium chloride in the powdery mass is thus driven off by the heat, and finely divided metallic platinum is obtained as the residue. This is in the form of a black, heavy powder, and is called by the chemist spongy platinum.

The spongy platinum is then rubbed into a fine powder between the fingers to break up any particles that may have aggregated into lumps, strict cleanliness being observed in the process in order to prevent the introduction of any greasy matter into the powder; and to ensure absolute cleanliness, it is advisable to finally wash the platinum powder repeatedly in very hot water, allowing it to settle each time thoroughly prior to each pouring away of the water. The metallic platinum has then only to be carefully dried to complete the result of obtaining the metal in a perfectly pure state.

The platinum may be precipitated from large solutions by suspending clean strips or sheets of wrought iron in the solutions from which the gold has been precipitated with sulphate of iron, but the solutions must be rendered acid to blue litmus paper, if not already so, by adding hydrochloric acid to cause that effect. Copper is not so satisfactory in liquid containing a salt of iron, and zinc must not be employed, as it would cause the precipitation of all the iron present in the liquid condition in the solution, its existence resulting from having used it in the form of sulphate of that metal in precipitating the gold. The iron sheets, while they slowly precipitate the platinum to the metallic state, leave the soluble iron in solution undisturbed. After the water is decanted off the metallic platinum powder (it is, however, advisable when this method is adopted, to treat the powder to a boiling solution of nearly concentrated pure sulphuric acid in order to dissolve any iron that may incidentally be present through inexperience in performing the operation) the resulting powder is then washed with water several times, and afterwards carefully dried, when the residue remaining will be pure platinum.

Recovering the Silver

The silver is contained in the water into which the sulphuric acid is poured from the first operation in the form of sulphate of silver. From this solution the silver is precipitated with strips of sheet copper in a finely-divided condition of metallic silver. After the water standing above the sediment is decanted or syphoned off, the silver is washed, dried, and melted into a button of fine silver in a fireclay crucible, with 1 oz. of bicarbonate of potash, to act as a flux to each 10 ozs. of the metallic silver powder. The liquid withdrawn from the silver will contain all the base metals in the form of sulphates of the respective metals, should they be copper, nickel, zinc, cadmium, or other common metals that may have been present in the original material, and can be thrown away with the liquid, as they are of no special value to the manufacturing jeweller in that condition, as the proportions are not likely to be large. All the solutions must be slightly acid—that is to say, they must be capable of turning blue litmus paper red, or the results will not be so satisfactory.

The silver may be more quickly obtained from this solution in the form of silver chloride, by adding to it spirits of salts or a solution of common salt, and may thus be separated from all other metals likely to be found in a platinum and gold alloy, as they do not form insoluble chlorides with either muriatic acid or common salt. The manufacturer, however, usually requires his silver in the metallic state, and to obtain it in this condition from the chloride, there are two methods in use for accomplishing it.

Method No. 1.—By first drying the chloride of silver, and then melting it in a fireclay crucible with about half its weight of soda-ash, which is simply dry carbonate of soda. Some furnace operators use bicarbonate of soda, but this is not so good a flux, as it contains much less soda. The chloride of silver need not be washed when precipitated by common salt, as a little salt in the residue will do no harm, for when chloride of silver is melted with carbonate of soda, the chlorine in the silver material unites with the soda and forms common salt, while the carbon of the carbonate, upon the action of the heat necessary to melt the silver, unites with the oxygen, as it is given off the fusing mass in the crucible, forming carbonic acid and oxygen gases, both of which escape into the flue of the furnace, and metallic silver results, which settles to the bottom of the crucible in the form of a button. A good-sized crucible is required to be used in melting chloride of silver, to avoid the boiling over of its contents. In some establishments more flux is used than the proportion I have stated, but it is advisable to use no more than is actually necessary, as a still larger crucible would be required; all that is wanted is sufficient to reduce the slag that is formed to a liquid condition, so that the tiny globules of silver, as they become melted, may collect together and fall through the slag to the bottom of the crucible, and there form into a solid mass of silver. If the chloride of silver is well dried, and well mixed with the dry soda-ash before putting into the melting pot, and the latter being of sufficient size to allow of the escape of carbonic acid gas as it arises, little difficulty will be found in the melting. The mass should be stirred from time to time with an iron rod, from partial fusion till the end of the operation, and when there is no more escape of gas, and the whole is in quiet fusion, the crucible is withdrawn from the furnace, allowed to cool till the silver has thoroughly set, when it may be broken, and the silver button easily separated from the slag or flux.

Method No. 2.—By reducing the chloride of

silver to the metallic state by means of zinc or iron scraps in a dilute oil of vitriol solution, after which the finely divided silver powder is melted down by means of a flux in a fireclay crucible into a solid lump. In adopting this method, the chloride of silver should not be dried or allowed to become dry, as it is not acted upon afterwards so readily if reduced to that condition. All that is required to be done is to pour off the water standing above the silver chloride, when it has been ascertained that all the silver has been "thrown down," and to add some pieces of scrap zinc and a little dilute oil of vitriol. This commences at once to act on the zinc, so that hydrogen gas is given off from the zinc, and the free hydrogen which is evolved in the nascent condition has the property of reducing the silver in silver chloride to the metallic condition. The oil of vitriol mixture consists of the following proportionate parts: oil of vitriol, 1 part to 16 parts of water, and sufficient of the dilute mixture is used to well cover the whole of the silver chloride. The acid should be poured into the water, and not the water into the acid. The whole contents of the vessel should be frequently stirred. The chloride of silver does not require to be washed before the zinc and oil of vitriol acid mixture are added to it. When scraps of wrought iron are used in place of the zinc, 1 part of acid to 8 parts of water are taken to prepare the mixture, as the iron is not acted upon so energetically with a weak mixture as is the zinc.

When the chloride of silver has all been reduced to a dark substance, it settles readily in the operating vessel, which operation usually takes several hours in a cold liquid. If the action of the acid ceases before all of the silver chloride appears to be reduced to the metallic condition, the liquid should be poured off and a fresh lot of acid and water added. When all the silver chloride has been converted, the clear liquid is poured off, and water added, the substance well stirred and allowed to settle. This is repeated several times to wash out any sulphate of zinc that may remain in the silver powder; the latter is then dried, and melted with 1 oz. of soda-ash to every 10 ozs. of the dried silver powder. The melting is performed in a fireclay crucible; the undissolved pieces of zinc, if any remain, being removed before any washing is done.

This method takes longer to perform than that of simply melting the chloride of silver after it is thrown down, and there is also a greater liability of losing some of the silver during washing operations.

The value of platinum is now so great that all operations in which it is used should be conducted with the greatest care, and particularly as regards the waste resulting from its manufacture. The process just described is pre-eminently the best for the separation of, and for the recovery of the three precious metals from a mixed waste material which contains them. The several operations in the process will, no doubt, assist the reader by recapitulating them as follows:—

(1) Melting each 1 part of the gold and platinum waste with 3 parts of fine silver to form the parting alloy.

(2) Removing the baser elements of alloy from the parting alloy by dissolving in slightly diluted pure sulphuric acid.

(3) Dissolving the gold and platinum residue in aqua-regia.

(4) Precipitation of the gold, after diluting the solution with water, with ferrous sulphate (copperas), washing the gold and melting with a flux in a clay crucible.

(5) Precipitation of the platinum with a strong solution of salammoniac (chloride of ammonium) and heating it red-hot to obtain metallic platinum.

(6) Precipitation of the silver by means of a copper sheet as metallic silver from the sulphuric acid solution, or as chloride of silver by means of muriatic acid, or a solution of common salt.

(7) Melting the silver, according to the condition formed by one of the methods described in the details of the general process of separating and recovering gold platinum and silver from their waste products.

CHAPTER XL

THE ASSAYING OF PLATINUM WARES

Platinum Only.—When the amount of platinum only is required to be known, the sample of selected metal must be entirely dissolved in aqua-regia, the solution evaporated at a low temperature, until it is very much reduced in quantity, to get rid of the nitric acid; the concentrated solution is then diluted with a small quantity of water, and filtered to remove insoluble substances, and next an equal bulk of alcohol is added to the filtrate; and the solution thus obtained is next treated with a saturated solution of chloride of ammonium, sufficient being added to "throw down" the whole of the platinum present in the liquid. When by the reactions occurring, the platinum will become precipitated as a yellowish sediment, leaving any other soluble metal in adulteration with it in solution. The solution should be stirred and allowed to stand for some time to enable every trace of the platinum to leave the liquid. Decant or filter off the liquid from the platinum residue, and wash the latter with diluted spirits of wine until the washings cease to be coloured. The resulting platinic ammonium chloride is insoluble in this menstruum, and has then only to be dried, and transferred to a suitable utensil wherein it may be heated, gently at first, as there may be danger of losing some of the platinum by the volatilisation of the ammonium chloride, and afterwards to a low red-heat. Allow to cool, and it will be ready to weigh as metallic platinum. The weight is that of the platinum, and the difference between this and that of the original weight of metal taken for testing gives the weight of the base metals, formerly in alloy with the platinum.

This process is more applicable to a high standard of platinum for hall-marking purposes, namely, $\frac{950}{1000}$ parts of pure metal. Silver, if existing in such a compound is soluble in a concentrated solution of ammonium chloride, as the amount in any case would be infinitesimal in the portion taken for assay, and could be poured off in the liquid form from the platinum residue. But when filtering is first adopted it remains on the filter as chloride of silver.

Another Method.—Take the required portions of scrapings from the platinum articles to be assayed; add three times their weight of fine silver; these materials are then wrapped together in the necessary quantity of lead foil, and cupelled to remove base metals. The resulting bead of metal, now composed of platinum and silver, is subsequently flattened and rolled to a very thin gauge size; it is next parted in slightly diluted pure sulphuric acid, when the silver will be dissolved out, and the platinum left as a dark residue, which is metallic platinum. This residue, after reboiling in fresh acid, if necessary, is washed by decantation, annealed, and weighed, to determine whether the articles really contained the required standard proportions of platinum. No hall-marking standard, however, up to this time, has been fixed, but the extensive use of platinum and its alloys in the manufacture of jewellery has created the necessity for establishing a standard for this metal.

Platinum and Gold.—Weigh out accurately a certain small quantity of the substance to be assayed, and reduce to a state of solution in aqua-regia, for when the platinum is simply alloyed with gold, it can be dissolved without any preliminary preparation. This can only be done by means of *aqua-regia*, with the mixture of hydrochloric and nitric acids already described, both metals being thereby dissolved. When the two metals have become completely dissolved continue the heat under the flask until the liquid is considerably reduced in quantity, adding a little hydrochloric acid occasionally, for the purpose of causing the *expulsion of the nitric acid*, the presence of which is injurious in the after part of the process. When the evaporation has been carried to a sufficient extent, 1 or 2 ozs. of water must be added and the liquid contents of the flask poured off into a glass beaker; add to it oxalic acid, of 20 per cent. strength, in solution, by degrees, until it ceases to have any effect on the liquid; place in a warm place, and allow to remain untouched for some hours, when the brown powder, which the oxalic acid has caused to be deposited on the bottom of the vessel from the fluid, will be quite pure metallic gold in a fine state of division. A small quantity of the clear supernatant liquid, before it is poured off, should be taken up in a pipette and placed on any clear white body, and a drop of the clear solution of ferrous sulphate added to it; if there be no change of colour, it is proof that the whole of the gold has been thrown down out of the liquid by the oxalic acid solution, but if there should be the slightest change of colour, more of the oxalic acid solution must be added, and the liquid heated, to produce a complete separation of the gold. The platinum alone now remains dissolved in the liquid, and in decanting this from the gold, the most scrupulous care is required not to disturb the gold powder. The decanted or filtered liquid is next mixed with its own bulk of spirits of wine, and a strong solution of salammoniac added, which precipitates the whole of the platinum into the double chloride of platinum and ammonium; this has only to be washed with diluted alcohol (in which the precipitate is insoluble), dried, and heated to a low red-heat, to get rid of the ammoniacal chloride left by the alcohol, and the residue on being collected together and weighed gives the amount of pure metallic platinum.

An alloy of platinum and gold can also be assayed by fusion in a cupel with the necessary quantity of fine silver and lead, which leaves a bead of platinum, gold and silver; the triple alloy is next dissolved in slightly diluted sulphuric acid, which dissolves only the silver, leaving the platinum and gold untouched, which are washed, and treated for their separation by the method described in the paragraph immediately preceding.

Platinum and Silver.—These metals may be assayed by means of slightly diluted sulphuric acid when the silver is in excess of or equal to the parting proportion of alloy, when the platinum is in excess the sample to be assayed will have to be cupelled with lead and the necessary quantity of fine silver added to form the parting alloy, when the sulphuric acid mixture will dissolve the silver and not the platinum, which has only to be washed, dried, and weighed to ascertain the proportion of platinum the object under examination contains.

Platinum and Base Metals. — Alloys of these metals present little difficulty in ascertaining the proportion of platinum. They are dissolved without any preparatory treatment, when the platinum does not exceed 25 per cent. of the weight of the mass, in boiling slightly diluted sulphuric acid which removes the base metals and leaves the platinum as a metallic powder; this has only to be washed with hot water, dried, and weighed to make an exact assay. But if the platinum exceeds 25 per cent. of the alloy under examination it is not possible to make an exact assay in this manner, and will require to be cupelled with lead, and adding the necessary portion of fine silver to reduce it down to the parting alloy, and then proceed in the way above described by means of sulphuric acid to effect the parting.

CHAPTER XLI

PLATINUM PRECIPITATING REAGENTS

HAVING now described most of the methods for the recovery of platinum from its solutions when appearing alone, and also from those in which other metals are present, and having thus imparted the means by which the recovery of gold, silver, and platinum may be obtained from all the "waste liquid residues" occurring in every kind of manufacturing establishment in which the precious metals are worked, I hope that this knowledge may occasionally prove useful to workers in their individual trades.

The following table gives some of the reagents by which platinum may be distinguished when in solution :—

Chloride of Ammonium ($AmCl_2$) precipitate platinum from its chloride solution as a yellow crystalline salt of the double chloride of ammonium and platinum (Am_2PtCl_6) which, on heating, is decomposed with the liberation of metallic platinum in a very finely-divided state.

Chloride of Potassium (KCl_2) precipitates platinum from its chloride solution as an insoluble double salt of potassium platinic-chloride (K_2PtCl_6), which is reduced to metallic platinum on the application of heat. Both these precipitates are produced more quickly on the addition of spirits of wine.

Sulphide of Hydrogen (H_2S) precipitates platinum from most of its solutions as a black powder of platinum sulphide (PtS), which is decomposed by heating to a red-heat, with or without the presence of air, and metallic platinum is left as the residue.

Carbonate of Potassium (K_2CO_3) in chloride solution precipitates platinum as a yellow precipitate ($PtCO_3$).

Carbonate of Ammonium (Am_2CO_3) in platinum chloride solution throws down a yellow precipitate and acts like the chlorides of ammonium and potassium.

Sulphide of Ammonium (AmS) produces the same precipitate as hydrogen sulphide, but with the former reagent the precipitate will be redissolved by adding excess of the liquid.

Carbonate of Soda (Na_2CO_3).—No precipitate.

Sulphate of Iron ($FeSO_4$).—No precipitate.

Oxalic Acid ($C_2H_2O_4$).—No precipitate.

Chloride of Tin ($SnCl_2$).—No precipitate, but a dark brown to red colouring is given to the liquid on its addition.

Copper, Iron, and Zinc slowly precipitate metallic platinum from its various solutions as black powders. The applications of these metals and their mode of treatment to the different solutions for which each is suitable, is given under their respective headings, and further detail will, therefore, I am sure, not be necessary for a complete understanding of the instructions provided for operatives engaged in the extraction of gold, silver, and platinum from their numerous and dissimilar solutions in a compact form, to render the operations more successful, and to obtain the respective metals with less difficulty than perhaps has generally been the case.

There are other reagents that can be employed as tests for platinum when in solution, but the tests most easily applied by workers in manufacturing establishments are provided in the foregoing table and in the general text.

CHAPTER XLII

RECOVERING GOLD, SILVER, AND PLATINUM FROM WASTE LIQUID RESIDUES OF EVERY DESCRIPTION BY ONE OPERATION

THE problem of treating the waste residuary liquids containing the precious metals is one of the most important factors to the successful running of a manufacturing jewellery establishment. To this end I have devoted very many years of close, personal attention and practical experimental research, the basis of all knowledge, and may justly claim to have now designed the most perfect contrivance known to workers in the precious metal industries, and which has probably reduced the loss of precious metals in liquid wastes to the minimum for the world.

The question as to the best and cheapest way to recover all the waste metal from the liquids of the workshops from which it may be recoverable, is therefore of considerable consequence financially, and one of the first things should be the adoption of a suitable apparatus, whereby the precipitation of the metal and the filtering away of the clear liquid, after all the precious metals have been removed, so that much better results may be achieved, and a far greater saving effected than has hitherto been the case.

The gold and silver waste-saving appliance which

I am about to describe is so much superior to all other contrivances for large establishments, that it cannot be placed in the same class as those which are in general use, as all descriptions of waste liquid products are treated in an ingenious arrangement of tanks by one operation, with eminently successful results, and this method has proved so thoroughly satisfactory, as tested by practical experience, as to leave no doubt that the whole of the precious metals have been extracted from the liquid at the end of the process. This apparatus is entirely self-acting, and it is guaranteed that the waters passing through it are completely freed from metallic constituents. With this contrivance there is absolutely nothing to do on the part of the employee except to fill the funnels at certain intervals with a chemical salt; this slowly dissolves away as the water percolates upon it, and it is only when it has all become dissolved that any addition of the chemical salt is required to be made. It is almost unnecessary to remark that the *waste liquid products* of precious metal workers cannot be treated successfully without the use of chemicals.

Another important feature of this contrivance is that employers can at any time inspect the working arrangements, and ascertain for themselves if the employee whose duty it is to attend to this work in the factory keeps up a constant supply of the precipitating salt in the funnels, and without the said person being aware of the fact. With this contrivance, therefore, any neglect of duty or inattention at the proper time is at once made apparent to the eyes of the employer, which is another distinct advantage.

This self-acting, waste liquid metal precipitating and filtering apparatus may be justly described, without question of doubt,

The Invincible Method.

The contrivance to which this method applies is the very latest invention, involving no loss whatever of the smallest particle of either gold or silver that may be present in the liquid which is passing through the apparatus. It is the most up-to-date method, and is so effective in recovering the precious metals from all kinds of liquid substances that it gives the very best returns that can possibly result from their treatment, and the advantages of the system over all others hitherto in practice, may be considered to be established by the recovery of more metal than ever before, and as the waste liquid residue is one of the most important things connected with every manufacturing jewellery firm, and one productive of a very great amount of loss if not carefully watched, it behoves manufacturers to inquire into the results of this system. The design on page 340 shows the form of apparatus required for a large factory employing 100 or more hands.

The design shown is a contrivance that has been especially constructed for use in large manufactories employing a large number of workpeople ; whilst the interests of smaller firms have not been overlooked, as a smaller series of tanks may be used, or the number reduced (see fig. 28) to meet their requirements, in proportion to the number of gallons of waste liquids that are being passed through the receptacles during the course of each working day, when, by adopting this method, it will be found more metal is recoverable by the use of this contrivance than that of any other which has preceded it.

The precipitating and filtering appliance for 100 workpeople consists of six tubs of 54 gallons each, arranged in series; old paraffin casks make very suitable receptacles for all descriptions of waste liquids, and the cost is very small. All the tubs should be uniform in size and each one stand about

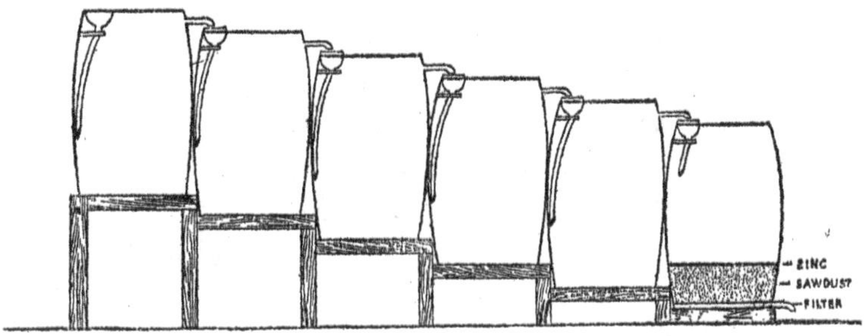

FIG. 28.—The Invincible gold and silver waste saver.

Scale: Funnels to be 1 inch above tops of tubs, outlet pipes to be 3 inches below tops of tubs, to allow the thistle tops to be 4 inches deep, and stand out of the liquid.

4 inches higher than the one which follows, for reason which I am now going to explain. As will be perceived, on looking at the drawing, each tank contains a long-legged funnel with a thistle-formed top to hold the precipitating salt (copperas), and the thistle part is required to project right out of the standing liquid, or the copperas would become too rapidly dissolved. The thistle part of the funnel should be, at the least, 4 inches deep and 4 inches across, for each of the six tubs, and it is an advantage in some cases for the thistle to the funnel in the first tub to be 6 inches each way. The height arrangement of the six tubs allows in each case for

the thistle parts of the funnels to stand out of the liquid, and also of their being 1 inch higher than the tops of the tubs; this prevents the copperas being unnecessarily dissolved. The running of the waste waters, from the series of conveyance pipes, on to the copperas will be sufficient for dissolving purposes. The large waste-water pipe entering the first tub should not be larger than 2 inches in diameter, and for the other tubs, 1 inch in diameter is sufficient; these should be bent downwards into the thistles of the funnels to prevent any overflow of liquid. The space in the tubs standing above the liquid and the outlet pipes is about 3 inches in each one, and this allows of the thistle ends of the funnels to be entirely out of the stationary liquid, when standing 1 inch above the tops of the series of tanks.

The sixth of the series of vessels is simply used as a filtering tub as security against floating gold or silver, with device for safeguarding any loss that might occur by all the gold and silver not being precipitated from the liquid into the metallic state in passing through the other vessels. The filtering apparatus consists of a circular board, fixed in the tub about 1 inch from the bottom, closely perforated, and on the upper surface is firmly secured a piece of baize, flannel or other suitable filtering material to act as the filter. The circular wooden board is closely perforated to allow the filtered water to pass freely through. On the top of this baize-covered board is placed a layer of about 12 inches in thickness of coarse sawdust, closely pressed together, and on the top of the sawdust is placed a circular disc of zinc, finely and closely perforated.

The zinc covering will decompose any liquid metal, and also prevent the sawdust rising up in the tub, and the baize underneath the sawdust will keep the latter from getting into the holes in the wooden board and stopping them up. It will also prevent any portion of the sawdust passing through and getting into the lower part of the tub into which the clear filtered water falls previous to running away into the drain. The tubs should be large enough to take the whole of the water used in each individual factory during the day; and, during the night and the time the works are not running, the filtering tub empties itself, and so becomes ready to take the next day's waste liquids as they pass in regulation order through the other five tubs. The liquids also slowly percolate away from the filtering tub during working hours, as the action of the contrivance is continuous. The process of filtering is uniformly brought about by the perforated zinc covering, which causes an equal distribution of liquid over the entire filter bed, which, after passing through the sawdust and the filtering material, is deposited into a clear space beneath, from whence it runs away, entirely freed of all metal, so that nothing whatever is lost. The zinc sheet will convert any liquid gold, silver, or platinum to the metallic condition, if any infinitesimal portion may by chance have reached the last tub, but this idea has been proved by practical experience, hypothetical.

The long legs of the funnels deliver the waste waters (after spreading over the copperas placed in the thistle parts) to the bottom of the tubs, carrying some portion of the dissolved copperas with it, which precipitates the gold in its downward course. It thus acts on the whole of the waste liquids in the right places, and causes the precipitates, through the large volume of water in possession above, to remain at the bottom of the tanks, the agitation in the liquid being mostly confined to the lower parts; the copperas is thus dispersed throughout the liquid in those parts where it is chiefly required, the lighter liquid then only rising to the tops of the tubs, and passing from there quite steadily through the outlet into the next funnel, and so on, until the last tub is reached, from which the clear water only is allowed to run away.

This contrivance carries on two operations simultaneously, namely, the precipitation of all the precious metals, and the filtering away of the clear liquid at the same time.

The copperas not only frees the waste liquids of all the precious metals by destroying their solvents, but clears them of organic matter as well, which settles to the bottom of each tub; the filter does not, therefore, become overloaded with sedimentary substances so as to impede the passing through of the liquid.

The last of the series of 54-gallon tanks, when the sawdust filtering cloth and zinc sheet are fixed therein, will hold about 40 gallons of liquid, and this is found sufficiently capacious in actual practice for the daily requirements of the largest firms, when taking into account the constant filtering away of the liquid from it during all the time, both day and night. No overflowing, therefore, of the liquid takes place from any of the tanks when this method is employed in recovering precious metals from waste-liquid residues. In case a larger arrangement is required than the one described, I need hardly remark that everything must be multiplied in proportion.

Liquid residues in goldsmiths' and silversmiths' establishments mean the used-up and exhausted solutions and waste liquids of every kind employed in connection with the various manufacturing processes. They consist of the exhausted gilding and silvering solutions and their rinsing waters; the exhausted colouring mixtures and swilling waters; the stripping solutions; the dipping and pickling solutions; the liquids in which both polished and finished work is washed out; the wash-hands waters; and all other kinds of liquid substances which accumulate in every one of the different workshops of the manufactory. Some of these waste-liquid products contain minute particles of gold and silver in one or other—or may be both—of the following forms:—Firstly, in extremely small portions in the metallic condition; and, secondly, in a dissolved condition in the liquid. If the precious metals existed only in the first of these forms, their recovery is not very difficult, and may be recovered by many of the different workshop appliances at present in operation for that purpose, for by carefully filtering, or by simply allowing the liquids to remain unmolested for some time in the collecting tank, all the particles of gold, silver, and platinum will, by their weight alone, sink to the bottom of the tank holding the liquids, and may then be covered by carefully withdrawing the watery liquid standing above the sediment at the bottom of the tank by means of a leaden syphon, drying the sediment, then burning it (including the sawdust and filtering cloth) and pulverising the resulting material to fine powder ready for the refiner's trial, or for fusing in a clay crucible with about three-fourths its weight of a good reducing flux, but it is more economical to entrust this work to the smelter.

If the precious metals existed in the second of these forms their recovery is far more difficult, for when in the liquid state, the atoms of the respective metals do not sink to the bottom of the collecting tanks as they do when in the metallic state, but will pass off with the watery liquid, if that be poured or syphoned from the tanks, as freely as the liquid itself, and, if the one operation of filtering the liquid be adopted, that will not mend matters, for the dissolved gold and silver will readily pass through the filtering material as easily as the watery liquid will; in such cases, should the waste liquids be well filtered before they are allowed to run away, the gold and silver will not all be saved, and unless the acid and alkaline liquids which hold the precious metals in solution are decomposed, no hope can be entertained for the recovery of that portion which has become dissolved in the various liquids. The liquids containing the precious metals in the soluble condition must be treated chemically, and some basic salt, or metal, must be employed, which will so act on the waste liquids as to cause the precious metals to leave their solvents and fall down as a sedimentary precipitate to the bottom of the tanks in an insoluble condition, and this must be done before any of the waste liquids, resulting from manufacturing processes, are thrown away, if all the precious metals contained therein are to be recovered.

There is very little doubt that some of the different kinds of liquid wastes of jewellers when mixed together, if allowed a long period of rest unmolested, would, in the ordinary way, become decomposed and give up most of the precious metals they held without assistance, but not all, and the application of this principle is totally unsuited (owing to the agitation constantly taking place in the receptacle) to manufacturing establishments where hundreds of gallons of liquid have to be dealt with during very short periods.

There are various methods adopted for the recovery of the precious metals by different sections of the goldsmithing and silversmithing trades from their waste waters, but the best of them can hardly be said to be completely successful. What has been a long felt want by those who have studied the question of waste is, the introduction of a simple contrivance well adapted to the treatment of all kinds of liquid wastes of precious metal workers, whereby every atom of gold and silver may be recovered from all sorts of liquids by one operation, and by an appurtenance in which not the least trace of gold or silver is left remaining in the liquid after treatment. Such an instrument has now been discovered by the writer, and the knowledge which has brought it about has come to him from a long series of investigations, and from a wide experience in dealing with many different kinds of mixtures in gold and silver working establishments; and as no other method or combination of methods has given anything like so satisfactory results that is to be obtained from this contrivance, it may justly be named *The Invincible Method.* Those who have tried this method are unquestionably agreed that it is the best solution of the "liquid-waste problem" yet introduced to the trade.

The practical application of the process to different classes of the liquids evolved under all ordinary workshop conditions has been carefully studied and worked out. The working of the contrivance is extremely simple, and shows the value of exact knowledge in its bearing on the subject, and the possession of such knowledge is only to be acquired by a large and varied experience in precious metal working establishments of different kinds.

The arrangement of the six tubs will allow for each tub to be easily separated and joined up again without expense, in order to recover the sediment at any desired period, for sale to the refiner.

The sediment which has accumulated in the different tubs, at clearing-up time may be scooped out in rotation order, and placed over the filter bed in the sixth tub, to drain the last remnants of the water away. Another plan may be adopted by tying or nailing the edges of a piece of washed or unglazed calico to those of a square frame of wood, or to a wooden hoop so as to leave a cavity to hold the sediments, and placing this over a cane-bottomed sieve or the top of an old beer barrel, with the top taken out, without putting them into the last tank. The sediment can easily be removed from such an apparatus when the water has drained from it, and further additions can then be made when the accumulations are large.

It is an excellent plan to coat the tubs with a protecting varnish, both inside and outside, otherwise the wood inside would become soft and absorb some of the gold and silver, and the iron hoops outside would be so acted on by the vapour of the acids in the liquid as to cause them to burst occasionally. A good mixture for this purpose may be composed of two parts of pitch to two parts of tar, and applied warm by means of a brush. Asphaltum varnish may be used as a substitute, but this requires two coatings, as it is too thin for one coating to last the usual period prior to cleaning up the residues with the view of sale to the refiner. The coating should be repeated periodically.

It is best to remove the sediment from each of the tubs at stated periods (usually every six months), and when all the water has been drained away, the material is then boiled down and burned in a specially constructed boiler furnace (see fig. 9), the coarse parts being afterwards well pulverised until all the residue will pass through an extra fine sieve, when it will be ready for trial by the refiners with the view of sale to the highest bidder.

It may be contended that no provision has been stipulated for in this contrivance for the precipitation of either silver or platinum, as the copperas is the only special precipitant provided, and that salt only acts as a reagent for the gold, and does not of itself precipitate silver or platinum. But when all the alkaline solutions and the three mineral acid mixtures are allowed to mix together, their reagents are already present for those solutions which hold the silver and platinum in the state of a double salt, become thereby decomposed, and the result is that all these double salts are reduced to single salts by the large quantity of oil of vitriol (which destroys the power of all other acids) employed in nearly every jewellery factory for boiling out and other purposes. Common salt, carbonate of soda, and carbonate of potash are without action upon liquids which hold silver and platinum in the form of double salts, but when such liquids are reduced to single salts by the sulphuric acid, which displaces the other acids through being much stronger, the silver is then easily precipitated by the common salt and muriatic acid of the colouring mixtures, as silver chloride, which settles down to the bottom of the tanks, and is held there as an insoluble substance by the organic matter removed from the waste water by the copperas. The carbonates of potash and soda also precipitate the silver as carbonate of silver, from very dilute solutions when rendered acid, as the waste waters always will be when everything is allowed to run into the same set of collecting tanks for the extraction of the precious metals. The carbonates of potash and ammonia and their caustic allies precipitate liquid platinum in all such waste liquids to the insoluble condition. So that all the metals of value are provided for in this contrivance, and experience has proved this to be a true question of fact, for as a result of several analytical tests being made, not an atom of precious metal has been found in the clear water after it has passed through the last tank. Gold and silver cyanide solutions, double tartrate of potash and silver solutions, double sulphite of soda and silver solutions, and all others, readily give up their silver to the action of this contrivance, for when aided by the oil of vitriol they are all reduced to single salts. Caustic soda and caustic potash, used for grease cleaning, throws down the silver as an insoluble oxide of silver sediment. The silver is thus easily precipitated by one or other of the numerous reagents for that metal which are regularly employed as auxiliaries in the manufacture of gold and silver jewellery. Some floating metal is always to be found in the waste waters, and this safeguarded and prevented from passing away into the drain by the filter bed fixed in the last of the series of tubs, while the circular piece of zinc reduces to the metallic condition any metallic salt of either of the three precious metals that may incidentally come in contact with it, and this action is promoted by the perfect steadiness of the liquid, and by the length of time it remains suspended over the zinc and filter bed without being disturbed in that part of the contrivance.

A large amount of foreign matter is removed from the dirty waters along with the precious metals, and this material has to be prepared for trial by assay before it can be sold. It is first burned in a suitable sweeps furnace (a special contrivance employed by sweep smelters), as nothing can be done with it until it is burned; it is then pounded in a mortar by means of a cast-iron pestle, and next riddled to separate the fine material; the coarse parts again pounded and the riddling repeated until the whole is reduced to a fine powder. There will be at the end of this operation a certain amount of uncrushable material left in the fine sieve; this will consist mainly of fine metallic shots, which may be melted down in a crucible into a button and treated in the same manner as lemel bar would be which has to be sold to the refiner.

The form of appliance for recovering gold, silver, and platinum by one operation from all descriptions of liquid substances is designed with six tubs or tanks for a large factory employing one hundred and more hands; but where there are many separate shops in the factory, and each one discharging its liquid residues through lead piping direct to the basement, it is advisable to set up a special receiving tank in the immediate vicinity of the precipitating and filtering appliance, into which all the waste liquids should first be allowed to run from their separate sources. An outlet pipe is fixed in this tank, about 8 inches from the top of it, to prevent overflowing, and to convey the liquid in a regulated flow to the first funnel in the appliance proper. This extra safeguard is only necessary where an unlimited amount of water and other liquors have to be introduced into the working arrangements of the business.

The main features of the contrivance are, the conversion of the soluble gold to the metallic condition by the copperas; the stillness of the liquid for the subsidence of the gold; and the device in the last tub for the filtering away of the clear water.

The conversion of the gold depends on the precipitant used; the subsidence of solid particles depends on the stillness of the waste waters, and on their own weight compared with the weight of an equal weight (not volume) of the liquid in which they are suspended, which would be much lighter, and to favour this process the most perfect stillness is required in the topmost part, and the water should leave the tops of the tubs, not from the bottoms, with the exception of the last, the filtering tub. The silver and platinum do not require special precipitating reagents to be employed, for when all the acid mixtures and all the alkaline solutions are allowed to mix together, their particular reagents will be found to exist already in the compounded watery liquid, from which the metals are thrown down in the insoluble condition, either as metal or as insoluble salts, which remain at the bottom of the tubs, while all light or floating material is prevented from finally passing away by the filtering device set up in the sixth tub. The large volume of water, standing steadily above the precipitated material in the five preceding tubs will prevent the precious metals from rising up from the bottoms of the tubs along with the upward movement of the water, owing to their being so much heavier than the water standing above, after separation has taken place.

Testing the Filtered Water

In order to ascertain whether the contrivance is working satisfactorily or not, it is necessary to sample the water which is being discharged through the filtering device occasionally, to make sure that all the precious metals are being extracted from the liquid, and to this purpose I have devoted considerable time in perfecting a testing method whereby that can be rendered certain. The solution of copperas is not the best mixture for the purpose of testing samples withdrawn from the running off water, not even when it is freshly made, and only the clear water-white liquid is taken as the testing medium.

Ordinary copperas sometimes makes a dingy solution, which will impart its own colour to the sample of liquid selected, and put into a glass test tube as a ready means of testing. You cannot, therefore, determine for certain whether all the gold has been removed from the waste waters by this means as well as you can by another reagent which I have devised for sampling running off waste liquids, which is far superior to copperas liquid, and an infallible test for gold in solution. It consists of a saturated solution of tin in hydrochloric acid, chemically called chloride of tin ($SnCl_2$). This testing mixture will at once tell you if there is the merest atom of gold left remaining in the liquid placed in the test tube, by at once changing its colour throughout from water white to a pale brown, the colour increasing in deepness the more gold there is in the liquid, to a purple, and then to an inky-black colour. A very dilute solution, containing only 1 gr. of gold in 12 gallons will, with a few drops of strong chloride of tin, if added to the clear liquid sample in the test tube, change its colour to a faint brown, while 1 gr. of gold in 3 gallons will change its colour to a brown black, and 1 gr. of gold in 1½ gallons will change its colour to a coal black. This test will also distinguish both silver and platinum. The former by giving a grey cloudiness to the liquid with precipitation, and the latter by giving the liquid a reddish brown colour without precipitation.

The chloride of tin reagent, on account of its being a weighty substance, does not, when the addition is made, remain on the top of the liquid being sampled, but quite freely penetrates through the whole of it, and acting in the above characteristic manner throughout. It is, therefore, a very delicate test for ascertaining the presence of gold and platinum when in solution. The commercial crystallised chloride of tin, sold by drysalters in that form, is unsuitable for the purpose of testing, for when dissolved in water it often becomes cloudy, and if a portion in that condition is added to the sample of filtered water under examination, a greyish colouring will also be imparted to it, and this may be taken by the inexperienced for silver still remaining in the waste waters, although not a trace of it may be present. An addition of hydrochloric acid to such a chloride of tin mixture will prevent its cloudy colouration, but even then that is not quite satisfactory for the particular purpose intended, as it will not be strong enough in tin to detect an infinitesimal quantity of gold.

The tin chloride is far better prepared in the factory by some person accustomed to making chemical solutions, and then it can be depended upon, if made in the following manner, free from water. Take—

Pure grain tin (Sn)	1 oz.
Pure hydrochloric acid (HCl)	4 ozs.

Dissolve the tin in the acid by means of gentle heat, in a dissolving flask; allow to cool, and then put the mixture into a glass-stoppered bottle, and preserve in a dark place out of the daylight, as it loses some of its virtue after a time, by being exposed to strong light. For the reason just stated, it is advisable not to make too large a quantity at one time, but if more is required than in the formula given, the quantities of the two ingredients must be multiplied proportionately, and not otherwise. Nevertheless, it is always advisable to have as much tin dissolved into the acid as the latter will take up, for you may have too little tin and too much acid, and that is not so good. Too much tin you cannot have, as the acid will not dissolve any more when it becomes thoroughly saturated with it.

Such a mixture acts more effectually on extremely weak metallic solutions, and enables the examiner to observe more clearly if any change of colour takes place, no matter how slight it may be. Chloride of tin will search right through the liquid selected for sampling, and detect in a most characteristic manner the smallest atom of gold, if one is still existing in any portion of the liquid under examination, by imparting to it the degrees of colour above stated according to the amount of gold there is in the soluble condition.

There are a few difficulties to be met with in testing waste water resulting from every description of liquid products being mixed together when even the proper tin chloride is employed, and I have devoted some considerable time to experiment to overcome these difficulties. Sometimes, when the "chloride of tin" is added to the waste waters resulting from the one-operation method, in sampling, a greyish colouring is imparted to the liquid, which at first sight might, as previously remarked, be taken for silver being converted into silver chloride, but I have tested this in numerous ways, and find such is not the case. The milky colour is produced when excess of alkaline carbonates, such as soda and potash, are found undecomposed in the running off water, or when sulphate of zinc is present, but these difficulties I have successfully overcome by the addition of a little hydrochloric acid to the sample liquid before testing with the chloride of tin. The hydrochloric acid will immediately clear the liquid, and change the substances into chlorides, so that a single drop of tin chloride may be seen making its way through the clear liquid to the bottom of the test tube, or other clear glass vessel in which the testing is being performed. Hydrochloric acid causes no reactions in colour to take place with the above-named salts or other substances when freely diluted with water, as all jewellers' liquid residues are wont to be. Hydrochloric acid does not react on sulphate of zinc, but simply changes its nomenclature, without effervescence or change in the colour of the liquid. It is therefore always advisable to add to the samples of waste waters, when testing for gold, a little hydrochloric acid before treating with the tin chloride, to make sure that there is no obstacle in the way of a satisfactory test being achieved.

I have given more instruction with regard to the fitting up and working of the invincible contrivance than to some others, but in view of its importance, less description would not have sufficed, for greater profit can be made out of the waste waters than was formerly possible, by using this appliance according to directions.

This special waste-saving appliance is not patented and sold by me with a view to profit. Anybody is at liberty to set up one at a cost of a few pounds, which will amply repay the cost in a few months' time by the extra savings effected.

Several of the leading jewellers of Birmingham

are at the present time using this appliance for treating their waste liquids, and all have testified to me that its use has resulted in a greater proportion of metal being recovered than ever before. In one instance more metal to the value of £100 was recovered in six months than ever before during a similar period. The firm believed that every atom of gold and silver was now being recovered by the action of the waste metal-saving appliance from their waste liquids. This means a saving of £4 per week over and above any previous corresponding period's receipts, which is a sufficient guarantee of the thorough reliability and efficacy of "The Invincible Method" for automatically treating the waste liquid products of the manufacturing branches of the precious metal-working industries.

CHAPTER XLIII

THE ELECTROLYTIC METHOD

EVERY industry that deals with precious metals should be marked by ever-increasing care with regard to the matter of the waste which accumulates in manufacturing valuable material, and among the most prominent is that of the liquid waste, since a great deal depends upon the principle adopted in regard to this waste saving as to effect a very different result in the financial success of the firm's business. As a matter of course, there are certain difficulties to be encountered in treating liquid wastes, and a little chemical knowledge is a necessary advantage in surmounting any casual difficulty which may arise. The precipitated material, after that has been brought about, requires care, for some of it is of a very light nature, and will rise up again into the liquid with only the slightest agitation. Now if this is allowed to happen, some loss will probably be incurred thereby, in drawing the water away from the precipitates if that method is adopted in removing the fluid standing above the precipitates, and I know in many instances sufficient attention is not given to this matter as there should be. "The Invincible Method," already fully described, has been designed with the view to prevent difficulties of that kind happening. As the liquid wastes occur in different quantities they are commonly treated by various methods, and in most of those I have inspected it is not possible, entirely, to prevent some loss of metal, which may easily be avoided by taking the proper precautionary measures. The apparatus shown below will complete successfully the many attempts which have been made to eradicate the evils above mentioned.

This contrivance is my own design, and consists

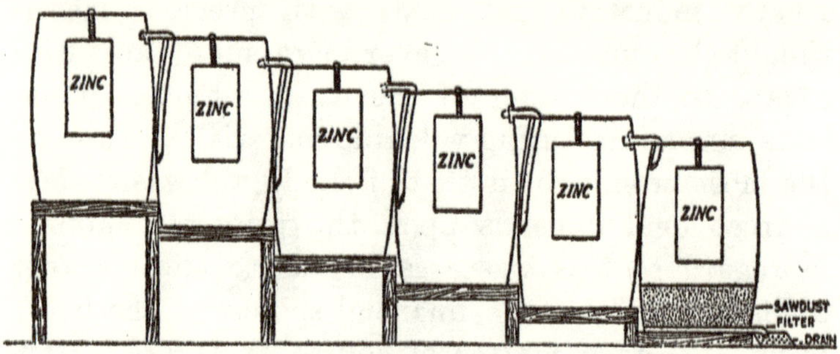

FIG. 29.—The electrolytic gold and silver waste saver.

of six large tubs or casks, each of which may have a 54-gallon capacity, or other volume in proportion to the size of the business, or to the number of gallons of liquid requiring to be dealt with in a given period of time. Each cask contains a piece of sheet zinc of fairly large dimensions, which exerts an electrolytic or chemical influence on the precious metals which are soluble in the waste liquids under treatment, by liberating them from all their solvents and precipitating them to the bottom of the series of vessels in the *metallic state*. Thus, the form of gold, silver, platinum and copper undergoes an entire change in the acid and alkaline liquids by the chemical action which is set up in the mixed liquors between the zinc and the acidulated chemical salts existing in the waste waters—when all are being treated by one operation — whereby the metallic constituents are separated and slowly fall downwards by the act of gravitation to the bottom of the tubs in the metallic condition. The chemical reactions are primarily caused by the greater affinity of the solvents in the wash-waters for the zinc than for the precious metals in solution. The zinc becomes gradually dissolved, and enters the liquid in the soluble condition, taking the places of all the metals which have been forced by the reactions caused by the immersed zinc, to leave their solvents in the waste liquids and return to the natural state; thus the metallic zinc goes into solution and remains in the liquid condition, while the precious metals go out of solution and return to the metallic condition.

The action of the contrivance is such, that all kinds of liquid wastes which are to be found in goldsmiths' and silversmiths' workshops may have the gold and silver thrown down completely, by immersing a sheet of zinc in each of the receptacles without having to use any copperas at all. The zinc sheets should be large, but not so large as to touch the bottoms of the tubs.

The theory of the process is, that the water and other constituents of the liquid are mainly decomposed; their oxygen unites with the zinc as it is acted upon by the acidified liquid, the hydrogen which is thereby evolved in the *nascent* form, passes through the solution, and being light, rises to the tops of the vessels and escapes in an invisible condition, the precious metals being, by these evolutions of nascent hydrogen in its most active form, separated from the liquid and conveyed to the bottoms of the vessels in their metallic state, their places in the liquid being possibly replaced by soluble zinc sulphate.

A further explanation may be given why metallic zinc immersed in wash-waters impregnated with acids and alkaline salts causes the precious metals, when they are present, to be "thrown down" out of the liquid, is that the zinc being a strong electro-positive metal, while those likely to be present in jewellers' waste solutions are electro-negative. The positive metal in such liquids always suffers first oxidation, and subsequent solution. The oxidation of the zinc prevents the precious metals from adhering to it, and whilst gradually dissolving away, nascent hydrogen gas is being constantly produced, and the gas precipitates the negative metals as metallic powders. The action of the zinc is not at all prevented by the alkaline constituents present in the liquid, unless insufficient acid mixtures are contained in the compounded liquids. The acidity of jewellers' mixed solutions is usually quite sufficient to keep the zinc sheets clean enough from oxidation for the object in view to be thoroughly accomplished. The soap-suds and other organic material proves no obstacle to the working of this contrivance. When water in negligent quantities is allowed to run to waste by careless people, and to enter the tanks, a little oil of vitriol must be added occasionally to the liquids in the tanks, as their acidity becomes weakened through too much water being allowed to flow into them; but strong chemical action is not necessary, as the zinc sheets will become too quickly dissolved, without any useful purpose being served. In the general way, it is advisable to have the wash-waters as dilute as possible; that is to say, not too strongly impregnated with corrosive acids of any kind, as the other constituents which jewellers' waste liquids contain, do their share in helping to set free the metallic elements of a compound soluble mixture into their component parts again.

The series of vessels, six in number, employed for the purpose of precipitation and filtration are arranged on a gradatory scale of about 4 inches each, in the manner shown in fig. 29, the construction of each being exactly the same, with the exception of the last, which has a support 1 inch from the bottom, on which is first placed a perforated disc of wood; upon the top of this comes a thick layer of coarse deal sawdust; and lastly, above this is fixed a piece of cloth filtering material, which is made tight against the sides of the vessel by being stretched upon a hoop. The filtering cloth may, if preferred, be attached to the perforated wooden disc at the bottom, and underneath the sawdust. Above the filter bed stands the liquid substance to be treated, and the filtered water falling into the bottom space of the vessel flows through the outlet pipe into the drain, and is carried off the premises freed of all its valuable metallic constituents.

To each of the other five vessels there is a 1-inch leaden pipe attached, extending downwards to a little more than half of the depth of the vessel, through which the liquid substance runs. It is in this manner emptied at nearly the bottom of each vessel in the order of rotation consistent with the plan of connecting up the apparatus. This avoids any disturbance of the liquid in the upper part of the vessel in contact with the zinc plates, except the chemical action, which by themselves create, for it is here that the particles of soluble material are being converted into metal, from which position the metal falls steadily to the bottom of the vessels. It cannot rise upwards owing to the steadiness of the water above, aided also by its volume, which prevents a more weighty substance reaching the tops of the vessels in the same manner as the water itself does in this contrivance.

The water rises upwards in each of the first five tubs, and when they are once filled with liquid, always remain full up to within 4 inches of the tops, at which distance the several outlet pipes are fixed. The liquid is thus in constant contact with the series of zinc sheets, and this ensures the precipitation of all the precious metals contained in the wash-waters subjected to this mode of treatment. The water when it reaches the end of the contrivance, having then passed through the filtering device, is allowed to drain away into the sewer.

The zinc sheets are suspended in the centre of the waste liquids from strips of wood placed across the tops of the tubs.

The leaden pipes through which the water passes from tub to tub have each a bent end of sufficient length when lodging on the tops of the tubs to pass through a waterproof hole in the tub immediately preceding. The pipes are made to rest against the sides of the tubs so that when the liquid passes through them and is discharged low down in the respective tubs, there is little agitation set up in the liquid by the flow of water, and this device allows for the volume of water above to always remain in an almost stationary condition, which gives the metallic constituents a safe opportunity to fall down to the bottom of the tubs and out of the liquid, as the waste water passes through the contrivance.

The water rises steadily upwards, but does not carry the metallic ingredients with it, for when these are converted into solid metal the substance is too dense to leave the bottom, and only the lighter material and unconverted metal rises upwards, and flows through the outlet into the next tub, and so on, the conversion becoming at each stage more complete, and when the end of the operation is finally reached, all traces of the precious metals will have been extracted from the waste liquid.

Gold, silver, and platinum is slowly precipitated from mixed "liquid residues" by the action of zinc, the reaction being one of simple displacement of the various metals. But when a mineral acid is not present in sufficient strength to act slowly on the zinc the cause of precipitation may be interrupted, owing to the formation of an indissoluble layer of hydrogen on the zinc sheets. When, however, the liquid residues are sufficiently acidulated this does not happen, for the hydrogen is then freely evolved, and the zinc gradually and slowly dissolves in the watery mixture. The greater affinity of the acids and alkaline salts for the zinc is the cause of this reaction, hence it forms an excellent reagent for completely throwing down all metals of a less electro-positive nature, the result being that the soluble metallic constituents of the liquid are decomposed and converted into their natural state; the oxygen of the watery solution attacks the zinc, forming zinc oxide, which in the presence of acid ultimately becomes dissolved into the solution. The nascent hydrogen gas given off from the surface of the zinc during these evolutions, in its endeavours to rise to the surface of the liquid comes in contact with the precious metals, liberating them from their solvents, which then fall to the bottoms of the tubs in the metallic state, the gas escaping in an invisible form; the action is continuous, and after the first charging up of the contrivance, inexpensive. The result is, therefore, that the work of precipitation is at no period delayed.

The larger the volume of free acid or alkaline salts in the different liquids used, the more capable are they of dissolving and holding the precious metals in solution, and the larger must be the quantity of water added to them, in the one-process operation, to cause the separation of the metals to be the more easily accomplished. This preparatory treatment is provided for by allowing all the swilling waters, and all the wash-hands waters, to flow into the same set of tanks, and by doing this all the different chemical mixtures will become sufficiently diluted to effect the object aimed at, without any special addition of water being made to them, which is imperative with some liquids when they are being treated separately as heretofore described.

Several methods have been tried and others suggested for dealing with the waste liquids of goldsmiths, silversmiths, and electro-metallurgists, with the view to recover therefrom all traces of the precious metals, some of which require too great a chemical knowledge, and are too complicated for the ordinary craftsman to master, and he wisely leaves them alone. But if those whose duty it is to look after the waste made in manufacturing personal ornaments from the precious metals were made acquainted with a few commercial wrinkles of real practicability respecting the treatment of waste residuary waters which would lead to their labours being more successful, no doubt they would be ready to embrace them, if the method suggested was simple and easily to be carried out. Now, the method I am describing is one which removes all soluble metals from solution and restores them to the metallic state, which, then, by the act of gravity, fall to the bottoms of the tanks, where the solid particles remain until they are required to be converted into money. They cannot pass over with the water and be lost. The method is so simple that the most dull and careless person cannot fail to understand its working advantages financially when the sedimentary residue is collected together, burnt, and sold to the smelter.

Zinc destroys the power of cyanide of potassium, as also those other salts of lesser powers used in the different branches of the precious metal-working industries. Cyanide gilding and silvering solutions, when spoiled and put into the general waste-water receptacles, it may be urged, will redissolve other salts of silver which may have been precipitated beforehand by some other reagents noted for throwing down the silver, and used in preliminary operations; several of such substances, it is well known, are soluble in cyanide of potassium; but practically this is not the result of my commercial experience in this, the one-process operation, for the mixed acids in the waste and exhausted mixtures, when all are being treated as one mixture, prevents any action of that kind from taking place, and, as both hydrochloric and sulphuric acid will release the dissolved gold and silver from their solution in cyanide of potassium when put into the general waste liquids, there is no danger to be feared of imperfect precipitation resulting from the adoption of their being added to the first of the series of tanks in the "electrolytic method" of arrangement without previous treatment, for the series of zinc sheets so effectually precipitates all the metals in whatever state of solution, that by the time the liquid reaches the last tank every trace of metal will have been removed from the watery mixture.

Carbonates of soda and potash precipitates the soluble silver in the waste liquids of jewellers as carbonate of silver, and this, in turn, is liable to be redissolved by strong cyanide solutions were it not for the reasons I have stated, namely, the mixed acids of other liquids have the properties to decompose the cyanides and other metallic solvents, and thus liberates the soluble metals from such solutions when compounded or mixed together. Then, these metals undergo further decomposition by the zinc plates, which causes their perfect reduction to the metallic condition, so that there is a series of double reactions constantly taking place in the precipitating tanks to throw down all the gold and silver in an absolutely safe manner from all liquid residues that no traces of those metals are left behind in the clear waters standing above the precipitates, the water running away from the last tank being almost as clear, when the operation is well performed, as clean spring water.

No other reagent than the zinc sheets is necessary to the thorough success of this process, provided it is carried out as directed, and a little attention given to it occasionally to ensure its working satisfactorily. All the usual expensive costs of other processes being thus saved in effecting the reduction of the gold and silver from the waste residuary liquids by this method, no injurious gases are evolved, no overflowing of the liquid takes place, and very little frothing occurs in the working arrangements of the contrivance, while the costs of acids and other chemical salts are saved by adopting this method of recovering gold, silver, and platinum from jewellers' waste-liquid products.

It is a most convenient process to adopt, for by it both large and small volumes of solution may be cheaply and perfectly treated when done with, the method being free from difficulties of any kind, and where large quantities of liquids have to be dealt with it is advantageous to recover the metal by a process which is both easy and inexpensive in performing the work.

It is not necessary to clean the surfaces of the zinc sheets at all, for the action of the acid (always present in jewellers' liquid residues) will dissolve the oxide formed on the zinc surfaces almost as fast as it is formed, and also prevent a layer of hydrogen adhering, and in that way a continually changing surface is being exposed to the action of the liquid, to bring about the desired results.

Zinc plates immersed in solutions containing gold and silver in any of their conditions will speedily reduce the metals to their natural state, and also clear the liquid of all murkiness by the time it reaches the last of the series of tubs which forms the outlet for the water after it has passed through the contrivance.

CHAPTER XLIV

MELTING SEDIMENTARY RESIDUES

To those unacquainted with the processes adopted in the workshop for the recovery of the precious metals from the "tub sediments," and sometimes a wish is desired to have these residues melted down on the premises instead of disposing of the resultant powder to the sweep smelter, in which case the following information will be useful to such firms who are desirous of putting into operation this practice for the recovery of their metal, in the solid compact form, from the worthless drossy material; but the competition among "sweep smelters" is now so keen that it is advisable to seriously consider whether it is the most advantageous to collect the gold and silver from these waters by individual manufacturers, considering the cost of materials required in the process and the time of the workman in running down all such material in order to separate the metal from the dross, or whether it would not be the better plan financially to sell all such residuary products to gold and silver smelters and refiners as being the more economical plan to adopt, and in the end most probably the least expensive course to pursue.

Smelters have every available source of dealing with such products to the very best advantage in the shape of the necessary plant and all other requirements for the purpose, which enables them to put through the fire very large quantities of material at one operation. This could not be accomplished by individual manufacturing firms, as they have neither the convenience nor the necessary plant on the premises; consequently their method will have to be one of melting the waste material in crucibles, assisted by means of a good fluxing mixture.

In the melting of most precipitated residues, in an attempt to recover the precious metals in solid compact form, a flux must be used, for unless a flux is used when the mass is raised to the point of fusion, a complete union of the metallic particles is prevented from taking place by the presence of so much foreign matter; and instead of every atom of metal being collected together into a solid button, there is usually obtained only a portion of it in the bottom of the crucible, and a large portion may be dispersed, in the form of shots, throughout the pasty material above. When, however, a suitable flux is used, all the foreign material is dissolved into a thin liquid, and this allows of all the metallic substances to fall out—or drain, as it were, from it—and gather together into a ball at the bottom of the crucible, as clean metal.

As a compound flux for use in melting tub sediments resulting from the *one-process operation,* I have found nothing better than one or other of the two mixtures here given. They are made up as follows :—

FLUX NO. 1.

Residue	10 ozs.	or	100 ozs.
Soda-ash	4 „	„	40 „
Common salt	2 „	„	20 „
Fluor-spar	2 „	„	20 „

FLUX NO. 2.

Residue	10 ozs.	or	100 ozs.
Soda-ash	4 „	„	40 „
Common salt	2 „	„	20 „
Salenixon	2 „	„	20 „

The treatment of the precipated residues after they have been removed from the tubs, is to put them through the sweeps furnace (see figs. 9 and 20). (A sweeps furnace may consist of any suitable boiler with as light a draught as possible to prevent the passing off of fine light particles during the burning.) The burned material, when that operation has been completed, is next subjected to riddling. The rough parts which do not pass through the sieve are put into a suitable apparatus, in which fine pounding can be rapidly and safely carried out without loss. For this purpose, when the quantities are small, an iron mortar with pestle is commonly used, which soon reduces the coarse parts to an impalpable powder. The whole of the pulverised material is finally sieved through a very fine-meshed sieve, when the substance left in the sieve will be found mainly to consist of metallic shots of various metals, and other hard, unpoundable material. This product is collected together and melted down separately in a clay crucible, assisted by the following flux, which never fails in its object :—

Sieve residue	10 ozs.	or	100 ozs.
Soda-ash	1½ „	„	15 „
Fluor-spar	½ „	„	5 „

This flux will melt anything. The button of metal thus obtained will be impure, and should be assayed and sold to refiners under mutual agreement.

The finely-powdered material which has passed through the fine sieve is the real tub sediment, and if you wish to run this down on the premises instead of trusting to the smelter and refiner of metals, it must be intimately mixed with one or other of the compound fluxes (also reduced to very fine powder) I have given for the purpose, and melted in a large clay crucible or in a clay-lined plumbago crucible. It cannot be melted in a plumbago crucible unless the insides of the crucible are protected, as the flux, when melted, becomes very fluid, attacks the kind of clay with which the graphite is mixed, and this, mixing with the contents of the crucible, prevents the material therein from fusing into a thin liquid, by raising the melting point.

If the slag, at the conclusion of the melt, is not sufficiently fluid to allow of all the particles of metal separating out, a little more flux must be added, in the same ratio of proportions of the different ingredients as those composing the original mixture, and when the mass is completely melted into a thin fluid, and a clay-lined plumbago crucible has been used in effecting the melting, pour the entire contents of the crucible into an open ingot mould which has been made quite hot beforehand. The slag will run into the mould with the metal and rise to the top. Allow the material in the mould to cool, then reverse the mould and its contents will fall out. The slag or flux can then be readily detached by a blow from a hammer. The crucible may be recharged with more of the residue to be melted and the melting continued while the first charge is cooling in the mould, so that no time is wasted in waiting, for before the second charge in the crucible is melted the mould will have had its contents removed and be quite ready to receive the next pouring.

But after all is said and done, it is perhaps the most economical to sell this kind of residue to the sweeps smelters; but even then, it is necessary to prepare the product before it can be sold to the most satisfactory advantage. The best method is to burn the residues of the tubs and grind them to a fine powder, so that a fair sample can be obtained for assay. Several assays may then be made by different smelters and a comparison made as to what difference is shown in the value. A word of warning is here given that, unless the residues are ground into very fine material and treated as before described, there will be very little chance of getting the full value. This failure to get the full commercial value is caused by the fact that the coarse material is filled with fine shots of metal, and when samples are obtained no notice is usually taken of these fine shots, as they are considered to be not evenly distributed throughout the mass of residue, and may result in a loss to the smelter who happened to select a portion containing more of these shots than the whole mass warranted; they are, therefore, placed on one side and not taken into calculation at all when the offer is being made; the only safe way, in selling to the smelter, is to grind as fine as possible and pass through an extra fine sieve so as to retain the fine shots, and then melt them down on the premises under the special flux I have provided for this purpose; have assays made, and sell to the highest bidder, as the metal cannot be used without refining. And now, with regard to the properties of the fluxes I have named for melting precipitated residues, a few details will be useful to the man at the furnace.

Soda-ash is dried carbonate of soda (Na_2CO_3); this acts as a powerful reducing agent upon many oxides, and reduces them to the metallic condition, while with those of a baser kind it acts as an oxidising agent, and then dissolves them into the slag. It readily forms with the non-metallic material of the residue a fusible slag, from which the metals separate.

Common salt is chloride of sodium ($NaCl$); this compound prevents the too violent ebullition of the contents of a crucible wherein organic matter is present. It prevents the formation of oxide, or reduces it if it has already been formed, and assists the metal to unite in a clean solid mass. It is cheap and easy to use.

Salenixon is bisulphate of potash ($KHSO_4$); this salt has solvent powers for iron and other metals, and acts by dissolving them into the slag which rises to the surface of a melting charge of this kind. It will search through any material which is refractory, and cause it to melt. Salt cake (the sulphate of soda, which is a by-product in making hydrochloric acid from common salt), which is equally cheap, will answer the same purpose.

Fluor-spar (CaF_2), or as it is often called, Derbyshire spar, is of great value as a flux in recovering precious metals from troublesome residuary products. It has great aversion to common metallic oxides, especially iron and zinc, which are quickly dissolved by it. It also acts on all kinds of silicious matter when that is present in a melting charge, and readily combines with earthy and dirty material, forming a liquid substance, from which all the contained metals quickly subside after the contents of the crucible have become completely melted.

In melting tub sediments by the crucible method, it is usual to make additions of the material to the crucible after the first lot has subsided.

INDEX

PRINTED IN GREAT BRITAIN BY NEILL AND CO., LTD., EDINBURGH.

DISSOLUTION AND RECOVERY OF DEPLETED URANIUM

By

Gary Czupryna
Rhonda D. Levy
and
Harris Gold

Revised and Reset from the 1987 Edition
for
Wexford Press
2008

TABLE OF CONTENTS

SECTION I
INTRODUCTION

OBJECTIVES

The primary objective of this project was to develop a selective solvent that can decontaminate impacted armor targets containing depleted uranium. The ideal solvent would reduce the radiation in the plates to a level that allows the disposal of the plate to commercial non-hazardous burial sites. The decontaminated plates can also be reused or recycled, if desired.

The best approach for the final disposition of the solvent containing the dissolved low level radioactive waste is to recover the depleted uranium from the solvent as a uranyl salt or oxide of acceptable concentration and purity for recycling by a refiner. If recycling is undesirable, the recovered uranium can be inexpensively disposed of in its volume reduced state as low-level radioactive waste. In either case, the solvent may be recycled. A secondary objective was devoted to identifying and characterizing technologies for removing DU from the solvent.

These objectives were met by carrying out the work in two experimental phases:

(1) Selective solvent dissolution studies relating the functional relationship between preferential uranium removal and process parameters including solvent composition, concentration, volume, temperature and immersion time, and

(2) The use of ion exchange, solvent extraction and precipitation technologies were investigated in order to evaluate the efficiency of precipitating agents and selective ligands to form readily extractable or insoluble uranium complexes that can be selectively removed from the acid solvent.

BACKGROUND

The impacted armor targets used in testing high density armor piercing ammunition containing depleted uranium (DU) are subject to disposal at low level radioactive waste. Depleted uranium, a low level radioactive by-product of the nuclear fuel enrichment process, is recognized as an outstanding material for use in kinetic energy armor piercing projectiles. Each branch of the U.S. Department of Defense has ordnance programs that utilize DU because of its superior armor penetrating capabilities and cost effectiveness. In currant production, DU penetrators are the principal ammunition for the U.S. Army's 105 mm 9-1 tank main cannon.

The various ballistic ranges have a great deal of armor that has been shot at with DU ammunition. The armor is normally in 4 ft x 4 ft (122 cm x 122 cm) sizes and varies in thickness from 0.5 in. (1.3 cm) thick to 4 in. (10.2 cm) thick. These plates have a varying number of holes in them; most of the holes are completely through the armor plate. In each case, the surface of the hole has been contaminated with DU. As the projectile passed through or even just impacted the target, the DU becomes molten and an adhesive bond is created on the surface. In some cases, the DU contamination penetrates into the armor plate.

Armor-piercing munitions are specifically designed to defeat armored targets through the primary target of a high-density, non-explosive penetrator. If depleted uranium (DU) is used as the penetrator material, its pyrophoric nature will cause the projectile to burst into burning fragments upon impact with armor. These fragments are dispersed into the air, the sand butts, and into the armor plate target. In the latter case, a eutectic alloy is formed with the mixed oxide, UO, and the iron from the steel target plate.

Depleted uranium is classified as a low-level radioactive material and, as such, must be handled in accordance with Nuclear Regulatory Commission (NRC) guidelines for low-level radioactive waste. The armor plates used in testing munitions containing DU are subject to disposal as low-level radioactive waste. The plates cannot be refined in their contaminated state and at the present time there are no operational alternatives to sending these plates to commercial disposal sites for burial.

Because of the cost associated with disposal of the entire armor plate as low-level radioactive waste and the limited use of secured commercial sites in the future, the U.S. Army is seeking to identify and evaluate feasible solvents that selectively remove fragments of uranium imbedded in armor plate as a result of firing DU penetrators into the target material.

The problem is to decontaminate the target plate to meet state and Federal requirements so that they can be safely disposed of off-site in unsecured landfills or can be returned to the ranges for other uses.

Mechanical and physical techniques for removing imbedded DU fragments that require isolating the penetrated area by cutting or melting is not only time consuming but also dangerous. Small uranium fragments are extremely pyrophoric, creating a potential fire hazard. The most deleterious health consideration comes from ingesting and inhaling fire uranium dust and therefore precludes fragment removal by cutting out the contaminated area. Dissolution of the entire armor target is possible, but the reagent cost of treating hundreds of targets and difficulties associated with uranium recovery make this method impractical.

SCOPE

The major effort of the Phase I program focused on developing a selective solvent that preferentially attacks the areas on the armor target contaminated with DU, leaving the armor plate unaffected and intact. A secondary effort was devoted to the removal of DU from the solvent. The volume of the spent solvent solution can actually exceed the volume of the treated armor plate. The greatest advantage of the process developed for the removal of the DU from the solvent is the volume reduction of DU requiring disposal. Once the DU is recovered, the treated solvent is available for reuse and can be returned to treat additional impacted targets.

The laboratory research plan involved screening candidate solvents for their ability to dissolve DU pellets containing 0.75 percent titanium in the same composition used for ammunition production). Solvents passing the initial screening criteria were tested for uranium selectivity by evaluating their ability to dissolve a DU-0.75 percent Ti pellet from a simulated test target. Testing on simulated targets provided valuable selectivity data, until actual target were obtained. The most promising solvent was evaluated using actual penetrated targets.

The feasibility of removing the dissolved uranium from the spent advent was demonstrated using precipitation, ion exchange, and solvent extraction technologies. The advantages and disadvantages of the techniques were assessed and are discussed in this report.

The program plan under the Phase 1 SBIR program consisted of three tasks described as follows:

Literature Search

A literature search was made to determine what solvents, combination of solvents, and combinations of solvents and oxidants (catalysts) will selectively attack (dissolve) uranium-titanium alloys, and other eutectoid alloys that any be formed upon projectile impact. Emphasis was placed on literature pertaining to uranium-selective chelating agents.

The range of operating parameters, such as solvent composition, solvent concentration and the reaction temperature at which selective dissolution of the uranium takes place, was initially defined based on the existing literature and past experience. Other characteristics of the solvent system, such as compatibility with the target material, flammability, and toxicity, were discussed when appropriate,

Based on the dissolution properties of the solvents, a list of candidate solvent systems was developed and have the potential of meeting the criteria for selective dissolution of uranium-alloyed fragments from steel plate targets.

Decontamination Studies

The solvents selected as potential selective candidates were subjected to dissolution studies with weapon grade uranium to evaluate their performance as solvents for the DU-0.75 percent Ti alloy. Candidates meeting the initial screening criteria weie evaluated to determine their selectivity characteristics for W alloy fragments imbedded in simulated test fragments. The most promising solvent was evaluated using the actual penetrated targets. The selectivity and rate of dissolution of the solvents were conducted according to the recommended methods of total immersion testing for ferrous metals described by Uhlig (1). When necessary, the testing procedures were amended to meet the technical objectives.

According to this procedure, the simulated and actual test targets are placed in the test apparatus and covered with the solvent. Chelate additives are also added, if appropriate. Test variables controlled included solvent composition, concentration, and volume; temperature and test duration.

Solvent selectivity was determined from elemental analysis of the solvent following the test.

Solvent Recycle and Uranium Recovery/Waste Reduction

The successful dissolution of DU from the target plates transfers the uranium disposal problem to the solvent solution. Removal of the DU from the solvent reduces the volume of disposal uranium and allows reuse of the solvent.

Three techniques considered for removing the DU from the solvent solution were precipitation, ion exchange and solvent extraction. The separation of uranium from solution by precipitation is based on the formation of sparingly soluble compounds. The purity of the uranium compounds is dictated by the resulting solvent containing dissolved uranium and the choice of precipitant. The recovery of uranium iron the selective solvents was greatly influcnced by pH. Increasing the pH of the acid solvent for uranium recovery neutralizes the solvent and precludes the opportunity for solvent recycle.

The feasibility of removing uranium from spent solvents using selective chelate ion exchangers was assessed using commercially available products. This was demonstrated by reacting liquid and solid ion exchangers by selectively removing uranium from a spent solvent without excessive extraction of other solution contaminants such as iron. Successful uranium recovery was achieved from an acid chloride/phosphate combination using the aminophosphonic acid chelating resin, Duolite ES 467. The effect of solvent composition and concentration on the capacity of the ion exchanger was assessed by comparing the uranium concentration in the solvent before and after treatment. The distribution coefficients for uranium and other metals dissolved by the solvent were calculated, from which the selectivity factors were determined.

SECTION II
LITERATURE REVIEW

During the process of penetration of armor targets, DU penetrators undergo severe fragmentation. Flash temperatures reached during impact of DU penetrators with armor plate have been shown to fall in the range of 3037 °C to 3093 °C over the entire range of impact test velocities. Such heat generation following impact causes the formation of molten uranium, uranium oxides, and eutectoid alloys with the steel plate. X-ray analysis confirms that airborne DU particles, formed when DU projectiles (99.25 percent DU, 0.75 percent titanium) impact armor targets, contain an extremely high iron content (2). It is reasonable to assume that the morphological characteristics of the uranium fragments imbedded in the armor targets will contain, not only DU, but also DU of a heterogeneous iron eutectoid composition.

The removal of DU from the target material is complicated by the presence of these uranium-iron eutectoid alloys. The dissolution of this eutectoid by a selective solvent was assessed using an impacted target contaminated with DU.

Rodden (3) and Katz and Rabinowitch (4) describe methods used for dissolving uranium metal and high uranium alloy. However, no detailed information was found concerning dissolving bulk depleted uranium-titanium alloys alone or from other bulk base metals. Most of the discussions in the literature are confined to the dissolution of uranium metal and its alloys for the purpose of analysis, in which uranium is the principal constituent. Other literature sources describe methods for the complete dissolution of alloys in which only small percentages of, uranium are contained in the metal sample, like reactor fuels.

In general, oxidizing acids, such as nitric acid and perchloric acid, as well as hydrochloric acid, react vigorously with elemental uranium. These acids are also likely to react with the homogeneous armor steel plate used for ammunition testing.

Mineral acids, like sulfuric and phosphoric acids, are slow to dissolve uranium. However, these acids used in high concentrations with the catalysts and accelerators, have the potential to dissolve the uranium fragments and also passivate the target material. Partial dissolution of the target is beneficial in order to remove ill traces of uranium and eutectoid uranium-iron alloy formed by projectile impact.

A list of acid solvent/solvent combinations that were considered as potential elective solvents is shown in Table 1.

NITRIC ACID

Probably the most common reagent for the dissolution of uranium metal and its alloys is nitric acid (HNO_3). Nuclear Metals, Inc. (Concord, MA), a manufacturer of the M744 armor piercing anti-tank round, dissolves DU-0.75 percent Ti, scrap for recovery by converting the depleted scrap to the metal oxide with 8 Molar (M) HNO_3, and with 0.5 percent hydrogen hexafluorophosphate (HPF_6) as an accelerator. A pure uranyl peroxide product is recovered by H_2O_2 precipitation at an adjusted pH of 2 to 4 (5).

HYDROCHLORIC ACID

Uranium reacts with hydrochloric acid (HCl) very rapidly to fora uranium (IV) chloride (UCl_4) and a black precipitate that is described as a hydrated (III-IV) oxide. In 1M HCl acid at 100°C, a 10 gram bulk sample can be consumed in one hour; in 12M acid. The evolution of hydrogen is explosively rapid (6). The addition of an oxidizing agent (e.g., sodium chlorate, hydrogen peroxide, perchloric acid) reacts rapidly with uranium (IV) to form a uranyl halide solution that can be precipitated and recovered.

AQUA REGIA

Aqua regia, which has a relatively high hydrochloric acid:nitric acid ($HCl:HNO_3$) volume ratio (4:1), is used for dissolving uranium alloyed with one percent or more zirconium (a Class V element along with titanium), molybdenum, ruthenium, and rhodium (4). This presents the problem of solubilizing the very reactive uranium matrix material in the same acid as that used for the very

TABLE 1. Reagents for Dissolution of Uranium and its Alloys.

Description	HNO_3	Aqua Regia	HNO_3-HF	HCl+ Ox	HCl-EtOAc	NaOH-H_2O_2	H_2SO_4-H_2O_2-HCl	H_3PO_4 (hot)
U	S	S	S	S	S	S	S	S
U-Zr	N	N	S	N	N	N	N	-
U-Nb	N	N	S	N	N	S	N	-
U-Fe	S	S	S	S	S	N	S	S
U-Cr	N	N	N	S	S	N	-	-
U-Ru	N	S	N	N	N	N	-	-
U-Mo	N	S	N	S	N	S	-	-
U-Fissium[a]	N	S	N	N	N	N	-	-
U-Si	S	-	S	-	-	-	-	-
U-Pu	S	S	N	S	S	N	-	-

[a] Alloys containing from 1 to 3 percent zirconium, molybdenum, ruthenium, rhodium, palladium and cerium.

inert alloying constituents. Alloys of this type have been satisfactorily dissolved in aqua regia when hydrofluoric acid is added in the appropriate amounts.

PERCHLORIC ACID

Uranium metal dissolves very rapidly and spectacularly in hot (160°C), 70 percent perchloric acid ($HClO_4$) (4). The character of the perchloric acid dissolves such to be desired from a safety aspect and was given a low priority as a dissolution method.

PHOSPHORIC ACID

Rodden (3) studied the dissolution of uranium in phosphoric acid (H_3PO_4)and reported that the bulk uranium metal is attacked by 85 percent phosphoric acid at a moderate rate to form a clear uranium (IV) phosphate solution. A plethora of information is available in the literature for the selective recovery of the resulting uranyl phosphate complex, $UO_2H_2PO_4$, from wet process phosphoric acid solutions by a number of ion exchange resins, such as Duolite ES467, Lewatit M504, and Levextrel OC 1023.

SODIUM HYDROXIDE - HYDROGEN PEROXIDE

It may be necessary to use a basic medium to selectively dissolve uranium fragments from the target material, if the target is soluble in most acid solutions. The dissolution in sodium hydroxide (NaOH) - hydrogen peroxide (H_2O_2) has the potential to meet this need. It has been reported that 10 gram samples of uranium metal and some uranium alloys react at a moderate rate with 50mL of 1M sodium hydroxide-5M hydrogen peroxide mixtures in less than one hour to form the uranyl peroxide complex (6). The advantage of this method is the solution containing dissolved uranium can be quickly converted to the soluble uranyl nitrate complex with the addition of nitric acid and subsequent precipitation of pure uranyl peroxide by adding additional hydrogen peroxide.

SULFURIC ACID

Uranium and most alloys in which uranium is the bulk constituent, do not react with sulfuric acid at moderate concentrations and temperatures. However, Larsen (6) has reported that dilute sulfuric acid (H_2SO_4) moved iron, magnesium, aluminum and other metals which have been bonded to, or mixed with, bulk uranium metal. Mixtures of sulfuric acid and hydrogen peroxide attack uranium slowly at 75°C. The addition of chloride in catalytic amounts increases the rate of dissolution markedly. It has been reported that 10 grams of bulk uranium can be dissolved in less than 30 minutes in 100mL of 6M sulfuric acid-1M hydrogen peroxide 0.0lM hydrochloric acid. Uranium (IV) has a strong tendency to form soluble uranyl sulfate anionic chelates. The complexes $[UO_2(SO_4)_2]^{-2}$ and $[UO_2(SO_4)_3]^{-4}$ can be recovered by ion exchange to yield a product of very high purity.

HYDROGEN CHLORIDE-ETHYL ACETATE

Hydrogen chloride (HCl) solutions in ethyl acetate (EtOAc) react with uranium at a moderate rate to form soluble uranium (IV) tetrachloride solutions. In a 3M solution, which is close to saturation, a 10 gram bulk uranium sample was reported to dissolve in a minimum of 10 hours (6). The hydrogen chloride-ethyl acetate dissolution of uranium is unique in that uranium metal is dissolved completely in a relatively mild reagent. This procedure has been used to selectively dissolve the uranium metal in metal oxide mixtures and separate certain intermetallic compounds by only dissolving the uranium matrix.

CHELATE ADDITION

Uranium forms stable complexes with liquid extractants and gel resins containing di (2-ethylhexyl) phosphoric acid (D2EHPA). The addition of chelates to selective solvents should promote the complexation cf uranium ions and accelerate the rate of uranium dissolution (4). Chelating agents with high stability constants, the most desirable for our intended use, are those whose donor atoms, such as fluorine and oxygen, are the most electronegative (7). The best known extractants that meet this criterion is the synergistic mixture of Di-2-ethylhexyl phosphoric acid (HDEHP) and trioctyl phosphine oxide (TOPO). Moreover, higher partition coefficients for uranium results it TOPO is replaced by phoshine oxides containing an ether oxygen at the beta position of the phosphorous atom. Examples are di-n-hexyl-2-oxa-decyl phoshine oxide [D(H)2-ODPO] and dibutyl (butyl-phosphonate) [DB(BP)]. In addition, the replacement of HDEHP by dialkyl phosphoric acids with ether chains may also synergistically improve the uranium/iron separation coefficients. Di (3-oxa-heptyl) phosphoric acid (HD3-OHP) is such an example (7).

The introduction of the ether oxygen in the above mentioned ligands make the phosphoryl oxygens less basic by reducing electron density (7). Consequently, the H^+-metal exchange is favored, and, given the acidity of the potential solvents should lead to an increase in partition coefficients.

When the ligand forming extractants are added to a compatible solvent like phosphoric acid, the ligand displaces the solvent molecules coordinated to the uranium metal. This reaction forms a chelate complex from the solvated uranium ion at the metal surface, exposing a fresh site on the target for the solvent to act on. For selective complexation of the depleted uranium ions in the presence of the metal target ions, chelating agents with high stability constants relative to those of the different solvated target ions are required. This will keep the equilibrium constant large enough to favor the sequestration of the uranium ion. The feasibility of improving uranium dissolution was evaluated using a synergistic mixture of HDEHP and TOPO.

SECTION III
DECONTAMINATION

The decontamination of impacted armor plates was conducted in three phases:

1. The dissolution characteristics of candidate solvents identified in the previous section were evaluated by measuring the percent dissolution of weapon grade DU-0.75 percent Ti pellets.

2. Selectivity of the two best solvents in (1) for DU-0.75 percent Ti was determined from simulated test targets.

3. The best solvent from (2) was tested against an actual DU contaminated penetrated target to determine what degree the radioactivity associated with the penetrator hole could be reduced.

TARGET MATERIAL DESCRIPTION

DU-0.75 Percent Ti Pellets

The pellets used to screen candidate solvents were provided by Nuclear Metals, Inc. (Concord, MA) and are the same source material used in the production of high density armor piercing ammunition. Each pellet weighed between 16 and 20 grams with approximate dimensions of 1.2 cm x 0.75 cm.

Simulated Target Material

Simulated target plates consisted of 6 in. x 4 in. x l in. (15.2 cm x 10.2 cm x 2.54 ca) ASTH A36 structural steel and weighed approximately 3 kg. The specifications for the actual target material for wrought-steel follows the military specification (MIL-A-12560G (MR)) homogeneous armor plate.

The chemical compositions of the simulated and actual target materials are shown in Table 2.

TABLE 2. Target Material Chemical Composition

Element	ASTM A36 (percent)	MIL-A-12560G (MR) (percent)
Carbon	0.17	0.10
Manganese	0.86	0.40
Phosphorus	0.011	0.025[a]
Sulfur	0.003	0.025[a]
Silicon	0.016	0.40
Nickel		0.50
Chromium		0.40
Molybdenum		0.15

[a] Combined phosphorus and sulfur content doe not exceed 0.04 percent.

Penetrated Targets

Pieces of impacted targets were obtained from the U.S. Army Ballistic Research Laboratory (Aberdeen Proving Ground, MD). The pieces measured approximately 5 in. x 5 in. x 1 in. (12.7 cm x 12.7 cm x 2.4 cm) and were torch-cut from impacted targets. Each piece contained a hole caused by a penetrator round. In the penetrations, the contamination is mainly fused uranium tightly bonded to the target material. The contamination on the surface was in the form of imbedded or fused fragments. Iron oxide, which formed with time on the metal targets, was contaminated from radioactive dust that became incorporated into the oxide.

A description of each sample is listed in Table 3.

SELECTIVE SOLVENTS

All selective solvents and other solutions or additives used in this project were prepared from reagent grade chemicals.

TABLE 3. Penetrated Target Characterization

Sample	Plate (cm)	Hole (cm)	Average Contamination (cpm)
A	2.5 x 13.0 X 14.0	9.0 x 8.4	48,000
B	2.5 X 14.0 X 15.0	4.0 X 4.5	1,100

ANALYTICAL PROTOCOL

Radioactivity measurements were made using a Geiger-Mueller (GM) counter equipped with a pancake tube having 0.2 millirea/hr window. All measurements were recorded in total counts per minute (cpm). No corrections for geometry were made.

Total iron in the acid solvent was determined by atomic absorption spectrometry. Measurements were made on a Perkin-Elmer 2380 Atomic Absorption Spectrophotometer at a wavelength of 248.3 nm.

Uranium analysis was made in contaminated solvent samples using an EG&G Model 264A Polarographic Analyzer/Stripping Voltammeter in conjunction with an EG&G Model 303A Static Mercury Dropping Electrode. Samples were prepared in a 1M sulfuric acid electrolyte. Measurements were made while screening at 10 mV/sec between the potential range of +0.03 mV through -0.50 mV.

LABORATORY PROCEDURES

All the screening and dissolution studies were conducted in an immersion test apparatus. The test apparatus employed a 3.0 L Pyrex reaction kettle (Figure 1). The reaction kettle was sealed with a Teflon O-ring and capped with four neck cover secured with a McCartier clamp. A DU-0.75 percent Ti pellet or target plate was placed into the reaction kettle and reacted with 300mL of solvent, the amount required to completely immerse the pellet or enough solvent to cover the target plate. The solvent was dispensed by

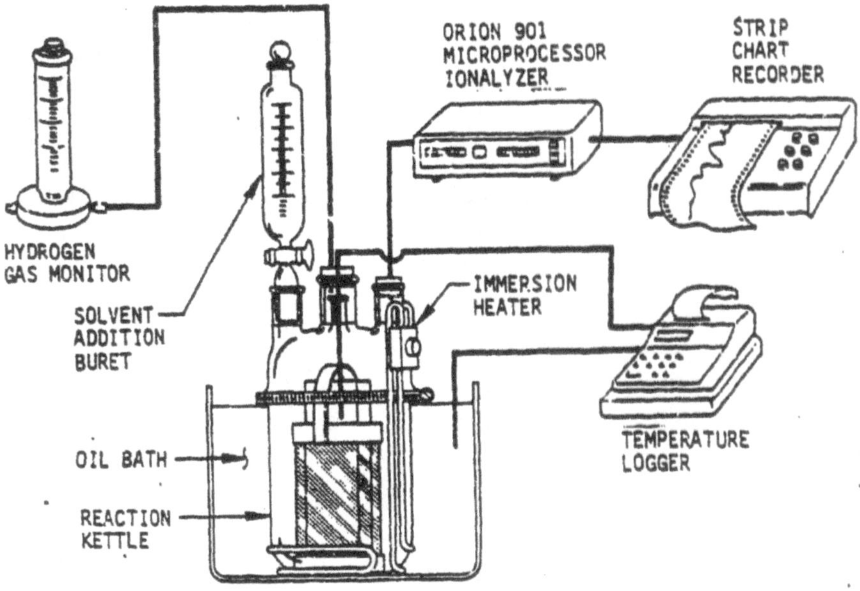

Figure 1. Reaction Kettle.

opening the stopcock of a 500mL separating funnel mounted on the reaction kettle cover. The solvent temperature was adjusted and maintained to the desired study temperature using a thermostated water bath. To allow intermittent inspection of the pellet/target during testing, the test sample was lowered and raised into the reaction kettle in polypropylene dipping baskets.

The temperature and pH of the solvent were monitored by probes immersed into the reaction bottle through the four-neck kettle cover. Qualitative observations regarding color changes of the solvent and gas evolution were recorded during testing.

In the first phase of laboratory testing the dissolution characteristics of candidate solvents identified in Section 2 were measured as a function of the percent weight loss of the DU-0.75 Ti pellet. The pellets were rinsed, dried, and weighed at the completion of each screening test.

In the second phase of testing, the selectivity characteristics of the two best solvent compositions were evaluated using DU-0.75 percent Ti pellets lodged into bore holes that were drilled into simulated target plates. This represents a large impacted depleted uranium fragment that may be found in actual penetrated armor targets. The dimensions and weights of the steel plate and depleted uranium pellet were taken prior to and immediately after testing. The steel plate containing the depleted uranium pellet was placed into the 3-liter reaction kettle and reacted with one liter of acid solvent. The plate was reacted at temperatures between 80 and 85°C for up to 75 minutes. The target plate was inspected every 10 minutes to determine the minimum reaction time seeded to completely dissolve the embedded DU pellet. The test was terminated after the pellet was completely dissolved or 75 minutes. After this test, the target plate and the DU pellet, if any existed, were weighed and the percent metal weight loss was calculated for both the uranium pellet and target plate.

In the third phase of the decontamination studies, the best solvent determined from the previous test was evaluated against a contaminated penetrated target. The most contaminated plate (Sample A, see Table 4), determined as the plate with the highest cpm value, was decontaminated with the best solvent using the procedure followed for selectivity testing. The contaminated plate was immersed into the 3-liter reaction kettle containing approximately 2 liters of solvent heated to 84°C. Periodically, the plate was raised from the kettle reactor, rinsed with deionized water and the average contamination (cpa) measured. This procedure was repeated until the cpm value was reduced by at least 95 percent.

RESULTS

DU-0.75 Percent Ti Pellets

In general, solvents that contain at least 4M hydrochloric acid (HCl) react very rapidly and completely with DU-0.75 percent Ti alloy to form uranium (IV) chloride (Table 4). When 4M HC1 is reacted with the DU-0.75 percent Ti pellet, there is a short quiescent period before the rapid evolution of hydrogen begins. Dissolution of the pellets with only HC1 forms a black precipitate

TABLE 4. DU-0.75 Percent Ti Pellet Dissolution Summary.

Solvent	Pellet Wt (g)	Percent Dissolved	Comments
12M HCl	18.345	100	-Rapid dissolution rate -Black precipitate -Forms uranium tetrachloride
4M HCl	19.143	100	Same as 12M HCl
3M HCl:1M H_3PO_4: ethyl acetate	20.375	93.5	-Moderate dissolution rate -Forms uranium tetrachloride -No black precipitation
4M HCl:7M H_3PO_4	15.952	100	-Rapid dissolution rate -Forms U(IV) chloride and sulfate complexes -No precipitate formation
7M H_3PO_4	17.206	0	-Non-reactive
4:1 - $HCl:HNO_3$	16.953	100	-Uranium IV chloro and nitrate complexes
1M NaOH:5M H_2O_2	18.869	2.2	-Slow dissolution rate -Uranyl peroxide complex
6M H_2SO_4:0.01 HCl: 1M H_2O_2	17.772	1.4	-Slow dissolution rate -Uranyl sulfate complex
4M HCl: 7M H_3PO_4 with 0.5M D2EHPA and 0.125M TOPO	17.676	100	-Rapid dissolution rate -Tetravalent uranium complexes -Organic complexes
6M H_2SO_4:0.01M HCl: 1M H_2O_2 with 0.5M D2EHPA and 0.125M TOPO	17.992	1.7	-Uranyl sulfate complex -Organic chelate does not significantly enhance rate

that is characterized as a pyrophoric and potentially explosive hydrated (III-IV) oxide. No amount of heating in the acid could solubilize the precipitate completely. In 4M HCl at 85 °C, a 19 gram sample pellet was 100 percent consumed in one hour; in 12M HCl, the evolution of hydrogen was explosively rapid. The pellet, when removed from the reactor kettle after partial dissolution, was covered with a heavy black coating. Most of this was easily scraped away, but next to the bulk metal it was very adherent. This coating is very reactive when dried, glowing red hot from sudden air oxidation.

Two reagents, phosphoric acid and ethyl acetate, when added to the hydrochloric acid dissolutions of the DU-0.75 percent Ti pellet, prevented the formation of the heavy black residue. Both are strong complexants of uranium. Phosphoric acid in high concentrations (7M) prevented the formation of the black residue associated with HCl uranium dissolution. A 16 gram sample of bulk DU-0.75 percent Ti was completely dissolved in one hour in a 4M HCl-7M H_3PO_4 acid solvent.

Hydrochloric acid solutions in ethyl acetate reacted with a DU-0.75 percent Ti pellet at a moderate rate to form a uranium tetrachloride solution. In a 3M HCl solution, which is close to saturation, a 20 gram sample pellet was 93.5 percent dissolved in one hour, but formed some residue. The addition of 1M H_3PO_4 resulted in a residue free dissolution.

Soon after the introduction of the pellet in the solvent, hydrogen evolution begins and the solution takes on the characteristic green color of quadrivalent uranium. The ethyl acetate-hydrochloric acid dissolution of depleted uranium is unique in that the metal can be dissolved completely in a relatively mild reagent. Its action is similar to hydrochloric acid but does not lead to the insoluble hydrated oxides.

Aqua regia (4:1 - HCl:HNO_3) was very effective in completely dissolving a DU-0.75 percent Ti pellet. A 17 gram sample pellet was 100 percent dissolved in one hour at 80°C. The dissolution resulted in the formation of uranium IV chloro- and nitro- complexes.

The dissolution of DU-0.75 percent Ti pellet in a sodium hydroxide-hydrogen peroxide mixture was tested to evaluate the effect of a basic medium. The literature (6) reports that uranium metal reacts at a moderate rate with 1M sodium hydroxide-5M hydrogen peroxide mixture to form a colored uranyl peroxide complex. However, a 19 gram DU-0.75 percent Ti pellet was only partially dissolved (2.2 percent) in one hour at 85° C.

Solvents containing sulfuric acid (H_2SO_4) do not significantly react with DU-0.75 percent Ti at moderate concentrations. Only 1.4 percent of an 13 gram pellet was dissolved in one hour at 75°C in 300 ml of a 6M sulfuric acid-1M hydrogen peroxide-0.01M hydrochloric acid mixture. The addition of HCl and H_20_2 in catalytic amounts was intended to increase the rate of dissolution, but proved ineffective. There was an initial effervescence at the pellet surface, indicated g the evolution of hydrogen gas and uranium dissolution, but all visible reaction ceased after 10 minutes, There was a visible darkening of the metal surface, which could be responsible for protecting the pellet from further attack. This is somewhat surprising because uranium (IV) is strongly complexed by sulfate, and uranium (IV) sulfates are soluble. The addition of the chelate 0.5M B2EHPA-0.125H TOPO did not have any significant effect on the rate of dissolution.

Based on the dissolution properties of the solvents for DU, two solvents were selected for further laboratory evaluation. The solvent systems

- 4M HCl : 7M H_3PO_4
- 3M HCl : 1M H_3PO_4:ethyl acetate

were judged most practical.

Simulated Target Material

A summary of the test conditions and results conducted on the simulated target are summarized in Table 5. The results indicate that 4M HCl:7M H_3PO_4 is an excellent solvent for selectively dissolving DU-0.75 percent Ti fragments from simulated target plates. In 55 minutes, a 17.36 g DU-0.75 percent Ti

TABLE 5. Simulated Target Summary.

Experimental Conditions		
Solvent Composition	4M HCl:7M H_3PO_4	4M HCl:1M H_3PO_4:EtOAc
Solvent Temperature (°C)	85	77
Solvent Volume (mL)	1000	1000
Test Duration (min)	55	75
Target Plate Dimensions (cm)	2.54 x 10.2 x 15.2	2.54 x 10.2 x 15.2
Bore Hole Dimensions (cm)	1.3 x 0.77	1.3 x 0.77
Target Plate Weight (g)	3117	3108
DU Pellet[a] Dimensions (cm)	1.2 x 0.77	1.2 x 0.77
Pellet Weight (g)	17.3604	18.6969
Test Results		
Target Plate Dissolution (%)	0.69	0.43
DU Pellet Dissolution (%)	100	Approx 50
Gas Evolution	Rapid	Moderate
Comments	Forms clear green U(IV) chloride and phosphate complexes. No residue.	Partial dissolution. Pellet fused to target.

[a] Pellet consists of DU-0.75 percent Ti.

pellet was completely dissolved in a pre-drilled bore hole. The dissolution is rapid and selective. The target plate remained intact with only 0.69 percent of its total weight dissolved. Vigorous gas evolution around the area of the pellet demonstrates the solvent's preference for DU-0.75 percent Ti over the target composition.

During decontamination, the acid solvent turned green in color, indicative of quadrivalent uranium chloride and phosphate complexes. No potentially hazardous precipitates were formed. The bore hole was inspected for residue and tested for radioactive contamination with a Geiger counter. No residue was found and only background radioactivity was detected

The 3M HCl:1M H_3PO_4:ethyl acetate solvent performed reasonably well. This solvent is less vigorous than the former, as indicated by the dissolution of approximately 50 percent of the uranium pellet in 75 minutes. Although this solvent is effective over time and produces no dangerous precipitate, its volatility precludes its use as a safe and economical decontamination solvent. Since this solvent offers no apparent advantages relative to the 4M HCl:7M H_3PO_4 acid solvent, no further testing was conducted.

Penetrated Targets

The 4H HCl:7M H_3PO_4 acid solvent was tested to evaluate the decontamination kinetics on penetrated targets. The target was decontaminated in the reaction kettle at 84° C. The target was periodically removed, rinsed in deionized water, surveyed to determine the residual contamination levels, and then reinserted into the reaction kettle. The results of this test are presented in Figure 2. The 4M HCl:7M H_3PO_4 formulation turned dark green during decontamination.

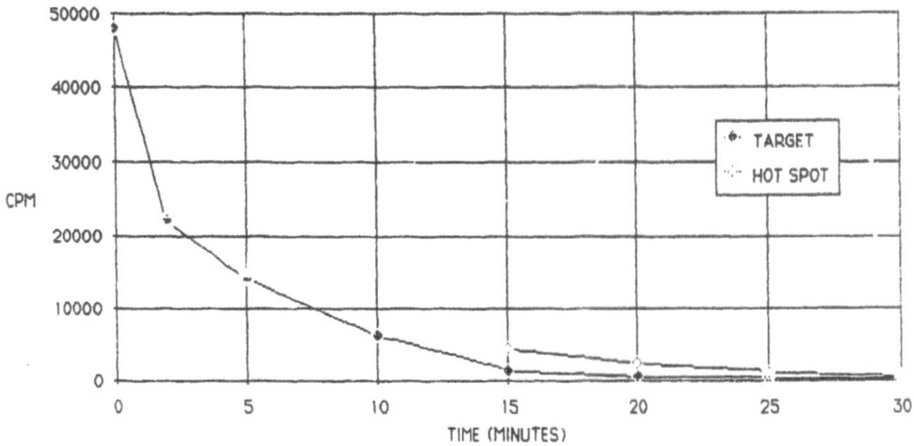

Figure 2. Decontamination of a Penetrated Target Using a HCl/H_3PO_4 Selective Acid Solvent.

A hot spot was identified during a 15 minute immersion inspection of the plate with a GH counter. The hot spot was located in a crevice located at the perimeter of the hole. The average decontamination of the hot spot was recorded at 4,400 cps and was monitored at subsequent inspections. It is unclear as to why the hot spot wasn't detected during initial characterization of the plate. A possible explanation is the uranium fragment was shielded by target material and was detected after a portion of the target was dissolved.

The 4M HCl:7M H_3PO_4 acid solvent effectively removed fixed uranium contamination from the penetrated area of the target including the hot spot. The average contamination of the hot spot was 4,400 cpn when it was detect and was reduced to 600 cpm (86 percent reduction) in the tine required to decontaminate the rest of the plate to 200 cpm (99.5 percent reduction). This suggests that regions which are contaminated with bulk uranium will require longer immersion times to reach the same level of decontamination as the rest of the target.

The acid solvent also removed rust from the surface of the target resulting in a dull gray appearance characteristic of the bare metal. The H_3PO_4 in the solvent inhibited excessive corrosion of the target plate, resulting in a 1.1 percent total weight loss. Target weight loss measurements are summarized is Table 6.

TABLE 6. Contaminated Target Dissolution

Solvent	Time (min)	Temp (°C)	Pre-Weight (g)	Post-Weight (g)	Weight Loss (percent)
4M HCl:7M H_3PO_4	30	84	3122	3089	1.1

DISCUSSION

The test results indicated both the DU-0.75 percent Ti pellet from the simulated target plate and the metallic radwaste from the actual penetrated target could be greatly reduced an 4M HCl:7M H_3PO_4 acid solvent. A synergistic effect occurred when the HCl and H_3PO_4 were combined for target decontaminated. This formulation removed rust and scale containing entrained uranium from the metal surface in addition to solubilizing uranium "hot spot" contamination. The decontamination solvent was selective and inhibited excessive attack of the steel target. The ferric iron that did result from partial target dissolution acted as an oxidant to facilitate uranium dissolution from the steel specimens.

SECTION IV
SOLVENT RECYCLE/WASTE REDUCTION

The feasibility of recycling (and reusing) the acid solvent buck to the decontamination process was demonstrated using ion exchange, solvent extraction and precipitation technologies. Recycling the solvent improves the efficiency of the decontamination process by maintaining a constant, but low level of contaminant concentration in the bath, minimizes operating cost, and decreases the volume of secondary waste. In addition, the DU is concentrated during the recovery process and is reduced to the smallest possible volume for disposal or purification/recovery.

ION EXCHANGE

Uranium normally forms cationic phosphate and chloride complexes in the IV and VI oxidation states in acidic phosphate and chloride solutions. However in the presence of Fe(II+), due to the partial dissolution of the target. U(VI) is reduced to U(IV). Excess phosphate in the solvent displaces the equilibrium towards the larger and more stable phosphate complexes, e.g., $UO_2(H_2PO_4)_2 \bullet H_3PO_4$

Resin Properties

The extraction of uranium from the 4M HC1:7M H_3PO_4 acid solvent was successfully demonstrated with the ion exchange resin, Duolite ES 467. Although not required as part of the Phase I program, FMI successfully demonstrated that the extracted uranium can be recovered from the resin as ammonium uranyl tricarbonate. Duolite ES 467, is a chelating resin that is commercially available from Rohm and Haas (Philadelphia, PA). The resin has a macroporous structure and its polystyrene matrix, which is cross-linked with divinyl benzene, contains aminophosphonic acid groups. Duolite ES 467 forms complexes with metallic ions and extracts uranium in the IV and VI valency states. Duolite ES 467 may be distinguished from the more standard aminodiacetic resins by its wider variation in stability in the presence of various cations and by the greater stability of the complexes formed with metallic cations of low atomic mass. A description of the chemical and physical properties of Duolite ES 467 is given in Table 7 (8).

TABLE 7. Properties of Duolite ES 467.

Matrix	Macroporous polystyrene
Functional Groups	$-NH-CH_2-PO_3Na_2$
Physical Form	Beige-colored beads
Bulk Density	About 0.75
Specific Gravity	About 1.12 (in Na form)
Particle Size (mm)	0.3 - 1.2
Moisture Content (percent)	60-65
Ionic Form (shipped)	Na
Minimum Total Exchange Capacity (g Cu/L)	45 (at pH 6)
Operating Temperature Limit (^{o}C)	65
Maximum Reversible Swelling (percent)	About 50 (H to Na form)
Solubility	Insoluble in water, dilute acids and bases, and common solvents

The characteristic reaction of uranyl species with the amino-phosphonic functional group of Duolite ES 467 is as follows:

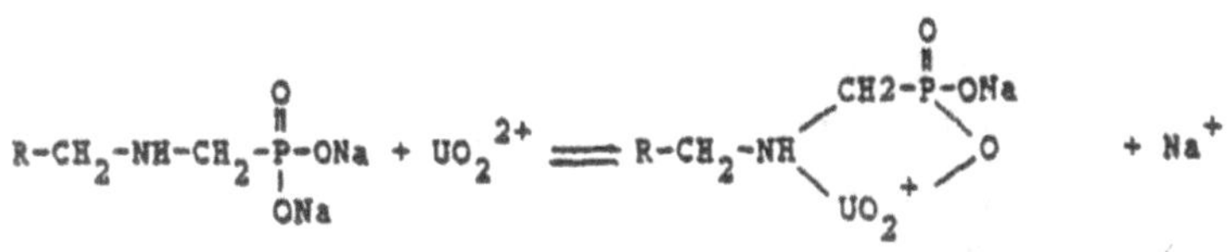

Test Procedures

An ion exchange column was prepared to remove dissolved uranium contaminants (cationic complexes of phosphate and chlorides) from the acid solvent. The column was charged with 122mL of Duolite ES 467 in the Na form.

The resin column was backwashed with 5 bed volumes (BV) of fresh solvent to remove any water that could cause uranium to precipitate from the acid solvent. The acid solvent containing depleted uranium at a temperature of 60° C was then permitted to flow through the column at an extraction flow rate of 8 BV/hr, or approximately 1 gpm/ft^3 resin. Flow wax provided by a peristaltic pump head driven by a variable speed drive.

As the acid solvent was treated, the exhaustion of the exchange resin was followed by noting the color change taking place in the resin bed. As the uranium complexes and the ferric chloride complex, which is present due to partial target dissolution, is adsorbed, the Duolite IS 467 darkens. Regeneration of the re«in was accomplished by washing the column with 5 BV of deionized water followed by 4.5 BV of 1M ammonium carbonate [$(NH_4)_2CO_3$] (8,9). The course of the regeneration can be followed by noting the color, as the refill becomes lighter. The flow rate for the regeneration cycle was 4 BV/hr, or approximately 0.5 gpm/ft^3 resin.

The temperature of the eluant was maintained at about 20° C to increase the kinetics of elution and the efficiency of the ion exchange column. Aliquots of the effluent and eluant were sampled at regularly timed intervals and analyzed for uranium and iron content.

Results

The Duolite ES 467 chelating resin gave very good sorption of uranium from the 4M HCl:7M H_3PO_4 acid solvent. Extraction data for the ion exchange resin is shown in Figure 3. In a parallel study, the equilibrium capacity of the resin was determined to be 0.33 equivalents (eq)/L. Because of the high acidity of the acid solvent, the efficiency of uranium removal is low and therefore the resin has to be regenerated often. Impurities in the acid, such as Fe (II+) and Fe(III+), have some effect on uranium uptake (9). The presence of Fe (II+) which reduces uranium to the tetravalent oxidation state, is also sorbed by Duolite ES 467. However, Duolite ES 467, is a powerful chelating resin in the strong acid solvent and selectively sorbs uranium over Fe (II+). The Duolite ES 467 aminophosphonic acid resin is readily eluted with an ammonium carbonate solution (Figure 4). About 90 percent of uranium is eluted at 20° C using 1M $(NH_4)_2CO_3$. Sorbed ion chloride complexes are selectively eluted with water prior to ammonia carbonate elution. This affords many advantages since it is possible to extract uranium from the acid solvent and recover the pure concentrate in the carbonate solution.

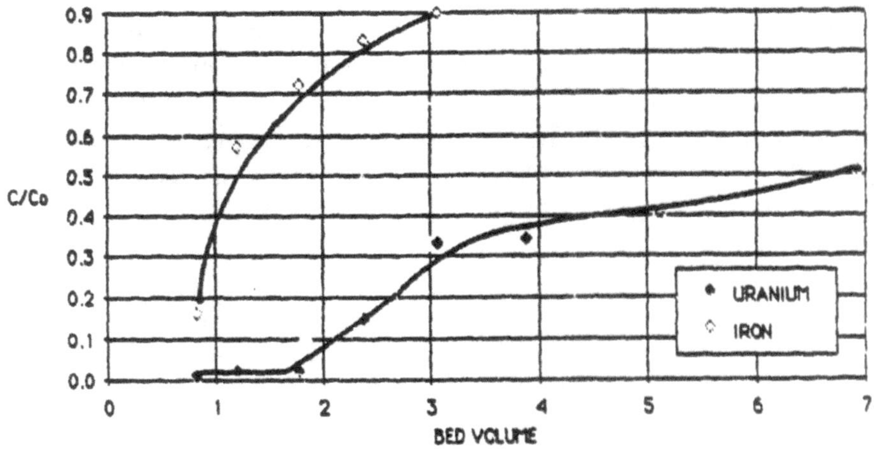

Figure 3. Extraction of Uranium Metal Complexes by Duolite ES 467.

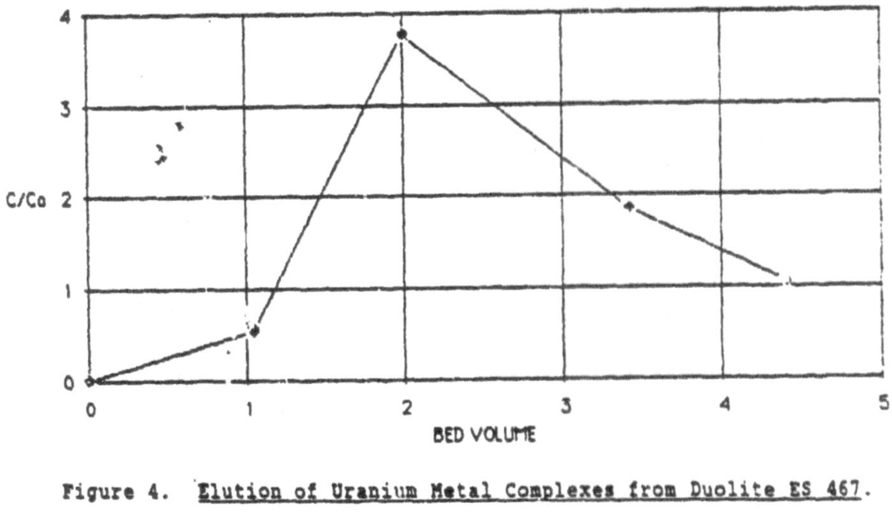

Figure 4. Elution of Uranium Metal Complexes from Duolite ES 467.

A conceptual flow sheet for recovering uranium using Duolite ES 467 is illustrated in Figure 5. The contaminant U (IV) in the 4M HCl:7M H_3PO_4 acid solvent can be sorbed in the extraction contactor at 60° C. After the sorption step (1), the resin is eluted with ammonium carbonate solution (2). The eluted resin is regenerated with a dilute NaOH solution (3). The ammonium uranyl tricarbonate eluant can be precipitated (4) directly with excess NH_3 And CO_2. The precipitate is filtered (5) and calcined (6) it 300° C to yield U_3O_8 "yellow cake" (8,9).

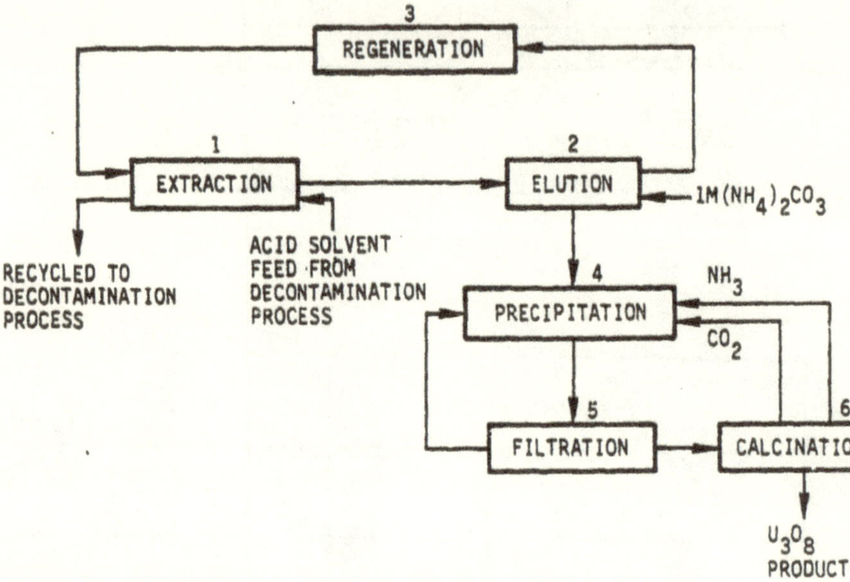

Figure 5. Flow Sheet for Uranium Extraction and Recovery by Ion Exchange.

This process route is simple and reduces the contaminated waste to the smallest volume possible. The recovery of uranium from the acid solvent allows the solvent to be recycled back to the decontamination process.

SOLVENT EXTRACTION

Extractant Properties

The potential of solvent extraction is well documented by the wide variety of solutions containing uranium which are amenable to recovery, purification, and concentration by liquid ion exchange. The extractants tri-n-octyl-phosphine oxide (TOPO) and di-2-ethylhexyl-phosphoric acid (D2EHPA), when used together in synergistic combinations and dissolved in an aliphatic solvent, exhibit very high uranium loading from sulfate leach liquors. Successful uranium recovery has also been demonstrated in acid chloride and phosphate solutions (10).

TOPO is basic in nature in that it coordinates to the uranium in a neutral metal complex, thereby causing the resulting organo-uranium complex to be organic soluble. The chemistry of D2EHPA resembles the chelate extractants and also show some similarity to neutral or solvating types ot extractants. The chemical structure of D2EHPA is represented as follows:

$$(RO)_2\overset{\overset{\displaystyle O}{\|}}{P}\text{-}OH$$

where $R = C_8H_{17}$. The following equation shows the behavioral characteristics of the two extractants when used together:

$$4(RO)_2\overset{\overset{\displaystyle O}{\|}}{P}\text{-}OH + UO_2^{2+} \rightleftharpoons [(RO)_2PO_2]_2UO_2[HO_2P(OR)_2]_2 + 2H^+$$

In the above formulation, two of the D2EHPA molecules lose a proton, much like a chelate extractant, while two other D2EHPA molecules solvate the uranium similar to solvating-type extractants.

The feasibility of using a synergistic mixture of TOPO and D2EHPA for the recovery of uranium from the acid solvent was demonstrated.

Test Procedures

The contaminated 4M HCl:7M H_3PO_4 acid solvent used to treat the actual impacted target plate contained 0.234 g U/L and 5.7 g Fe(total)/L. The uranium was extracted from the acid solvent with 0.5M D2EHPA:0.125M TOPO dissolved in reagent grade kerosene. The ratio of organic/aqueous was unity, which permitted good separation of the two phases. The two phases were contacted in a 100 mL Erlenmeyer flask. Mixing was accomplished with a magnetic stirrer. It was experimentally determined that equilibrium was established in less than 5 minutes, but for convenience 30 minutes was selected. Separation of the aqueous and organic phases was carried out in a 100 mL separatory funnel. After complete separation of the two phases, samples of the extracted acid solvent were taken for uranium and iron analysis.

Results

In a simple experiment on uranium removal in a contaminated selective acid solvent, the effect of the D2EHPA and TOPO synergistic mixture is shown in Table 8.

TABLE 8. Results of Uranium Removal by Solvent Extraction.

Element	Acid Solvent Concentration (g/L) Before Extraction	After Extraction	Removal (percent)	Distribution Coefficient [a]
U	0.234	0.062	73.5	2.78
Fe (total)	5.70	2.30	59.6	1.49

[a] Molar ratio of the element in the organic and aqueous phases at equilibrium.

The distribution coefficient, D, is defined as the molar ratio of the element in the organic and aqueous phases at equilibrium. The relative selectivity of the solvent for uranium relative to iron is measured by the separation factor; the larger the separation factor the more selective the solvent is for uranium. The separation factor, S, which is defined as the ratio of the distribution coefficients for uranium and iron, was found to be S = 1.87 for the experiment.

Discussion

Extraction of the uranium is very rapid. The uranium distribution coefficient, $D_U = 2.78$, obtained using the TOPO/D2EHPA extractant, was higher than the distribution coefficient for iron. $D_{Fe} = 1.49$, emphasizing the selectivity of the extractant for uranium. The percent removal of uranium is dependent on the D2EHPA concentration (with a constant TOPO:D2EHPA ratio). Consequently, the extraction efficiency of 73.5 percent can be increased by increasing the extractant concentration. Although not demonstrated in these experiments, the uranium is effectively stripped from the TOPO:D2EHPA extractant by contacting with ammonium carbonate solutions (10,11.) . The stripped solution can be precipitated and calcined to yield U_3O_8 as described previously for ion exchange.

Some of the major drawbacks of a solvent extraction process for decontaminating impacted targets include the possible buildup of solids at the aqueous/organic interface, the entrainment of organic solvent in the acid solvent solution, and the high cost of TOPO.

In conclusion, the method based on solid ion exchange using Duolite ES 467 resin will offer several advantages over the liquid extractant system, the major ones being the ease of separating the reactants from the products and the ease o£ recycling the extractant and the 4M HCl:7M H_3PO_4 acid solvent.

PRECIPITATION

The recovery of uranium from the contaminated 4M HCl:7M H_3PO_4 solvent is based on the formation of sparingly soluble compounds. A list of sparingly soluble uranium compounds is given in Table 9. The separation of a cure product from a contaminated 4M HCl:7M H_3PO_4 acid solvent cannot be achieved because the precipitate often contains considerable amounts of the precipitant itself, or co-precipitation of interfering elements, like iron. However, precipitation is often very effective in removing up to 97 percent of the solution uranium content.

TABLE 9. Solubility of Sparingly Soluble Uranium Compounds.

Compound	Solubility Product	Solubility
$UO_2NH_4PO_4$	$4.36x10^{-27}$	-
$(UO_2)_3(PO_4)_2$	$4.73x10^{-47}$	-
UO_2KPO_4	$7.76x10^{-24}$	-
UO_2HPO_4	$2.14x10^{-11}$	-
UO_2SO_3	$2.56x10^{-9}$	-
$UO_2NH_4AsO_4$	$1.71x10^{-24}$	-
UO_2KAsO_4	$2.52x10^{-23}$	-
UO_2NaAsO_4	$1.35x10^{-22}$	-
UF_4	-	10^{-4} mole/L(25 °C)
$U(OH)_4$	$(1.1±0.7)x10^{-52}$	-
$UO_2(C_6H_5N_2O_2)_2 \cdot NH_4C_6H_5N_2O_2$[a]	$(5.8±2.5)x10^{-10}$	-
UO_2HAsO_4	$3.17x10^{-11}$	-
$UO_4 \cdot 2H_2O$	-	$7.78x10^{-10}$ mole/L
$U(C_2O_4)_2 \cdot 6H_2O$	$4.3±0.4)x10^{-22}$	$5x10^{-2}$ g/L(25 °C)
$U(HPO_4)_2$	$3.1\ x10^{-18}$	-
$NH_4[UO_2(OH)_2VO_3]$	$1.7\ x10^{-13}$	-
$NH_4[UO_2(OH)(VO_3)_2] \cdot 1.5H_2O$	$2.9\ x10^{-14}$	-
$NH_4[UO_2(VO_3)_3] \cdot 3.5H_2O$	$9.6\ x10^{-15}$	-

[a] Uranyl ammonium cupferronate

Test Procedure

The contaminated 4M HCl:7M H_3PO_4 solvent solution resulting in the decontamination of the impacted target was precipitated with sodium hydroxide (NaOH) and ammonium phosphate [$(NH_4)_2HPO_4$] solutions. Each solution, at a concentration of 2M (molar) was titrated into the contaminated 4M HCl:7M H_3PO_4, solvent until the precipitation reaction was evident. Sufficient excess of precipitant was added to drive the precipitation reaction to completion. Constant stirring was maintained with a magnetic stirrer. The progress of the reaction was monitored by sampling aliquots of the acid solvent for analysis. The final pH value was recorded following the last addition of precipitant.

Results

The addition of sodium hydroxide and amonium phosphate to the 4M HCl:7M H_3PO_4 acid solvent resulted in the precipitation of uranium as the insoluble uranyl ammonium phosphate, $UO_2NH_4PO_4$ and uranium hydroxide, $U(OH)_4$, with solubility products of $(1.1 \pm 0.72) \times 10^{-52}$ and 4.36×10^{-27}, respectively. A summary of the final pH values and percent uranium removal are given is Table 10. Removal efficiencies of 97.6 percent for NaOH and 94.5 percent for $(NH_4)_2HPO_4$, demonstrate effective removal of uranium in the 4M HCl:7M H_3PO_4 contaminated solvent.

TABLE 10. Results of Uranium Removal by Precipitation.

Precipitant	Final pH	U Concentration (g/L) Before Precipitation	After Precipitation	Removal (percent)
NaOH	10.9	215	5.2	97.6
$(NH_4)_2HPO_4$	6.5	215	11.9	94.5

Discussion

Precipitation has proven to be an effective technique for removing uranium and other contaminants from contaminated 4M HCl:7M H_3PO_4 acid solvents. Uranium compounds that precipitate out can be removed in a subsequent treatment step and disposal as a low level waste. The recovery of uranium from acid solvents is greatly dependent on pH. Increasing the pH of the acid solvent for uranium recovery neutralizes the solvent and precludes any chance of recycling the solvent. However, precipitation remains a viable waste reduction method pending a cost analysis of the other recovery methods mentioned here.

SECTION V
CONCLUSIONS AND RECOMMENDATIONS

CONCLUSIONS

All of the objectives of the Phase I SBIR program have been achieved, namely:

- To develop a selective solvent that can decontaminate impacted armor targets containing depleted uranium for disposal or recycle.

- To identify and characterize technologies that can remove depleted uranium from the solvent for solvent recycle and uranium recovery for easier disposal.

This program has demonstrated the technical feasibility of using a 4M HCl:7M H_3PO_4 acid solvent to dissolve depleted uranium fragments from contaminated metal targets for disposal or recycle as non-radioactive material. Decontamination testing on targets characterized as containing residual entrained penetrator fragments is necessary to determine if all of the radioactivity associated with penetrator holes in the metal target material can be removed.

The HCl:H_3PO_4 acid solvent completely and very rapidly dissolved a DU alloy pellet inserted in a pre-drilled bore bole in a simulated metal target, while the target plate remained intact with less than 1 percent of its total weight dissolved. In less than 30 minutes, the acid solvent reduced the average contamination of a hot spot on a real penetrated target by 86 percent, while reducing the overall contamination of the rest of the plate by 99.5 percent; the phosphoric acid in the combined solvent inhibited excessive corrosion of the target plate resulting in about a one percent total weight loss.

The 4M HCl:7M H_3PO_4 decontamination acid solvent has many advantages over existing chemical decontamination methods. The most noteworthy are:

- Hydrochloric acid is one of the most active solvents for uranium alloy dissolution.

- Phosphoric acid is a powerful sequestrant that removes radioactive oxide scales and rust on the target and passifies the target surface with a film of iron phosphate.

These characteristics make the addition of expensive corrosion inhibitors (e.g., formaldehyde and gluconic acid) that may interfere with uranium recovery and solvent recycle unnecessary.

Removing the depleted uranium from the acid solvent results in recycling of the acid solvent and in concentration of the depleted uranium for easier disposal. Three technologies for removing the depleted uranium from the acid solvent were demonstrated experimentally, namely ion exchange, solvent extraction, and precipitation.

Ion exchange using an amino phosphonic acid chelating resin, Duolite ES 467, appeared to he the most feasible technically for removing and concentrating the depleted uranium from the 4M HCl:7M H_3PO_4 acid solvent. A conceptual flow sheet, based on preliminary test data, suggested that uranium could be extracted from the acid solvent while at the sane tine recycling the acid solvent. The resin could be elated with ammonium carbonate solutions, ultimately recovering stable U_3O_8 ("yellow cake") for hazardous waste disposal. This technique can completely separate the depleted uranium from the target generating a small volume of contaminant for disposal as low-level radioactive waste.

Organic extractants containing the synergistic mixture D2EHPA:TOPO was also evaluated for the recovery of uranium from a contaminated 4M HCl:7M H_3PO_4 acid solvent. About 75 percent of the uranium and 60 percent of total iron content was recovered from the acid solvent. However, the major drawbacks of a solvent extraction system involve the possible buildup of solids at the aqueous/organic interface, entrainment of the organic solvent in the acid solvent, and the high cost of the extractants.

Precipitation also presents an attractive method of recovering uranium from the acid solvent, since the uranium product is obtained directly with a consumption of reagent that is limited to the amount required to form the uranate salt and the stoichiometric equivalent of the other precipitated metals, namely Ti and Fe. Increasing the pH of the acid solvent for uranium recovery neutralizes the solvent and precludes any chance of recovering the solvent. If uranium is recovered from the contaminated acid solvent by simple hydrolytic precipitation, neutralization of the moderately high acid concentrations needed may involve a considerable chemical expense. Further testing and cost analysis are required to indicate the feasible limit of this procedure.

RECOMMENDATIONS

Since the major cost of the recovery process is that of the acid solvent, the economic success of the process depends on keeping the frequency of solvent replacement or acid make-up at a tolerable level. Although solvent recycle was demonstrated by removing uranium from the contaminated solvent in the laboratory by ion exchange, it will be necessary to demonstrate that the decontaminated 4M HCl:7M H_3PO_4 acid solvent is strong enough to decontaminate additional targets before solvent disposal is necessary. The effectiveness of this chemical decontamination process, measured by meeting the disposal criterion for DU contaminated materials and acid solvent recyclability, can be more accurately determined in a small pilot plant located at a plant site where the process can be run continuously. The results of this pilot scale test can be used to determine the working parameters, equipment, safety, cost and other factors that nay be identified as pertinent for the design and operation of the target decontamination facility.

Treatment of metal targets containing large residual bulk penetrator material needs evaluation to determine if complete target decontamination can be achieved. Sufficient attention should be given to incorporating chelate ion exchange into the pilot plant testing to demonstrate continuous depleted uranium removal from the acid solvent for solvent recycling and to demonstrate depleted uranium recovery and concentration for easier hazardous waste disposal.

SECTION VI
REFERENCES

1. Uhlig, H. (ed.), The Corrosion Handbook, John Wiley and Sons, Inc., New York,New York, 1948.

2. Patrick, M.A. and Cornette, J.C., "Morphological Characteristics of Paticulate Material Formed from High Velocity Impact of Depleted Uranium Projectiles with Armor Targets," DTIC Report AD-A067 121.

3. Rodden, C.J., Analytical Chemistry of the Manhattan Project, McGraw-Hill Book Co., New York, New York, 1950.

4. Katz, J.J. and Rabinowitch, E., Chemistry of Uranium, McGraw-Hill Book Co., New York, New York, 1951.

5. Shlifer, A., Personal discussion concerning the use of nitric acid for the dissolution of uranium metaland its alloys, Nuclear Metals, Concord, MA, January 1986.

6. Larsen, R.P., "Dissolution of Uranium Metal and Its Alloys," Analytical Chemistry, Vol. 31, p. 545, 1959.

7. Li, N.N. and Nauratil, J.D. (eds.), Recent Developments in Separation Science, Vol. III, CRC Press, Boca Raton, FL, 1986.

8. Duolite ES 467 Technical Bulletin, Diamond Shamrock Corporation, Redwood City, CA.

9. Naden, D. and Streat, M. (eds.), Ion Exchange Technology, John Wiley and Sons, New York, New York, 1984.

10. Blake, C.A., Brown, K.B. and Coleman, D.F., "The Extraction and Recovery of Uranium (and Vanadium) from Acid Sulfate Liquors with D-2-ethylhexyl Phosphoric Acid and Some Other Organosphosphorous Acids," Atomic Energy Commission Report ORNL-1903, May, 1955.

11. Bunus, F.T., Domocos, V.C. and Dumitrescu, P., "Synergic Extraction of Uranium from Phosphate Solutions with Di (2-Ethylhexyl) Phosphoric Acid and Tri-n-Octylphosphine Oxide," Journal of Inorganic Nuclear Chemistry, Vol. 40, pp. 117-128, 1978.

ALSO SEE:

https://e-reports-ext.llnl.gov/pdf/304820.pdf

to read the updated edition of this report.

www.ingramcontent.com/pod-product-compliance
Lightning Source LLC
Chambersburg PA
CBHW031928130626
46609CB00015BA/241/J
* 9 7 8 1 9 3 4 9 3 9 1 0 9 *